8th International Workshop NDT in Progress

Prague, Czech Republic
12 – 15 October 2015

ISBN: 978-1-7138-5224-7

Printed from e-media with permission by:

Curran Associates, Inc.
57 Morehouse Lane
Red Hook, NY 12571

Some format issues inherent in the e-media version may also appear in this print version.

NDT.net
Rheintalstr 3
53498 Bad Breisig
Germany

Phone: + 49 2633 4756670
Fax: + 49 3212-1030513

info@ndt.net

Additional copies of this publication are available from:

Curran Associates, Inc.
57 Morehouse Lane
Red Hook, NY 12571 USA
Phone: 845-758-0400
Fax: 845-758-2633
Email: curran@proceedings.com
Web: www.proceedings.com

TABLE OF CONTENTS

A New Approach of Microwave Resonant Methods for Bone Replacement Biomaterials Testing 1
D. Faktorová, A. Savin, M. Pápežová, R. Steigmann, F. Nový, O. Bokůvka

A Robot Inspection System Allows the Detection of Defects in Adhesive Bonds Between CFRP Components by Using Active Thermography, Leading to Reduced Cycle Times 11
M. Busch, B. Faupel, D. Weyrich

New Areas of Application of Acoustic Emission Method 18
P. Mazal, L. Komenda, F. Vlašic, V. Bukacek, M. Jana

Propagation Property of the Leaky Surface Acoustic Wave on the Monocrystalline Silicon Using Large Aperture Line-Focused PVDF Transducer 28
D. Lee, S. Kim, Y. Noh, M. Son, Y. Kim

Shortest Ways Finding in Discrete Bodies. 32
M. Chlada, Z. Převorovský

Study on the Lamb Wave Propagation in the Monocrystalline Silicon Wafers Using Large Aperture Line-Focused PVDF Transducer 38
J. Oh, D. Kim, I. Jung, K. Han, Y. Kim

Influence of Transducer Coupling in Ultrasonic Testing 46
L. Carbol, I. Kusák, J. Martinek, P. Vojkůvková

Acoustic Emission Crack Detection in Injection Molding 53
J. Grum, T. Kek, D. Kusic, A. Hančič

Ultra Sniffer New – LEAK Detection Method 61
R. Brockmann

In Situ Calibration of Acoustic Emission Sensors 66
J. Kober, Z. Převorovský

Introduction to the AFF Universal File Format and Its Use for NDT Data 71
M. Dráb, M. Hanak

Modelling of the Influence of Iron Oxide Layers on Microstructural Damage of Boiler Tubes 76
J. Angelo

Progress in NDT Methods and Topics During the Last 15 Years 84
Z. Převorovský

Modal Analysis as High Sensitive NDT Method of Compressor Blades 91
M. Witoś

Non-Destructive Testing of Electronic Components Overheating Using Infrared Thermography 106
K. Petrosyants, I. Kharitonov

Charging Wirelessly the Robot Inside Pipeline 111
M. Chaari

Damage Detection of Reinforced Concrete Elements Retrofitted with FRP by Using a Wireless Measurement System 112
T. Triantafillou

Some Aspects of Signal Processing for Nonlinear Pulse Ultrasonic Mixing Spectroscopy 118
K. Hajek

Nonlinear Time Reversal for Non Destructive Testing of Complex Medium : A Review Based on Multi-Physics Experiments and Signal Processing Strategies 126
S. Dos Santos, M. Lints, N. Poirot, Z. Dvořáková, A. Salupere, M. Caliez, Z. Převorovský

Characterization of the Anisotropy and Directional-Dependent Nonlinearity of the Human Skin In-Vivo Using Ultrasound 136
D. Tokar, Z. Převorovský

A Reference Free Method for Damage Detection and Location Based on Geometric Similarity 145
M. Salmanpour, Z. Sharif Khodaei, M. Aliabadi

Interpolation Analysis of Ultrasonic Guided Wave Signals in the Temperature Domain 155
R. Liou, J. Michaels

Laser-Ultrasonic Measurement of Local Elastic Moduli 165
E. Cherepetskaya

Ultrasonic Guided Wave Propagation and Detection of High Density Core Region in a Honeycomb Composite Sandwich Structure Using Embedded Piezoelectric Transducers 170
S. Sikdar, S. Banerjee

A Multichannel Ultrasound Pulser with Arbitrary Signal Excitation Unit for Scanning Acoustic Microscopy 180
A. Juhrig, S. Kümmritz, M. Wolf, E. Kühnicke

Complementary Near Field Techniques for Assesment of Fiber Reinforced Composites 185
A. Savin, R. Steigmann, N. Iftimie, M. Pápežová, D. Faktorová

Eddy Current Array No Contact Screw Inspection 196
C. Patrick

Material Characterization of Layered Structures by the Simultaneous Determination of Sound Velocity and Thickness 213
S. Kümmritz, M. Wolf, E. Kühnicke

OEM opBox ver 2.3 221
N/A

Author Index

A new approach of microwave resonant methods for bone replacement biomaterials testing

Dagmar FAKTOROVÁ [a], Mária PÁPEŽOVÁ[a], Adriana SAVIN[b*], Rozina STEIGMANN [b,c],

František NOVÝ[d], Otakar BOKŮVKA[d]

[a] Faculty of Electrical Engineering, University of Žilina, Univerzitná 1, 010 26 Žilina, Slovak Republic
[b] Nondestructive Testing Department, National Institute of R&D for Technical Physics, Iasi, Romania
Phone +40232430680; fax: +40232231132, email asavin@phys-iasi.ro
[c]Faculty of Physics, University Al.I.Cuza, 11 Carol I Blvd, Iasi, 700506, Romania
[d]Faculty of Electrical Engineering, University of Žilina, Univerzitná 1, 010 26 Žilina, Slovak Republic

Abstract
This work was to find out to what extent the knowledge of dependence of impedance on the crack depth observed by means of microwave frequencies can be used as nondestructive techniques for biocompatible materials which are used in bone replacement. The measured and calculated results are given in the graphic form and are compared each other in common graphs which provide a good overview about basic quantities and simultaneously give an initiative for orientation on practical applications in the process of defect presence finding in materials used in bone replacement.

Keywords: microwave, biomaterials, resonant methods

1. Introduction

Nowadays the significant need for the development of quantitative non-destructive testing (NDT) methods to measure bone implants stability and homogeneity is connected with exponentially increasing of biomaterials employing for improved life [1], [2], [3]. This work presents the investigation of the possibility to use microwave (MW) resonant methods both for in vivo and in vitro testing of homogeneity of biocompatible materials [4], [5].
Metallic implants interact with the surrounding biological environment [6], [7]. This interaction depends on the properties of bone surface and the shape of bone.
Using MWs as NDT method, comparatively with ultrasound [8], eddy current [9] and thermal imaging [10], allows the crack detection under various coating [11] without coating removal prior to testing. Other advantage of MW [12] is the obtaining of higher resolution images with deeper penetration than the thermography and eddy current techniques [13].
The experiments in this work were realized on flat stainless steel sample and have shown that the defect acts like the loss waveguide and the reflected signal amplitude is strongly dependent on the depth of the defect [12]. The depth of defect in the volume of investigated biocompatible material can be calculated from the resonant frequency at which the signal is maximum reflected.

1. Theoretical basis and applied formulae

Maxwell equations provide the basis to solution, and for the experimental part, the waveguide technique was chosen to make use of the same theoretical basis.

Every component of electromagnetic field satisfies the same equation with three coordinates and for the transversal electric field **E** having a sinusoidal character with the angular frequency ω

$$\frac{\partial^2 \dot{\mathbf{E}}}{\partial x^2}+\frac{\partial^2 \dot{\mathbf{E}}}{\partial y^2}+\frac{\partial^2 \dot{\mathbf{E}}}{\partial z^2}+\frac{\omega^2}{c^2}\dot{\mathbf{E}}=0 \tag{1}$$

where $\frac{\omega}{c}=\frac{2\pi}{\lambda}$ is the phase constant for the TEM waves and λ is the wavelength in free space. Supposing that the change of the **E** in dependence on coordinate x has the form

$$\frac{\partial^2 \dot{\mathbf{E}}}{\partial x^2}=-\beta^2 \dot{\mathbf{E}} \tag{2}$$

where $\beta=\frac{2\pi}{\lambda_g}$ is the propagation constant and λ_g is the wavelength in the waveguide, we get

$$\frac{\partial^2 \dot{\mathbf{E}}}{\partial y^2}+\frac{\partial^2 \dot{\mathbf{E}}}{\partial z^2}+\left(\frac{\omega^2}{c^2}-\beta^2\right)\dot{\mathbf{E}}=0 \tag{3}$$

From the condition for **E** on the waveguide surfaces it can be shown that

$$\lambda_g=\frac{\lambda}{\sqrt{1-\left(\frac{\lambda}{\lambda_c}\right)^2}} \tag{4}$$

where λ_c is the cut-off wavelength.

The complex impedance $\dot{Z}$ which characterizes the conditions in waveguide is $\dot{Z}=\frac{\dot{E}}{\dot{H}}$.

Transversal electric (TE) waves are used and therefore the representation only for these ones is given. The characteristic impedance of the waveguide is

$$\dot{Z}_0=\sqrt{\frac{\mu_0}{\varepsilon_0}}\frac{1}{\sqrt{1-\left(\frac{\lambda}{\lambda_c}\right)^2}} \tag{5}$$

with $\sqrt{\frac{\mu_0}{\varepsilon_0}}$ the characteristic impedance of the free space. Using the signal reflected by defects, the measurements and calculations are based on the waveguide technique, where the complex reflection coefficient $\dot{\rho}$ can be measured and it is given as $\dot{\rho}_E=\frac{\dot{\mathbf{E}}^-}{\dot{\mathbf{E}}^+}$ with $\dot{\mathbf{E}}^+$ and $\dot{\mathbf{E}}^-$ the intensities of reflecting and incident waves respectively.

Taking into account the expressions of $\dot{\mathbf{E}}^+$ and $\dot{\mathbf{E}}^-$ by means of β

$$\dot{\rho} = |\dot{\rho}_0| e^{j(\phi_0 + 2\beta x)} \tag{6}$$

where ϕ_0 is the phase in x = 0 and $|\dot{\rho}_0|$ is absolute value in the same point. Standing wave ratio (SWR) is

$$s = \frac{|\dot{\mathbf{E}}_{min}|}{|\dot{\mathbf{E}}_{max}|} \tag{7}$$

can be also measured and from the E_{min} position (d_{min}) it is possible to determine the phase

$$\phi = 2\beta d_{min} - \pi \tag{8}$$

But with regards to the definition of ρ̇

$$s = \frac{1-|\dot{\rho}|}{1+|\dot{\rho}|} \tag{9}$$

respectively

$$|\dot{\rho}| = \frac{1-s}{1+s} \tag{10}$$

The impedance can be calculated from measurements as

$$\dot{Z} = \dot{Z}_0 \frac{1+\dot{\rho}}{1-\dot{\rho}} \tag{11}$$

The complex impedance $\dot{Z}$ is calculated

$$\dot{Z} = Z_0 \frac{1-|\dot{\rho}|^2}{1+|\dot{\rho}|^2 - 2|\dot{\rho}|\cos\phi} + jZ_0 \frac{2|\dot{\rho}|\sin\phi}{1+|\dot{\rho}|^2 - 2|\dot{\rho}|\cos\phi} \tag{12}$$

where $\varphi = 2\beta d_{min} - \pi$ is the phase angle of $\dot{\rho}$, d_{min} is the distance of the first minimum of electromagnetic wave from the load. s and d_{min} are directly measurable on the MW slotted line and in the similar way λ_g can be determined, complex impedance of the investigated sample can be calculated using eq. (12) .

In the case when SWR has low value that it is impossible to measure it on one measuring range, it can be determined measuring w in the minimum of standing wave (Figure 1).

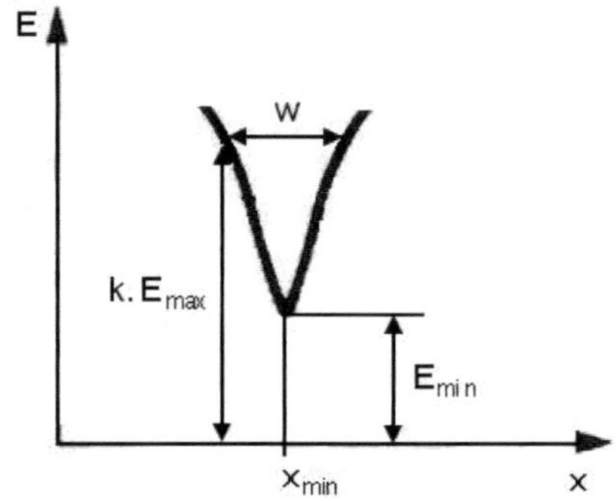

Figure 1. The method for small SWR measurement.

and SWR became

$$s = \frac{\sin\left(\frac{\pi w}{\lambda_g}\right)}{\sqrt{k^2 - \cos\left(\frac{\pi w}{\lambda_g}\right)}} \tag{13}$$

These equations allow to evaluate the measurements and report experimental results by plotting representation.

2. Experimental set-up and results

The experimental set-up (standard laboratory MW equipment [14]) is represented in Figure 2.

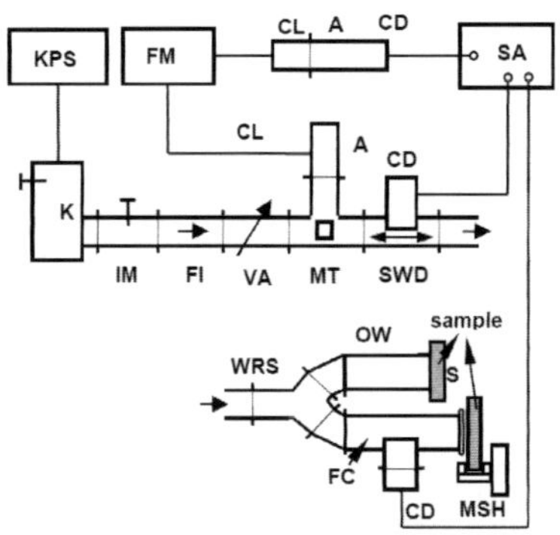

Figure 2. Experimental set-up: K – klystron, KPS –klystron power supply, IM- impedance match, VA – variable attenuator, MT – magic T, A – adapter, CL – coaxial line, FM – frequency meter, WRS – waveguide rotation change–over switch, FI – ferrite isolator, SWD – slotted section, FC – ferrite circulator, CD – crystal detector, OW – open waveguide, SA – selective amplifier, S – sample, MSH – movable holder.

The reflex klystron modulated with 1 kHz signal was used as a source of MW signal. The measurements were carried out from the ranges X and G band frequencies on the wave TE_{10}. The measured quantities were detected on the selective amplifier on the end of the line. The switch enables measuring both SWR and direct reflections in the same connection.

The measurements of SWR were taken with the switch position to the open waveguide (OW). OW was terminated with metal samples with the artificial slots representing cracks of the different depth and width. The samples with the defect depths from 5 to 20 mm were at disposal and the SWR was measured for every depth at each frequency by the standing wave detector. The measured and calculated values are plotted in Figure 3.

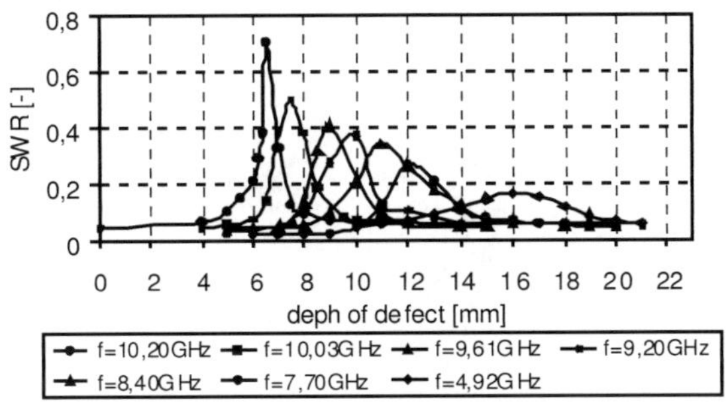

Figure 3. Dependence of SWR on the defect depth for seven frequencies.

Another point of view is in Figure 4, where it can be seen how the various depths of defect affected the value of standing wave ratio for various frequencies.

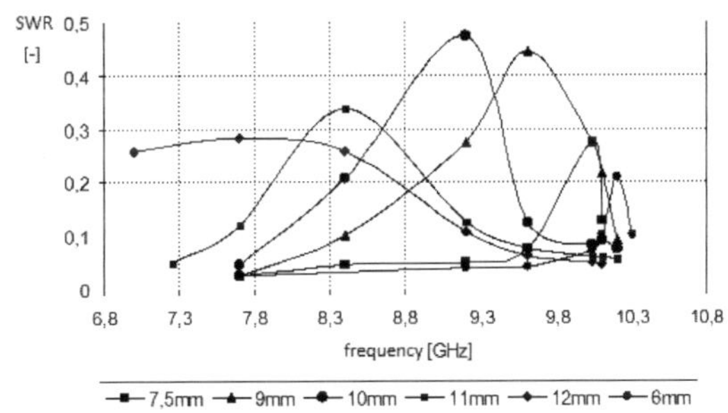

Figure 4. Dependence of SWR by frequencies.

The successive curves represent values of waveguide terminating impedance in the waveguide–defect contact position although they show quasiresonant course. Expressions of $s, \dot{\rho}, \dot{Z}$ are related. Figure 3 shows that individual samples at particular frequencies behave as a quarter–wave transformers and further increased continuously the defect depth and SWR was represented in Figure 5.

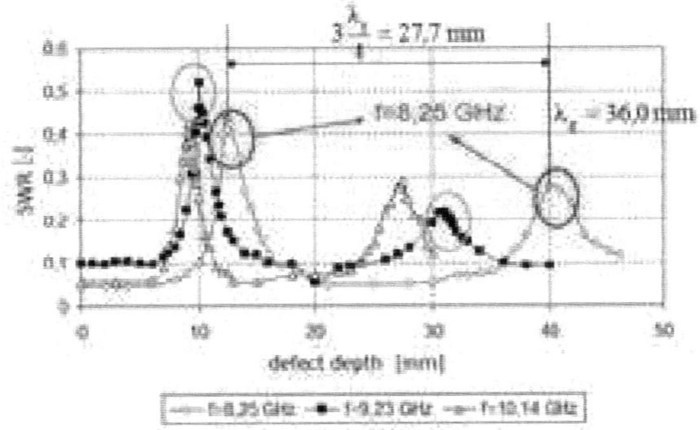

Figure 5. Dependence of SWR on the defect depth for several frequencies.

It can be seen from the all three curves (for frequencies 10.14 GHz, 9.23 GHz, 8.25 GHz) that the quarter–wave transformer effect really manifests itself at individual frequencies at three multiple of $\lambda_g/4$.

For the more complex assessment of the measured results from the point of view of quantities with which the MW technique operates, the values of impedance were calculated (11), and their dependences on the defect depth were represented at the frequency 9.23 GHz, (Figure 6).

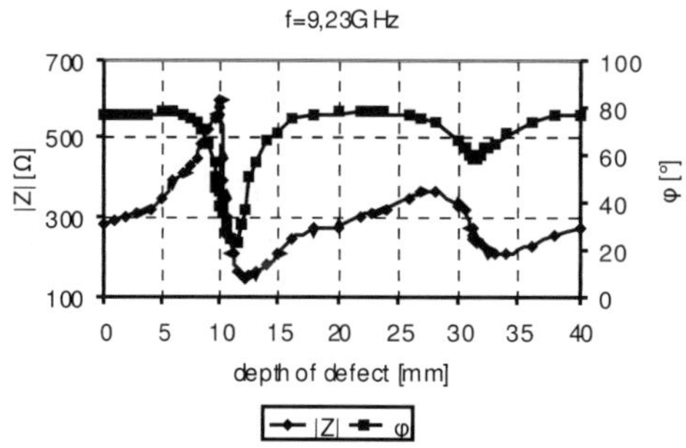

Figure 6. Dependence of amplitude and angle of impedance on depth of defect.

An illustrative image about impedance course for the defect quarter–wave transformer affords Figure 7, where closed curves belongs to the defect depths $\lambda_g/4$ and $3\lambda_g/4$, in concordance with Figure 5.

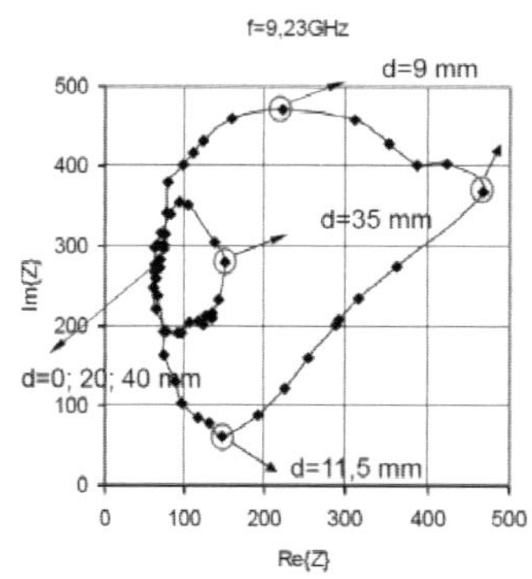

Figure 7. Dependence of impedance angle on the defect depth.

In order to obtain information about the influence of defect width on the reflected signal, the amplitude of the reflected signal with the moving probe position were measured. The results for different defect widths are presented in Figure 8. It can be observed that the sensitivity increases with the increasing of the defect width. The least recordable defect width was from the interval <0,05mm ÷ 0,1mm> what was confirmed by repeated measurements, too.

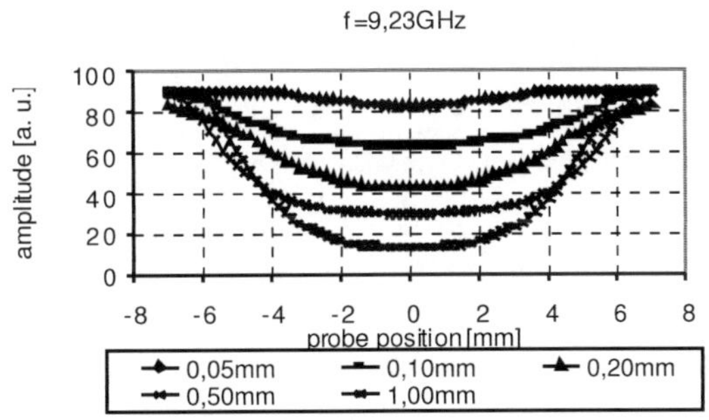

Figure 8. Dependence of signal amplitude on probe position.

In Figure 9 are represented the experimental results for influence of probe position above defect on defect impedance absolute value and angle.

Because the sample with defect acts like the complex impedance, the information about its amplitude and angle is required in the process of NDT. The next measurement was directed to the assessment of the open waveguide loading impedance character. The probe was moving above the artificial defect in the sample, similarly with the previous measurements. Measurements were done at the frequency 9.23 GHz. Figure 9 and 10 show that the existence of defect in the sample is emphasized by modification of the amplitude and the phase of complex impedance.

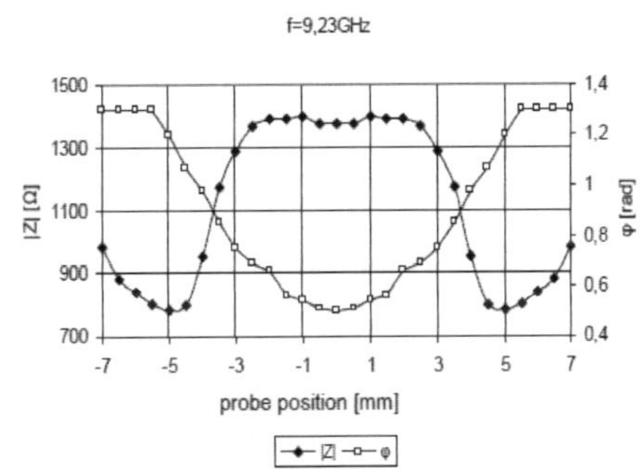

Figure 9. Dependence of impedance absolute value and angle on probe position.

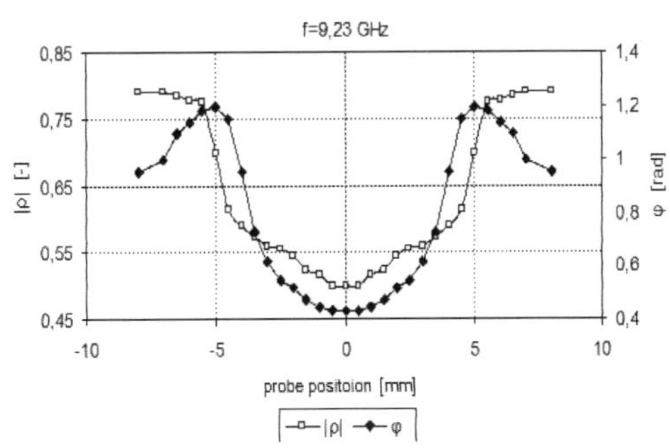

Figure 10. Dependence of complex reflection coefficient absolute value and angle on probe position.

After calculating the real and imaginary part of defect complex impedance, the Lissajoux curve was represented for the various position of open waveguide sensor moving above defect, Figure 11. By comparison of Lissajoux curves in Figure 7 and Figure 11, it can be seen the various character of Lissajoux curves for changes in depth of defect and changes due to sensor displacement in the vicinity of defect. The changes in the depth of defect lead to close curves and in the opposite side the moving of open waveguide sensor above defect leads to the open curve.

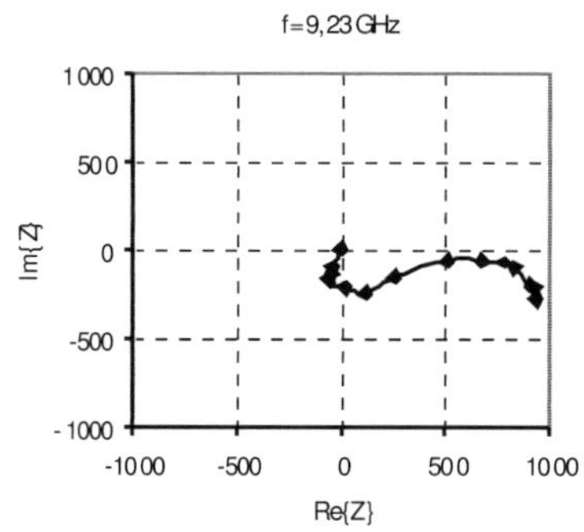

Figure 11. Lissajoux curve for various positions of open waveguide sensor.

With the open waveguide measurement it could be possible to obtain information about the defect orientation. Changing the angle between the waveguide H–plane and the straight line passing along the defect the reflected signal amplitude was measured and the dependence is presented in Figure 12.

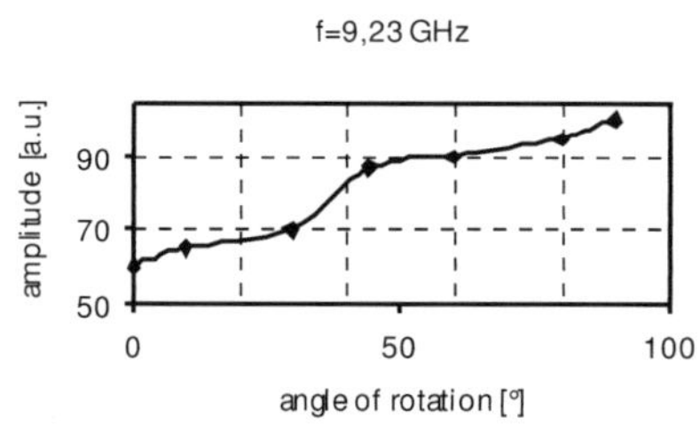

Figure 12. Dependence of signal amplitude on angle of rotation.

The orientation of defect is very important information, which can lead to the prediction of defect grooving in various applications. The measurement of defect orientation was done in connection with ferrite circulator (Figure 2). In experiments we used the defect with the width 0.1 mm. It can be seen from Figure 12 that the defect rotation has considerable impact on amplitude MW signal reflected from defect with varied orientation. This conclusion proves the above investigated statement that the defect acts like the complex impedance and can influence its amplitude and phase.

Conclusions

This work was toward MW technique use through nontraditional way and the attention was given primarily to the experimental verifying of MW in defects detection in biocompatible metals. Cracks were tested from the point of view the waveguide techniques and on this base, under certain conditions the defect can manifest itself as a quarter – waveguide transformer. This property allows detecting it as a quasiresonant effect and it can be stated what frequencies appertain to the individual defect depths.

It is possible to find the open defect in metal sample with open waveguide probe and using the information about amplitude and phase of complex impedance changing, in order to estimate the geometric properties of the defect in metal sample. MWs can be used for finding out crack presence in metal biocompatible materials, its depth, width, and orientation and in cooperation with other method they can be used as effective tool for material testing.

Author Contributions

All authors have equal contributions to this paper.

Conflicts of Interest

The authors declare no conflict of interest.

Acknowledgements

This paper is partially supported by projects VEGA 1/0846/13 – grant of The Ministry of Education, Science, Research and Sport of the Slovak Republic, and by a grant of the Romanian Ministry of Education, Research and Innovation CNCS – UEFISCDI, project number PN-II-ID-PCE-2012-4-0437.

The author would like to thank MSc. Pavol Žirko director of High School for Agriculture and Fishing in Mošovce for technical help at realization of experiments.

References

1. S. Franko, E. Babusova, M. Badida, Thermography and possibilities of its application in practice, Annals of DAAAM for 2011 & Proceedings of the 22nd International DAAAM Symposium, 23-26th November 2011, Vienna, Austria, ISSN 1726-9679, ISBN 978-3-901509-83-4, Katalinic, B. (Ed.), pp. 1233-1234, Published by DAAAM International, Vienna, Austria.
2. A.Savin, M.L.Craus, R.Steigmann, Resonant Ultrasound Spectroscopy used for diagnosis of some medical prosthesis components made from zirconium oxides, ASME 2014 12th Biennial Conference on Engineering Systems Design and Analysis, Volume 1: Applied Mechanics; Automotive Systems; Biomedical Biotechnology Engineering; Computational Mechanics; Design; Digital Manufacturing; Education; Marine and Aerospace Applications, Copenhagen, Denmark, July 25–27, 2014, ISBN: 978-0-7918-4583-7, ESDA2014-20597, pp. V001T03A010, 5 pages.
3. J.Delgado, J.R. Blasco, L. Portoles, J. Ferris, E. Hurtos, G. Atorrasagasti, FABIO project: Development of innovative customized medical devices through new biomaterials and additive manufacturing technologies, Annals of DAAAM for 2010 & Proceedings of the 21st International DAAAM Symposium, 20-23rd October 2010, Zadar, Croatia, ISSN 1726-9679, ISBN 978-3-901509-73-5, Katalinic, B. (Ed.), pp. 1541-1542, Published by

DAAAM International Vienna, Austria.
4. H.-J. Lee, K.-A. Hyun, H.-I. Jung, A high-Q resonator using biocompatible materials at microwave frequencies, APPLIED PHYSICS LETTERS, 104, 023509, 2014.
5. D. Zhao, G. Rietveld, G. M. Teunisse, A multistep approach for accurate permittivity measurements of liquids using a transmission line method, IEEE Trans. on Instrum. Meas., 60, (2011), 2267-2274.
6. S.Hloch,J. Foldyna, P. Monka, D. Kozak, D. Magurova, (2013) Advances in (Un)Conventional Engineering of Biomaterials and Nursing Care, Chapter 13 in DAAAM International Scientific Book 2013, pp. 297-316, B. Katalinic & Z. Tekic (Eds.), Published by DAAAM International, ISBN 978-3-901509-94-0, ISSN 1726-9687, Vienna, Austria.
7. P.A. Dearnley, A brief review of test methodologies for surface-engineered biomedical implant alloys, Surface and Coatings Technology, 198, 1–3, (2005), 483-490.
8. P. R. Armitage, C. D. Wright, Design, development and testing of multi-functional non-linear ultrasonic instrumentation for the detection of defects and damage in CFRP materials and structures, Composites Science and Technology, 87, (2013), 149-156.
9. N. Yusa, H. Hashizume, R. Urayama, T. Uchimoto, T. Takagi, K. Sato, An arrayed uniform eddy current probe design for crack monitoring and sizing of surface breaking cracks with the aid of a computational inversion technique, NDT & E International, 61, (2014), 29-34.
10. E. F. J. Ring, K. Ammer, Infrared thermal imaging in medicine, Physiological measurement, 33.3 (2012), R33.
11. M. Pastorino, A. Massa, S. Caorsi, A Global Optimization Technique for Microwave Nondestructive Evaluation, IEEE, Transaction on Instrumentation and Measurement, Vol. 51, (2002), pp. 666-673.
12. S.Kharkovsky, R. Zoughi, Microwave and millimeter wave nondestructive testing and evaluation - Overview and recent advances, Instrumentation & Measurement Magazine, IEEE, 10, 2, (2007), 26-38.
13. R. Zoughi, S. Ganchev, Microwave NDE – State – of – the –Art Review, NTIAC 95 – 1, Austin, TX, (1995).
14. D. Faktorová, Application of short circuit waveguide method for evaluation of inhomogenities in dielectric sample, ElectroScope, ISSN 1802-4564. - No. 6, 2008, 4 p.

A robot inspection system allows the detection of defects in adhesive bonds between CFRP components by using active thermography, leading to reduced cycle times

Matthias Busch[1], Dennis Weyrich[1], Benedikt Faupel[1 2],

ZeMA Zentrum für Mechatronik und Automatisierungstechnik[1]; Saarbrücken, Germany, htwsaar[2], Phone: +49 681 85787 537, Fax: +49 681 85787 11, e-mail: m.busch@zema.de

The bonding of large industrial components made of carbon fiber reinforced plastics (CFRP), as they are used in the aviation industry in case of structural components, requires an inspection system which is able to verify these bonds. In this contribution, the advantages of a robotic inspection system, which will be achieved through continuous testing, will be introduced. Of particular interest here are the bonds with which, for example shell elements and reinforcements (so-called stringers) are bonded together. In order to make a qualitative statement about the bond, it is measured using a non-destructive testing method, the active thermography. For elongated part geometries or large components, an advantage can be achieved by a continuous testing. In this variant of testing with active thermography, the positioning time for the measurement system is eliminated and by spatial shifting, the contrast of failure measurement, a much shorter testing time will be achieved. The inspection system consists of an excitation source (infrared quartz radiator) and an infrared camera for the far-infrared spectral range (FWIR) which are both attached to a lightweight robot.

Keywords: active thermography / carbon fiber reinforced plastic CFRP / inspection system

The competitiveness compared with low-wage countries shows a great challenge for enterprises resident here. These exist and are convincing less by the price of a product rather than by high-class standards and high loyalty of delivery. A zero-defect production or manufacturing processes can be hardly reached, therefore, 100 % testing of the manufactured products is often necessary to do justice to the high quality standards and to reach a competitive advantage. An advantage can also be reached by the increased application of carbon fiber reinforced plastics (CFRP) in different industrial branches, like in the automotive or in the aviation industry, by weight savings. By the application of new materials the joining technology is also of essential interest. The claim of the security relevance becomes particularly obvious if for joining, the adhesive technology is used. The advantage of this joining technology is, in spite of low weight to reach high demands for the rigidity and stability. The application of non-destructive testing methods in this area is gaining in. If new materials like CFRP and the joining technology infers it is necessary to verify the quality of the bonded joint. A procedure which admits reliable statements about the adhesive bond is necessary moreover. A non-destructive testing method which meets the requirements for these demands is the active thermography.

1 Thermal Sensors

A part of the IR radiation going out from an object is displayed with optics which shows the transmission τ on a sensor element and is absorbed by this (Figure 1: action principle of thermal sensors, $\Delta\phi_0$radiant flux of the object, $\Delta\phi_S$ radiant flux of the sensor, τ transmission of the optic, ΔT temperature difference, ΔU output voltage of the sensor). Thereby a temperature difference ΔT arises in the sensor element, which is leading to a change of an electrical quantity, though. Depending on the type of sensor there is a changing value of electrical resistance (ΔR, bolometer), electrical charge (ΔQ, pyroelectric sensor) or electrical voltage (ΔU, thermoelectric sensor).

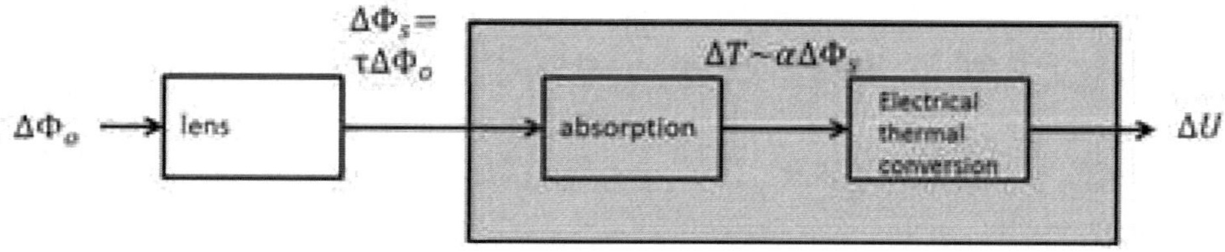

Figure 1: action principle of thermal sensors

Thus the temperature change is very big in the sensor element; this must be isolated thermally very well to not be influenced by neighboring elements. Figure 2 shows the construction of one pixel of a microbolometerarray like it is installed in the infrared camera which is used for this work.

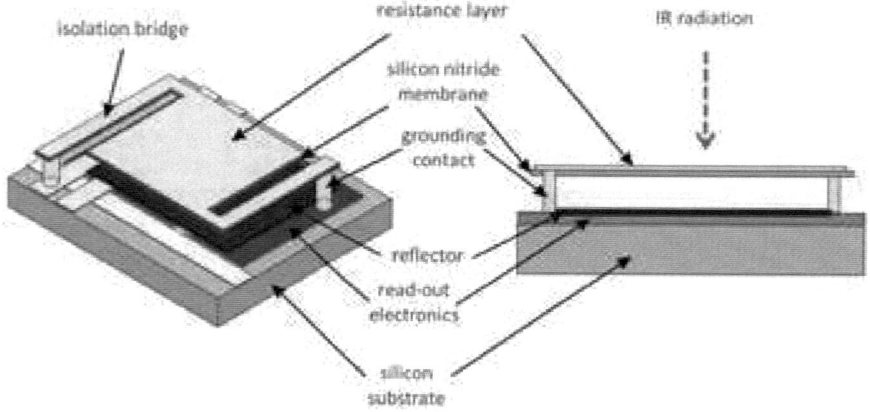

Figure 2: micro bolometer element

A pixel is performed as a MEMS element (Micro-Electro-Mechanical Systems) with an upset resistance layer. The resistance bridge above the selection circuit is about 2,5µm. The incident infrared radiation is absorbed by the micro bridge. This is leading to a change of temperature ΔT and therefore a modification of resistance ΔR. To ensure the thermal isolation, the micro bridge is located into vacuum. Besides the construction (leg supports and bridge) enables a lower thermal conductance to the selection circuit, which can be assumed as thermal mass[1].

1.1 Thermal resolution

The thermal resolution of a sensor element can describes with the NETD (Noise Equivalent Temperature Difference). It declares the smallest resolvable temperature difference of an object. This value is mostly given for objects with a temperature of 27°C, whereas the aperture of the lens is assumed by one.

1.2 Geometric resolution

The spatial or geometric resolution describes the capability of a system to separate nearby structures of an object. The geometric resolution is limited by the Fraunhofer-diffraction. The smallest illustratable measuring point is the Airy-slice, which arises by the Fraunhofer-diffraction on a perforated disc.

1.3 Temporal resolution

The temporal resolution describes the capability of a system, to detect temporal subsequent temperature changes of an object; it is described by the time constant τ. In a micro bolometer the time constant is defined by the behavior of the micro bridge[1].

2. active thermography

To identify failures in adhesive bonds between CFRP-components, there are various thermography processes which were developed in the last years. This various processes can be distinguished by the excitation which is used. Firstly, on all active thermography processes, a thermal flow will be generated and the subsequent balancing process within a certain period will be observed. Additionally, the various processes can be distinguished in qualitative and quantitative thermography processes. The so called active thermography is characterized by generating a thermal flow into a component by different excitation sources. Different thermal conductivity within the component is leading to temperature differences at the surface of the component which can be measured. As contactless excitation sources for generating a thermal flow, mostly optical sources in visual spectrum (halogen lights, flash lights or laser) or infrared spectrum (quartz- or carbon radiators) are used. Components made of metal can also be excited inductively by eddy current. Furthermore, a thermal flow can be generated by ultrasonic. Depending on the position of the infrared camera and the excitation source relative to the surface of the component a distinction can be made between reflection (Figure 3) and transmission setup.

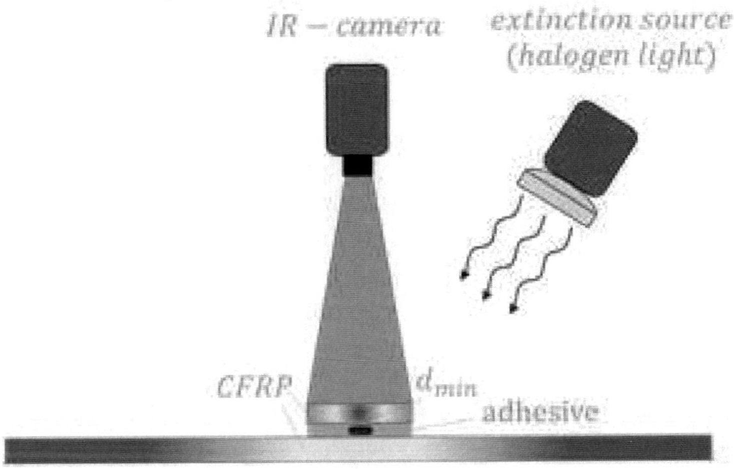

Figure 3: active thermography in reflection configuration

Choosing a suitable setup occurs appropriate to the accessibility and the depth of a relevant defect. For the analysis of the recorded data with regard to possible defects, three methods come into consideration. These methods are called pulsed thermography, lockin thermography and pulsed phase thermography[2].

1.1 pulsed-thermography

Pulsed thermography is based on a high-energy pulse which is bringing the component into a thermal instationary status. For analysis the procedure of excitation and decay behavior is observed. The excitation for pulsed thermography is a high-energy pulse which is as short as possible. The thermal excitation can occur either inside or outside of a component. The heat sources for pulsed thermography should have a preferably wide and homogeneously distributed thermal flow density.

2.2 lockin-thermography

Just like the pulsed thermography, the lockin thermography is used to detect defects in components. The lockin thermography is using continuous sine signals of a defined frequency for excitation. The modulated response signal is recorded by an IR-camera and analyzed relating to pulse and phase. The lockin thermography is using digital Fourier transformation for the lockin-technique by four measurement points each period for every pixel of the infrared image. The phase data is of particular interest because it is almost independent of external perturbation[3].

2.3 pulsed-phase-thermography

The pulsed phase thermography was introduced to combine the advantages of the pulsed thermography and the lockin thermography:

– pulsed thermography: rapid data acquisition with flexible setup and with a high-energy pulse suitable for a larger surface

– lockin thermography: a better resolution for the depth and a smaller influence of optic and infrared perturbation

The pulsed phase thermography is using the Fourier transformation according to the method of the pulsed thermography. The results of the pulsed thermography on the time level remain entirely and are only replenished by the results of the frequency domain[4].

2 continuous inspection with thermography

To inspect adhesive bonds between reinforcement components and shell elements made of CFRP with non-destructive testing, a continuous inspection process is appropriate. With this alternative approach of testing the positioning time for the inspection system is eliminated and with the displacement of the failure contrast measurement, a distinctly shorter inspection time can be reached. This is in case of large component assembly leading to smaller cycle times, lower costs and establishing a competitive advantage. Figure 4 shows the setup for active thermography in reflection configuration for continuous testing. Here, the component which has to be inspected is passed over by the inspection system with the velocity v, which is adapted to a light weight robot with a defined path. As a result the excitation source with the intensity I_{exc} and the effective excitation width e is moving in a defined distance to the infrared camera over the component. A continuous heating of vertical component areas occurs.

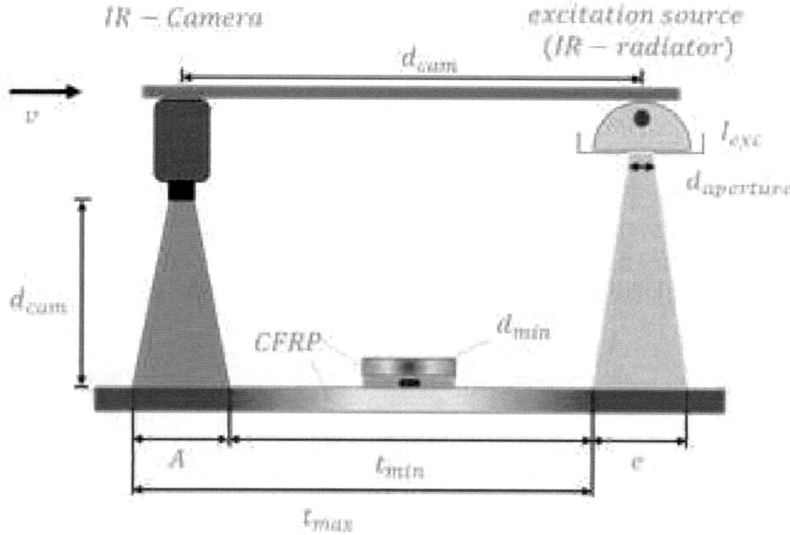

Figure 4: process parameters of the continuous inspection

First of all, the amount of parameters must be restricted to identify the optimal process parameters by design of experiments. Figure 5 shows two characteristic cooling curves which represent the temporal temperature behavior of an intact and a failure containing component area. If defects are contained in the adhesive layer, a failure contrast is visible after a certain time t_1 which is reaching its maximum at the time t_{max} and disappears until t_2.

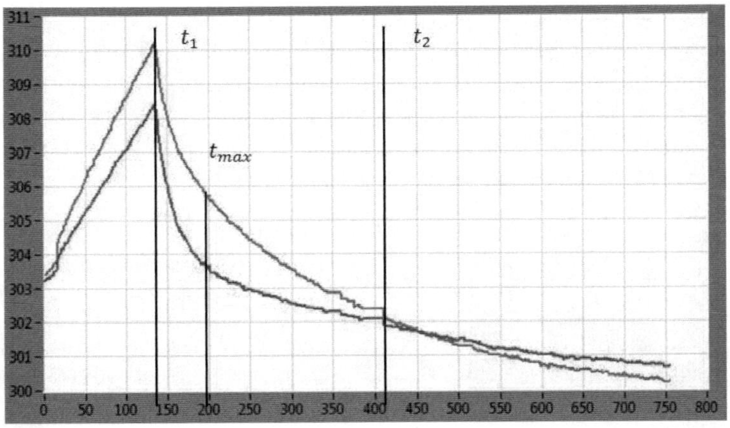

Figure 5: characteristic cooling curves of an intact and a failure area

Depending on what defect the adhesive layer consists, different characteristic times after a thermal excitation occur. The smallest measurable defect in the adhesive layer d_{min} is limited by the geometric resolution. Due to this reason, at a given velocity only a particular temporal sector of the balancing process can be detected. To optimize the probability of defect detection, the excitation source is positioned ahead of the IR camera in such a way that the most significant measurable failure contrast occurs while the camera with the detection area A is into the position just above the component. The best distance between excitation source and IR camera can be defined by the time t_{max} and the velocity v of the light weight robot. Therefore the cooling curves are measured in static test setup [5].

3 results

3.1 specimens

In order to enable a quality assessment about defects in adhesive bonds between CFRP components, the performance and the provability of the inspection system has to be specified. To achieve this objective, specimens with different defects were produced. At this point of the contribution it should be noticed that a standardized specimen such as for ultrasonic testing is available does not exist yet. Figure 6 shows how to simulate different kinds of defects into the adhesive layer:

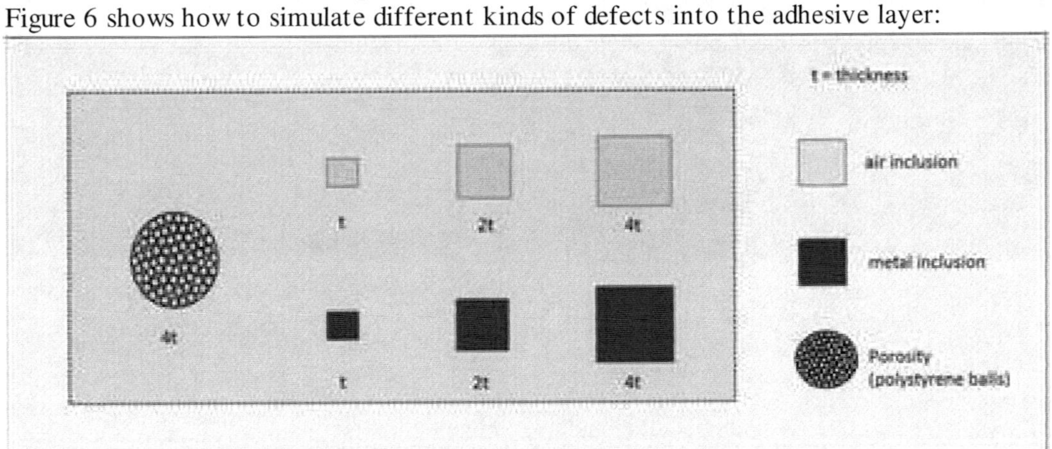

Figure 6: template to establish specimens containing defects

To gain a better access for measuring, a single adhesive layer for each kind of defect was made.

3.2 parameter setting of the continuous inspection

After optimal process parameters are specified, the geometric setup of the inspection system can be build up. Based on the smallest measurable defect, firstly the appropriate image field of view must be chosen. The smallest measurable defect must be mapped on three pixels at least to enable a safe distinction of the measured noise. In a subsequent step, a suitable excitation source must be selected. This excitation source must be comparable in intensity and resulting pulse length to the source which is used in the static test setup. The length of the pulse can be adapted by an aperture which is attached in front of the excitation source. The area which is available for data acquisition can be calculated by the detection area A and the velocity v.

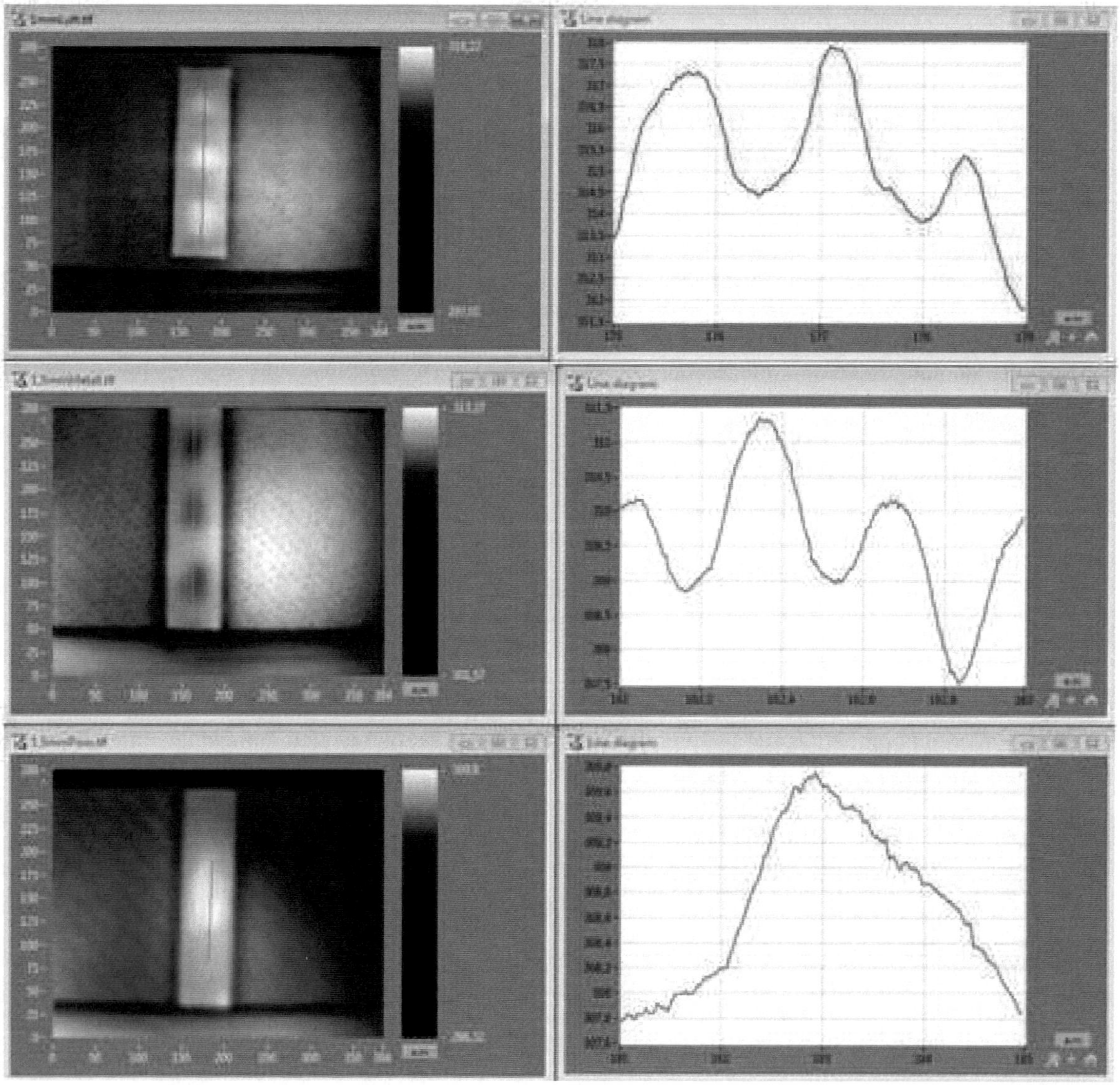

Figure 7: results of continuous testing, the upper CFRP layer was 1.5mm

4 summary

In this contribution, the advantage of an inspection system which is using active thermography to detect defects in adhesive bonds between CFRP components is shown. The advantage is due to the continuous testing process which is enabled by adapting the inspection system on a light weight robot. Cycle times are reduced because the positioning time of the robot and the thermal balancing process within the component are eliminated. Additionally the automation of the entire method is simplified.

References

1. Thermische Infrarot-Sensoren für die Thermografie - Stand und Tendenzen; LIFIS ONLINE, 2011
2. Maldague, X.: Non-Destructive Evaluation of Materials by Infrared Thermography. London : Springer-Verlag, 1993
3. Wu, M. Sc. D.: Lockin-Thermographie für die zerstörungsfreie Werkstoffprüfung und Werkstoffcharakterisierung, Universität Stuttgart, Dissertation, 1996
4. Maldague, X. ; Marinetti, S.: Pulsed Phase Thermography. In: Appl. Physics 79, 1996
5. Thiemann, C. Zäh, M.,: Zerstörungsfreie Prüfung durch aktive Thermografie im kontinuierlichen Bauteildurchlauf: Thermographie-Kolloquium, 2011

New areas of application of acoustic emission method

Pavel MAZAL [1], Lukas KOMENDA [1], Frantisek VLASIC [1],
Vladimir BUKACEK [2], Miroslav JANA [2]
[1] Brno University of Technology, Faculty of Mechanical Engineering; Technicka 2, Brno, Czech Republic
Phone: +420 541 143 229; mazal@fme.vutbr.cz
2 Policské strojírny, a.s., Bariny 1145, Policka, Czech Republic

Abstract

Acoustic emission laboratory at the Institute of Machine and Industrial Design of Brno University of Technology is long focused on the use of acoustic emission testing (AT) for diagnostics of damage development in cyclically loaded materials and machine parts. In addition to these relatively traditional applications already however this laboratory workers devote other non-traditional possibilities of using acoustic emission method. In this paper there are presented the first interim results of the project, which is focused on applications of AT in function diagnostic of pneumatic devices. There are compared the signals obtained from the fully functional pneumatic cylinders with signals from cylinders with various types of artificially created damage. The second part briefly presents the first results of the acoustic emission application in other very non-traditional areas. The attention is paid to the usability of AT for identification and localisation of undesirable discharges in gas-insulated conductors for high-voltage substations and for increasing of accuracy and objectivity of the tests for sensitiveness determination of explosives to friction.

Keywords: Acoustic emission, non-destructive testing, pneumatic cylinder, gas-insulated conductors, explosives

1. Introduction

Acoustic emission method (AE) passes through the very rapid development especially in the field of instrumentation. Modern analysers enable highly sophisticated scanning of signals on the surface of the tested bodies. Somewhat worse situation is in the exact identification of sources of acoustic emission signal. Nevertheless, the AT method has been routinely used for the diagnosis of pressure vessels, storage tanks, pipelines and power equipment. Reasons are of course quite specific characteristics of AE method that is able to identify and locate the presence of active defects which can be relatively easily found and verified by other NDT methods in this case.

2. Pneumatic cylinders

The first research of use of AE as a method for defects detection in pneumatic circuits starts by research of Dickey, Dimmick and Moore [1]. In their paper is described relation between amplitude of AE signal and rate of gas leak in valve. Research in the paper [2] is focused on noise measurement of gas valve. Results of this research describe correlation between signal of acoustic emission and leakage rate in frequency domain. In those papers was just performed potential of acoustic emission as a non-destructive method for pneumatic industry with no particular results connected to development of specific methodology of measurement.

AE as a method of NDT for pneumatic industry have been developed during last decades. Main purpose of this method is to analyse and gas leakage on a specific parts of the system such is pneumatic pistons and cylinders. In the paper [3] is described utilization of AE for diagnostic pneumatic systems of power plants. In this study measurement of leakage defects in pneumatic systems was undertaken by three methods like AE, noise measurement and vibro-diagnostic. Comparison of signals from each measurement proofed that results obtained by AE fully describe defect on testing specimen. In other studies like the papers [4] or [5] was measured by AE used for detection of leakage in valves with aim of development a universal method for description of leakage defects. In the paper [4] was performed research of influence of leakage rate (from a defect) on RMS value of the AE signal. This measurement was provided by sensors of AE. An effect of various internal pressures on a signal of AE was tested as well. The results of the research proof that RMS of AE signal rise with higher leaking rate as well as with higher internal pressure.

In the study [5] was aim to analyse the signal of internal leaking of valves so that their rate could be calculated by analytical method. Effect of different sizes of valves on a signal of AE was tested and the amplitude of the signal was determined from the signal. Analysis of the obtained signal was performed by Fourier transformation. Moreover, in this study was suggested procedure for creation of methodology of determining leakage rate from the signal of AE. This procedure contains parameters as size and type of valve and internal pressure. By applying those parameters on analytical processing of the AE signal the leakage rate should be obtained.

However experiments in papers were undertaken in laboratories and under specific conditions. Therefore, particular tool for detection of leakage of pneumatic circuits have not been developed.

The AE laboratory at the Institute of Machine and Industrial Design of Brno University of Technology has long focused on the use of AE for diagnostics of damage development in cyclically loaded materials and machine parts (standard fatigue, contact fatigue, bearings etc.). In addition to these relatively traditional applications already however this laboratory specialists devote other non-traditional possibilities of using acoustic emission method [6, 7, and 8].

2.1 Experimental procedure

The research of AE as a method of non-destructive testing for parts of pneumatic circuit is undertaken at Brno University of Technology, Institute of Machine and Industrial Design. This research is focused concretely on pneumatic cylinders as a project in cooperation with company Policske strojirny a.s. The aim of this research is to develop system for non-destructive damage detection of specific types of pneumatic cylinders by use of AE.

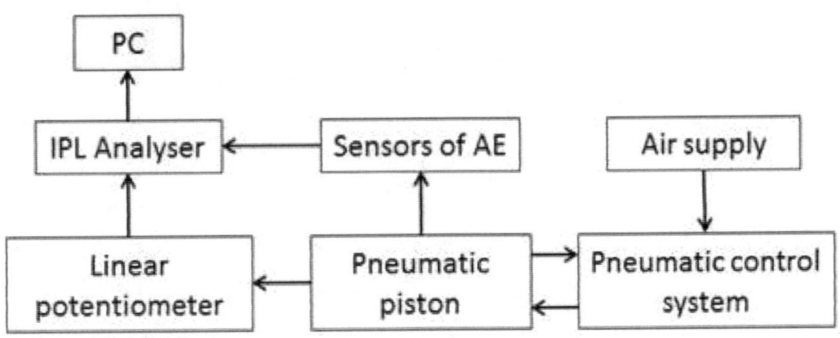

Figure 1. Schema of the experimental stand for testing of pneumatic cylinders

An experimental stand was created for simulation of pistons behaviour in operation. For analysis of cylinder's damage during operation a measuring chain was created. This chain usually comprises three or four sensors of AE, linear potentiometer and 4-chanel IPL analyser. A system DAKEL was used for signal processing and post-processing. During the experiment pneumatic cylinders were operated by pneumatic control system. A schema of the experimental stand is shown in Fig. 1.

Firstly cylinders without damage were used for experimental measurement to obtain signal of AE which correspond to proper operation of cylinders. After that, defects on each cylinder were created. There were 14 cylinders, each with different kind of a defect. During operating cylinders were measured also by sensors of AE in aim to analyse which defect is possible detect by signal AE. Sensors were placed into three positions on the piston's body to find the best position for signal detection. Linear potentiometer was used for measuring of the position of the cylinders (Fig. 2).

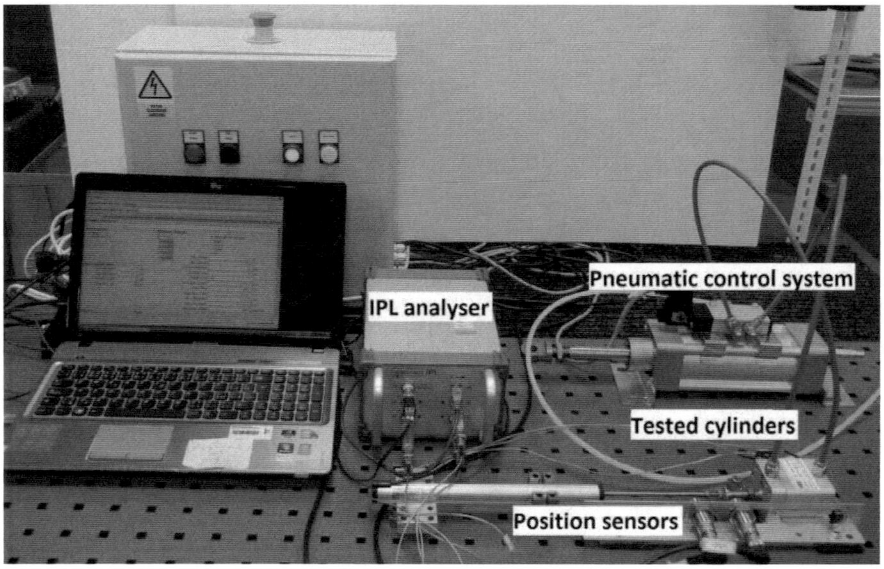

Figure 2. Assembly of experimental equipment

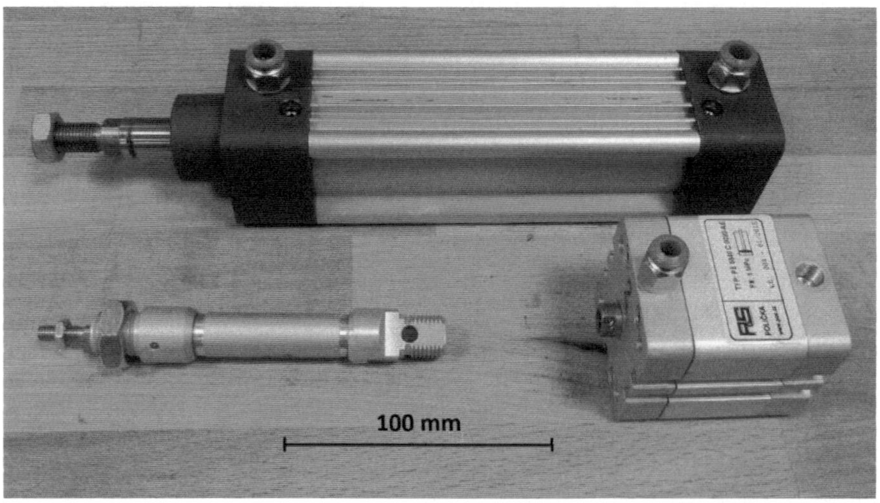

Figure 3. Types of tested pneumatic cylinders

2.2 Experimental results

For data analysis was the most effective method spectral analysis. By this method was possible to identify particular frequencies and distinguish signals from cylinders with damage and without. Envelope method was also used for signal analysis but this method was not as accurate as spectral analysis. Further presented results were obtained on the cylinders PB 6040C (producer Policske strojirny). Structure of cylinders is shown in Fig. 4.

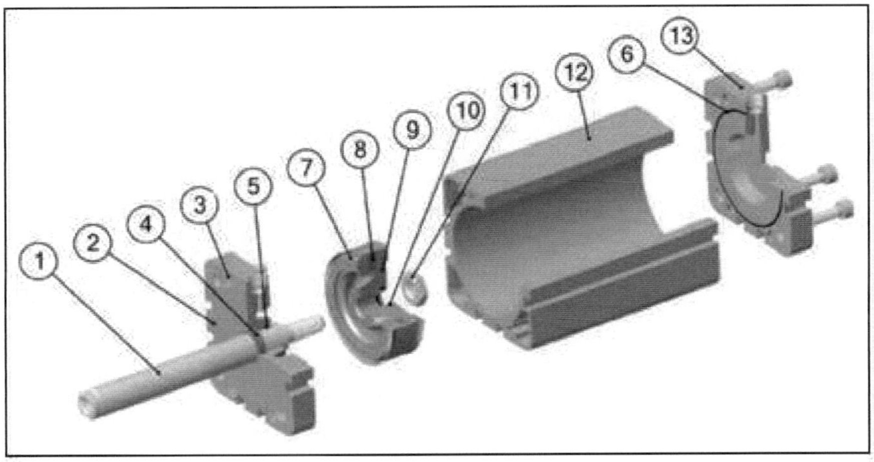

1 Piston rod
2 Front cover
3 Bolt
4 Sealing of the piston rod
5 Rod conductor
6 O-ring
7 Piston sealing
8 Magnet
9 Piston
10 O-ring
11 Nut
12 Cylinder body – tube
13 Back cover

Figure 4. Scheme of the pneumatic cylinder PB 6040C

The range of measuring frequencies of the signal is naturally set by AE sensors and its size is approximately from 50 kHz to 400 kHz. According to damage of cylinders some dominant frequencies were lost or some were gained in the AE signal.

Figure 5 shows the record of one AE sensor placed in the middle of cylinder's body which has no defect. From the graph is visible that AE sensor detects signal only if the position is in ejected mode. Cylinders were operated with frequency 1Hz. Each event is characteristic by it's dominant frequencies which are repeatable in range from 20 to 150 kHz. This measurement was repeated for other 14 pistons of the same type and results of AE signals were same.

As following, measurement for cylinder with defect was undertaken. For this measurement was used same cylinder as before but with created defect. In particular O-ring on position 10 had been cracked (Fig. 4). Moreover linear potentiometer was used to measure position of the piston rod during operating. Results plotted in Fig. 5 show that signal events occur when piston goes from one position to another. Signal from potentiometer is plotted on the right side of the graph in Fig. 5 and symbolize movement of the cylinder. Defect created on piston cause move of range of dominant frequencies to higher magnitudes. Those frequencies have also repeatable characteristic for each movement of the piston like in Fig. 5 where are pictured 3 operational circles of the piston.

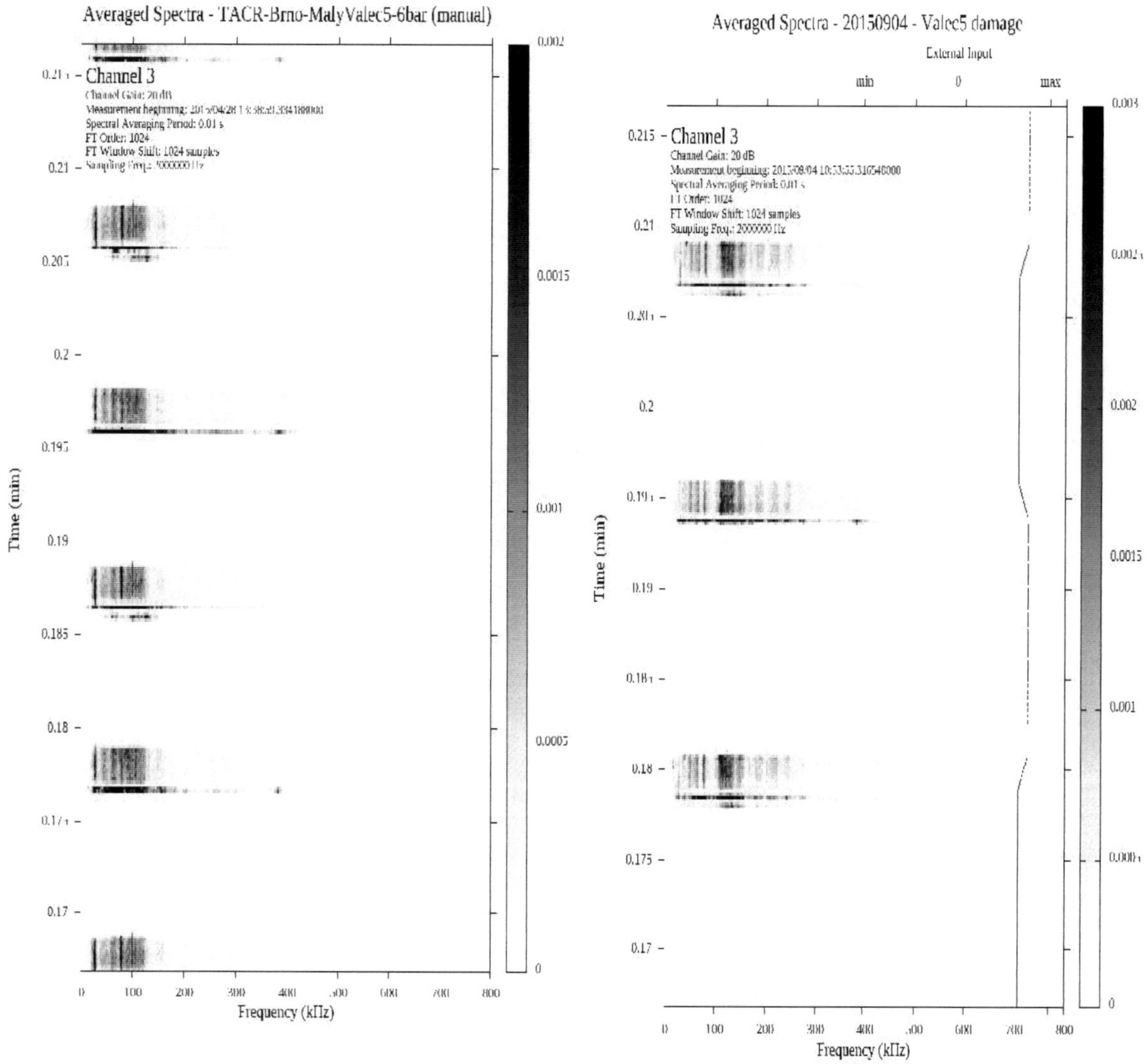

a) b)

Figure 5. Spectral analysis in particular time - differences between signal spectrums obtained from signals of pistons without defect (a) and with defect (b).

In order to highlight differences between signals spectrums obtained from signals of pistons without defect and with defect (Fig. 5) the spectral analysis in particular time were plotted. From results on graphs in Fig. 6 and Fig. 7 is visible the difference caused by defect created on the cylinder. Domain frequencies were moved from area under 100 kHz to area above 100 kHz.

2.3 Conclusion

In this part of our paper were presented results of testing pneumatic cylinders by acoustic emission. By this method was verified that cylinder of same type have same characteristic of AE signal during operating and this can be used as a pattern for a comparison of signals from defected cylinders. The evaluation of the influence of different types of cylinders disturbances on the nature of signal changes is currently underway.

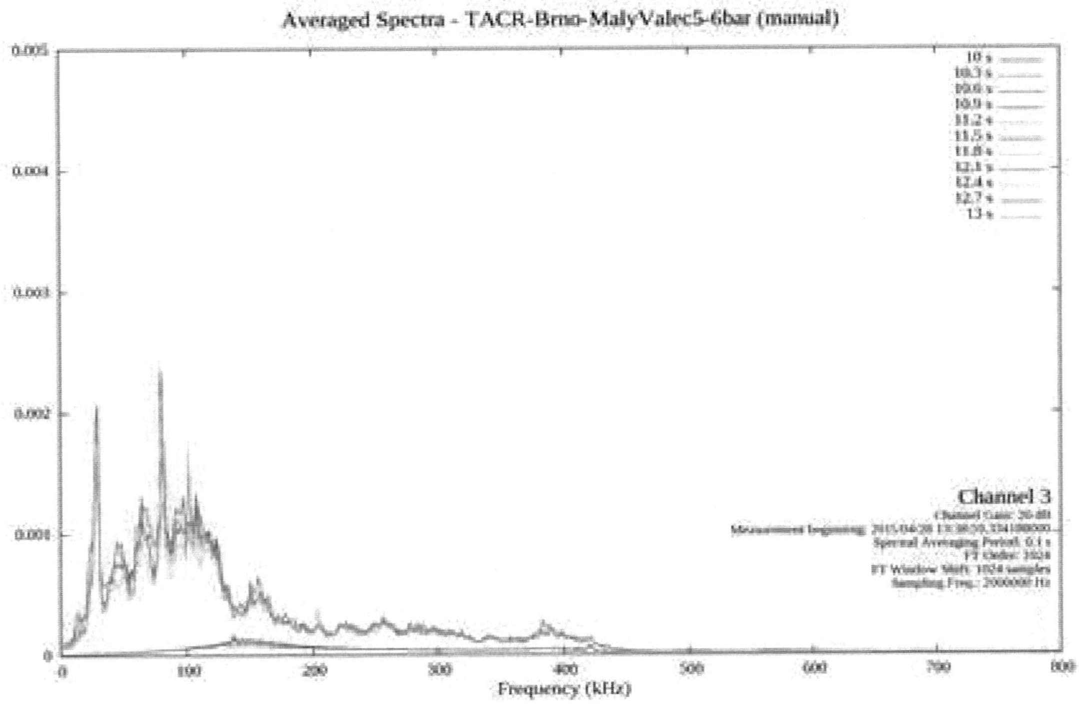

Figure 6. Signal spectrums from piston without defect

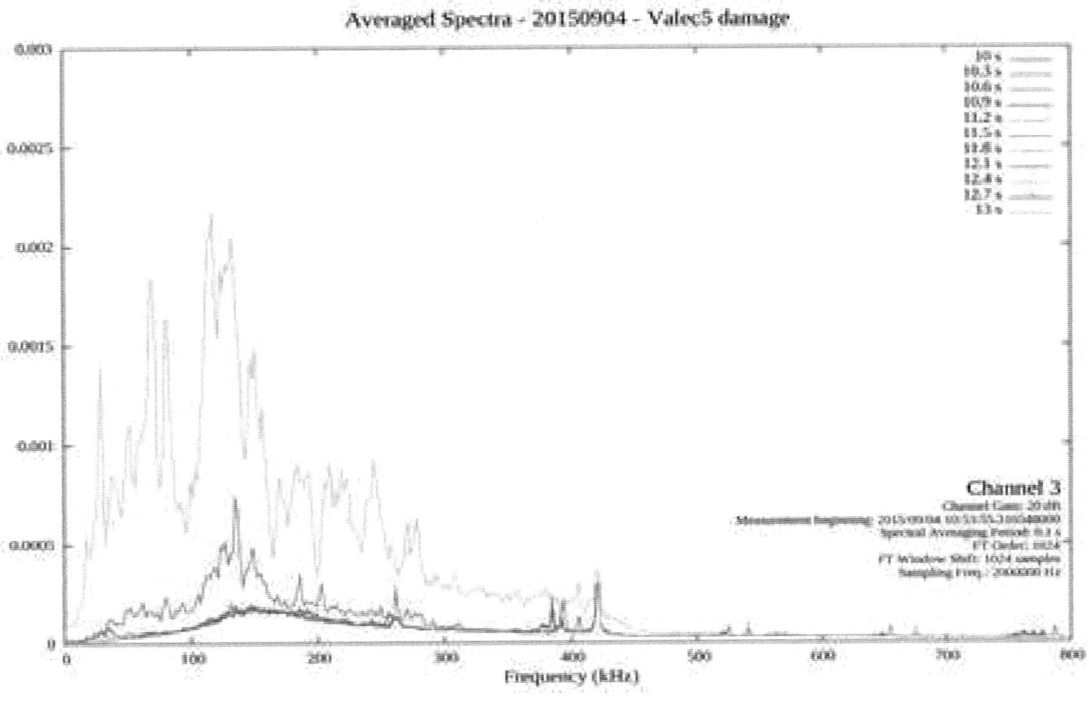

Figure 7 Signal spectrums from piston with defect

3. Breakdown of gas insulated lines of very high voltage

The main objective of this new area of researches in our laboratory was the measurement of the output during the tests to determine whether it is possible to use AE for detection of the partial discharges occurring in increasing voltages up to 650 kV in the entire system of gas insulated lines. Undesirable electrical discharges occur within the conduits on various impurities or bumps on the surface in this case. The output control is done on the assembly of several connected parts. In the event of a discharge it is necessary to disassemble the entire unit and all parts must be inspected. To accelerate repairs would be very useful to identify the faulty component and locate the place of discharge.

3.1 Experimental results

Several sensors were placed onto surface of power line and signal during operating was obtained. The measurement was provided by system DAKEL-XEDO, which is used for analysis of signal AE and digital record of signals.

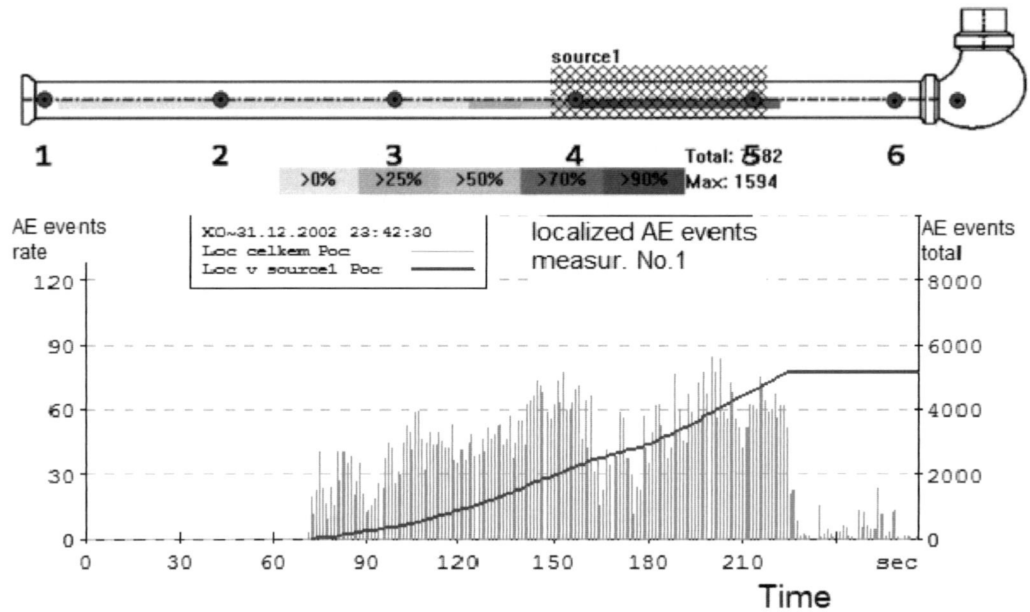

Figure 8. Demo of location map (top) and time curves of AE localized events (bottom) at output test no.1.

Measurements showed that by using more AE sensors fitted to the surface of power line is possible accurately localize position of breakdown. Due to the testing of a long pipe (about 10 m) in which the AE registered discharge was carried out localization of discharge. Repeated measurements showed formation of discharge between the transducers Nos. 4 and 5. Measurements were not successful between parts mutually insulated. This result, however, is useful for reliable determination of defective parts in the whole assembly.

3. 2 Conclusions

a) Acoustic emission method can be used for basic identification of metal parts of EHV lines in which occurs partial discharges,

b) When using multiple sensors on components it is possible to locate the place of the discharge,

c) Eventual discharges into supporting insulators were not been registered at the current setup of the AE system. The cause is probably an effective damping material of these insulators. On the other side this feature increases the reliability of discharges identification in other parts.

4. Tests of resistance of explosives to friction

We also received a very specific requirement to verify the possibility of evaluating the resistance to friction of explosives. These tests are carried out according to EN 13631-3 - Explosives for civil uses - Part 3: Determination of sensitiveness to friction of explosives. The test is conducted on a simple device in which the explosive is placed on the ceramic plate (Fig. 9). Tested explosive is placed into a porcelain plate with a roughened surface with grooves. Standard evaluation is done only sensually and is largely dependent on the experience of the test engineer. The aim of the project was to verify the possibility of an objective evaluation of these tests using AE method.

4.1 Experimental results

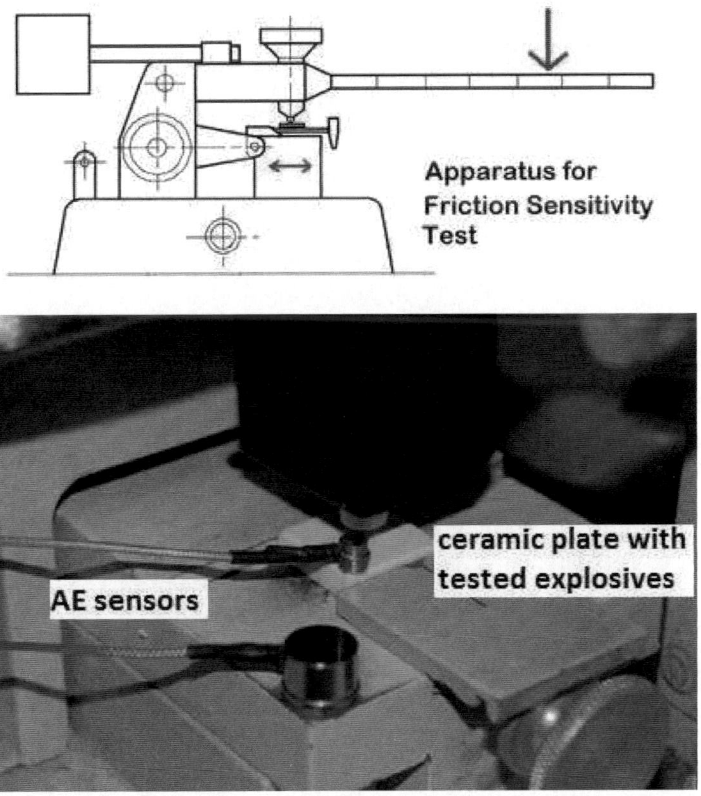

Figure 9. Scheme of the test equipment and detail of the location of AE sensors

AE sensors were placed directly on a porcelain plate and on the clamping plate of the testing machine. By comparing the individual emission events were observed differences in the nature of the signal, when an explosive responded compared to a negative result. During successful tests were detected oscillations of high amplitude and high frequency in the second part of the hit represents the retraction and higher signal amplitude at the beginning of the hit (see Fig. 10).

It was conducted only a limited number of tests and therefore it is necessary to validate that findings of experiments with an optimal setting of apparatus based on preliminary measurements and make a deeper analysis of the signal in the frequency and time-frequency domain. Overshoots will also be evaluated via the detection thresholds and distribution of signal energy. For practical use, it is necessary to find the optimal placement of sensors and resolve the problem with their fastening on the surface the test ceramic plate.

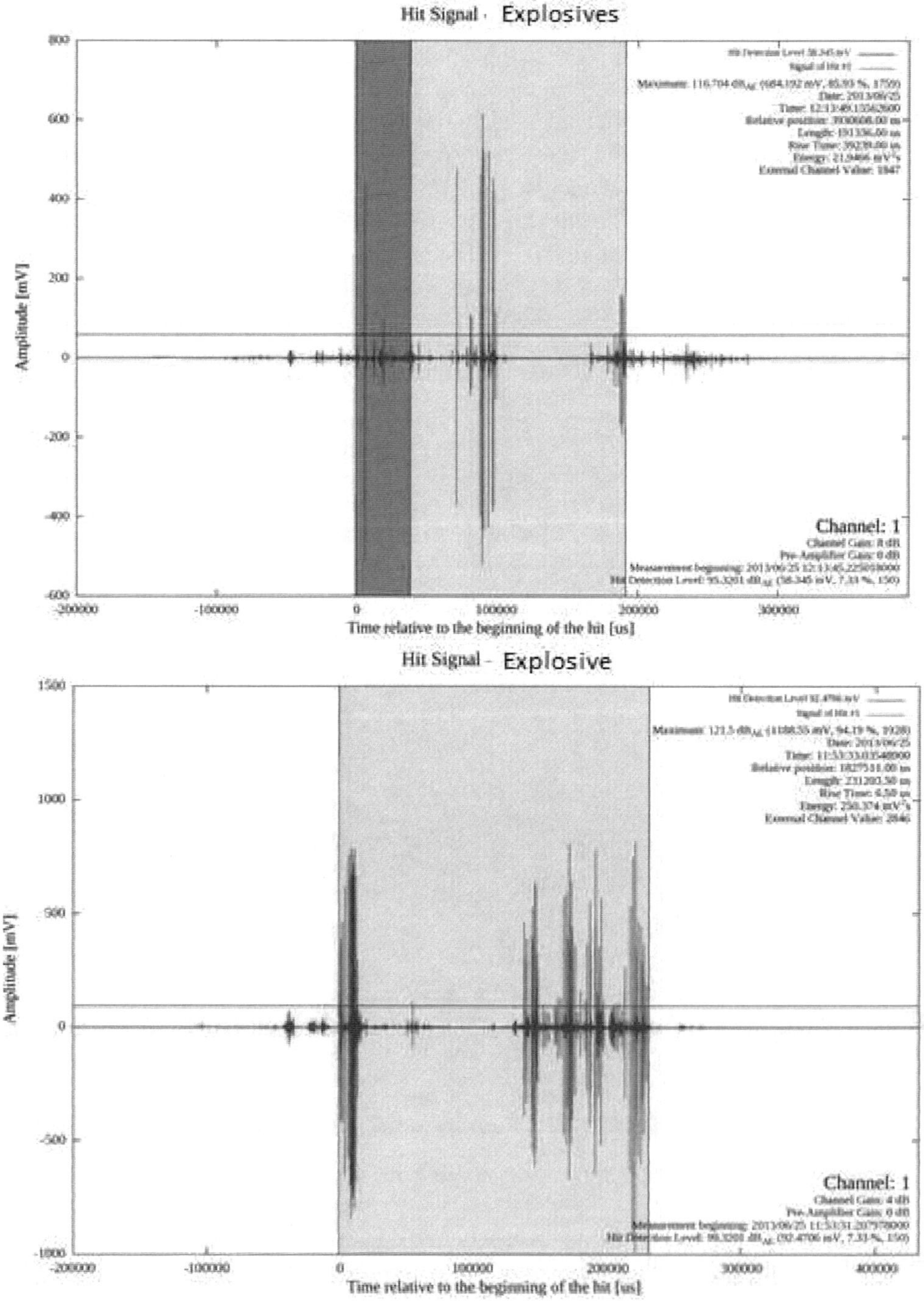

Figure10. Comparison of AE signal hits for not responding (above) and reacting explosives (below).

5. Conclusions

Acoustic emission method was used in three areas of non-standard technical diagnostics, which were made in the Laboratory of AE at FME BUT recently: a) applications of AE for function diagnostic of pneumatic devices, b) usability of AE for identification and localisation of undesirable discharges in gas-insulated conductors for high-voltage substations, c) tests for determination of sensitivity of explosives to friction. In connection with the actual solution of the project of TA CR, we will focus on identifying the area of development of damage of pneumatic cylinders. In collaboration with other workplaces will be developed the professional apparatus, which will be included in the final inspection at the manufacturer. Other developed instruments will be used for external diagnostics among users of pneumatic cylinders in the production lines and transportation facilities.

The presented results confirm the wide range of new industrial applications acoustic emission method. For the application in routine practice, the basic problem is the high sensitivity of the method and the need for suppression of disturbing signals. Certain problem is then also a need to ensure the long-term repeatability of the measurement conditions.

Acknowledgements

This work is an output of research and scientific activities of the Technology Agency of the Czech Republic project No *TA04011374*" The new system is non-invasive diagnostics of pneumatic and hydraulic components" and was also supported by NETME Centre (New Technologies for Mechanical Engineering), Reg. No. CZ.1.05/2.1.00/01.0002 and, in the follow-up sustainability stage, supported through NETME CENTRE PLUS (LO1202) by financial means from the Ministry of Education, Youth and Sports under the „National Sustainability Programme I".

References

[1] DICKEY, J.; DIMMICK J. and MOORE, P.M.: Acoustic measurement of Valve Leakage Rates. *Material Evaluation*. 1978, No. 36, p. 67-77.

[2] PLLOCK, A.A.; HSU, S.Y.S.: Leak Detection Using Acoustic Emission. *Journal of Acoustic Emission*. 1982, No. 4, p. 237-243.

[3] CHEN, X.; LI, B.: Acoustic emission method for tool condition monitoring based on wavelet analysis. *Int J Adv. Manuf. Technol.* 2007, No. 33, p. 968–976. DOI: 10.1007/s00170-006-0523-5.

[4] LEE, S.; PARK, J. H.; YOO K. B.: Evaluation of Internal Leak in Valve Using Acoustic Emission Method. *Key Engineering Materials*. 2006, p. 661-664.

[5] KAEWWAEWNOI, W, A.; PRATEEPASEN, P.; Investigation of the relationship between internal fluid leakage through a valve and the acoustic emission generated from the leakage. *Measurement*. 2010, No. 43, p. 274-282.

[6] MAZAL, P.; NOHÁL, L.; KOMENDA, L.: Diagnostics of damage development of bearings and gears by acoustic emission method. International Journal of Microstructure and Materials Properties. 2014. 9(3/4/5). p. 433-447. ISSN 1741-8429.

[7] MENTL, V., VLASIC, F., BARTKOVA, D., MAZAL, P.: Application of AE for Identification of Differences in Fatigue Damage of Selected Materials for Power Plants. Key Engineering Materials (print). 2014. 627(313). p. 313 - 316. ISSN 1013-9826.

[8] MAHMOUD, H.; VLASIC, F.; MAZAL, P.: Simulation of operational loading of pressure equipment by means of non-destructive testing. In METAL 2015, Brno, Tanger Ltd. 2015. p. 321 - 327. ISBN 978-80-87294-58-1.

Propagation Property of Leaky Surface Acoustic Wave on Monocrystalline Silicon Crystal using Large Aperture Line-Focused PVDF Transducer

Sung-Jun KIM, Yoon-Ho NOH, Min-Cheol SON, Dong-Jin LEE and Young H. KIM

Applied Acoustics Laboratory, Korea Science Academy of KAIST; Pusan, Korea
e-mail: sjty97@naver.com , lionnoh@naver.com, smc1515115@naver.com, brain0626@naver.com, yhkim627@kaist.ac.kr

Abstract

Nondestructive evaluation of crystal structure is one of important tasks. The ultrasonic wave speed is related to the elastic properties of solid specimen, and it is varied with the propagation direction in anisotropic solids. In the present work, three monocrystalline silicon of different crystal structures, [100], [110] and [111], are employed as anisotropic media. The speeds of Leaky Acoustic Surface Wave (LSAW) on the monocrystalline Silicon were measured by using large aperture line-focusing PVDF transducer and angular profile of LSAW speeds was obtained by rotating the specimen. Finally obtained angular profile of LSAW speeds were compared with elastic moduli of anisotropic plates. As a result, it is found that the angular profile of LSAW speeds offers information of crystal structure of silicon.

Keywords: Large aperture line-focusing transducer, Leaky acoustic surface wave, Silicon

1. Introduction

Nondestructive evaluation of crystal structure is one of important tasks. The ultrasonic wave speed depends on the density and the elastic properties of solid medium. In addition, the propagation speeds are varied with the propagation direction in the anisotropic solids [1].

Leaky surface acoustic wave (LSAW) propagates along the surface of solid specimen at the interface between a solid and fluid. Speed of LSAW depends on the elastic properties of solid. Anisotropic solids such as single crystals have the anisotropic elastic properties, so that speed of LSAW depends on propagation direction. Large aperture line-focusing PVDF transducers are useful to measure LSAW and showed the great potential to be applied to materials characterization [2, 3].

In the present work, the propagation property of LSAW has been investigated by using line-focusing PVDF transducer with large aperture. [100], [110] and [111] monocrystalline silicon was employed as anisotropic solid specimens. The speed of LSAW was measured from the arrival times of directly reflected echo and LSAW signal, and its angular dependence was obtained by rotating the specimen. The elastic moduli of specimen were compared with obtained angular profile of LSAW speeds.

2. LSAW measurement principle

Figure 1 shows ray representation of line-focusing PVDF transducer. Path I is the directly reflection from the surface of specimen, and Path II involves LSAW, that is incident at the critical angle for Rayleigh wave and propagates along the surface and finally incidents as a compressional wave by mode conversion at symmetric point.

As varying the defocusing length z, the time interval of surface wave and reflect wave's separation are changed, and the speed of the surface wave can be determined by calculating the slope of the time difference versus defocusing length z difference. Then the velocity of the leaky surface acoustic wave (v_{lsaw}) can be evaluated by following formula.

$$t_w = \frac{2(R-z)}{v_w},$$

$$t_s = \frac{2(R-z/\cos\theta)}{v_w} + \frac{2z\tan\theta}{v_{lsaw}},$$

$$v_{lsaw} = \frac{v_w}{\sqrt{1-\left(1-\frac{\Delta t v_w}{2z}\right)^2}}$$

Where t_w and t_s are elapsed times for path I and that for path II, respectively.

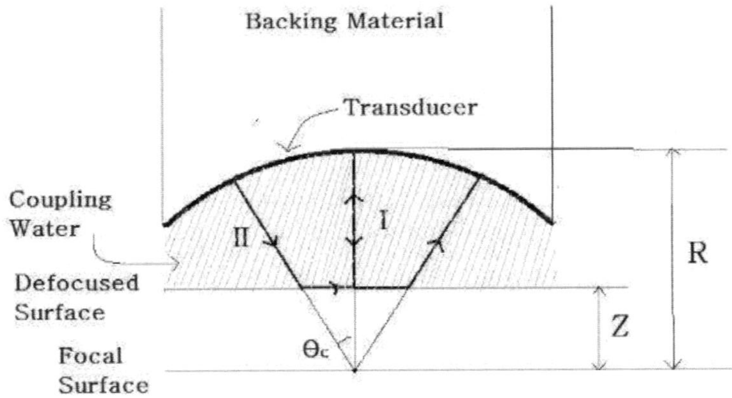

Figure 1: Two ray paths in ultrasonic pulse-echo signal.

3. Experiments

A PVDF transducer with radius of 25 mm was fabricated by using PVDF strip with thickness of 28 μm, width of 11 mm and length of 52 mm. Epoxy backing material has acoustic impedance of 3.8 MRayl and transmittance between PVDF and backing materials is 99%, so that reflection in bonding layer is minimized. Peak frequency of the fabricated transducer was about 5 MHz.

Panametrics NDR-5800 was employed as an ultrasonic pulser-receiver. Three dimensional position of transducer was controlled by x, y and z stage and NI Motion Controller MID-7604. Pulse-echo signals were captured by NI PXI-5124 200 MHz digitizer.

4. Results and discussion

4.1 Pulse-echo signal as varying defocusing length

Pulse-echo signals obtained for various defocusing length in 1 mm step are shown in Figure 2. Two echoes of directly reflected wave and LSAW were clearly observed in pulse-echo signals, and two echoes moves with the different slopes as decreasing defocusing length. The slope of LSAW is 3.49 mm/μs, and the speed of the LSAW is determined as 2.43 mm/μs.

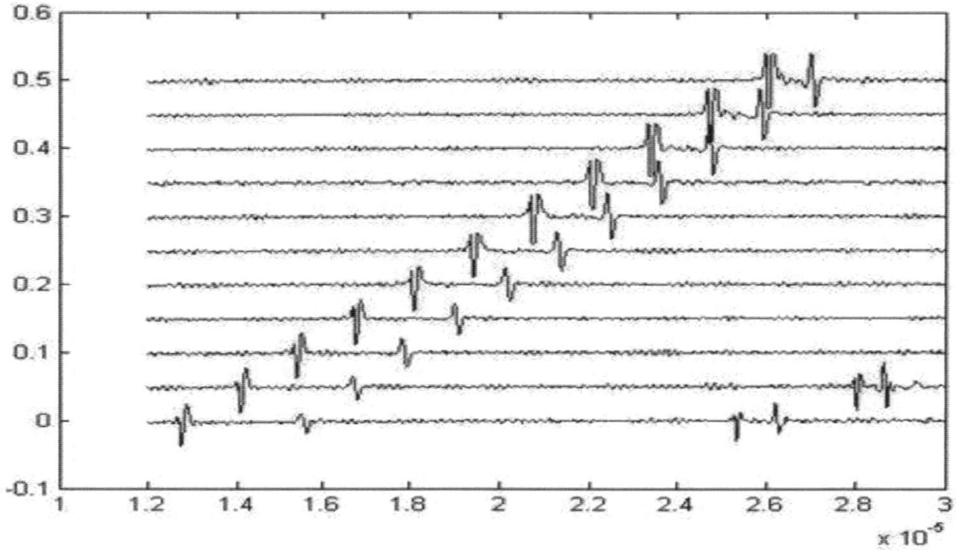

Figure 2: Pulse-echo signals for different defocusing length in a steel specimen.

4.2 Comparison of speeds of LSAW and shear modulus of monocrystalline silicon

Figures 3, 4 and 5 show angular plots of pulse-echo signals, elastic moduli of [100], [110] and [111] monocrystalline silicon, respectively. Angular Pulse-echo signals were obtained by polar plot of pulse-echo signals. Peak positions were detected and marked in polar plot in second graphs and wave speed (inner circle) and slowness (outer circle) are shown in third plot. The rightmost graph shows Young's and Shear moduli of monocrystalline crystal.

From the experimental result, it was confirmed that the shape of angular plot of echo signals, peak positions of that signals are similar to Young's and Shear moduli of monocrystalline crystal. This is because shear modulus is proportional to the speed of a shear wave, and the speed of the Rayleigh wave is 80~90% of the shear waves speed. It implies that angular profile of LSAW speeds offers some information on the crystal structure of silicon.

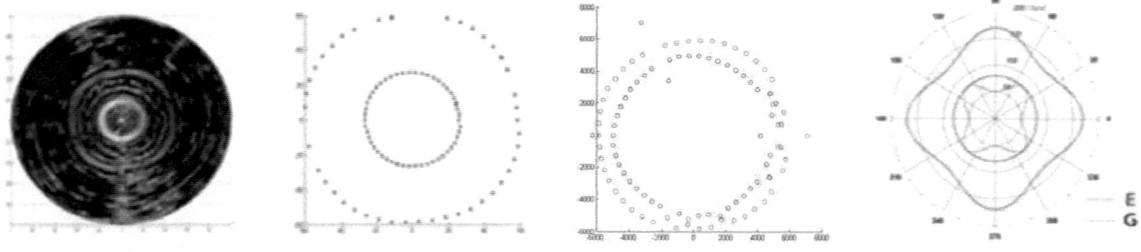

Figure 3: [100] monocrystalline silicon: Angular plot of pulse-echo signal in grey scale, peak positions of pulse-echo signal, speed and slowness curve graph, Young's and shear moduli of monocrystalline silicon.

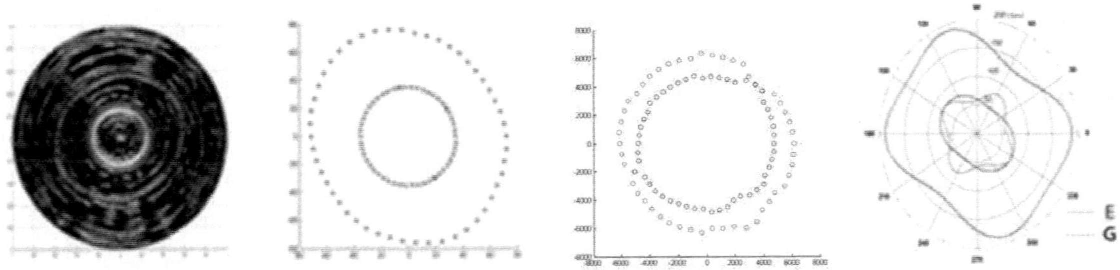

Figure 4: [110] monocrystalline silicon: Angular plot of pulse-echo signal in grey scale, peak positions of pulse-echo signal, speed and slowness curve graph, Young's and shear moduli of monocrystalline silicon.

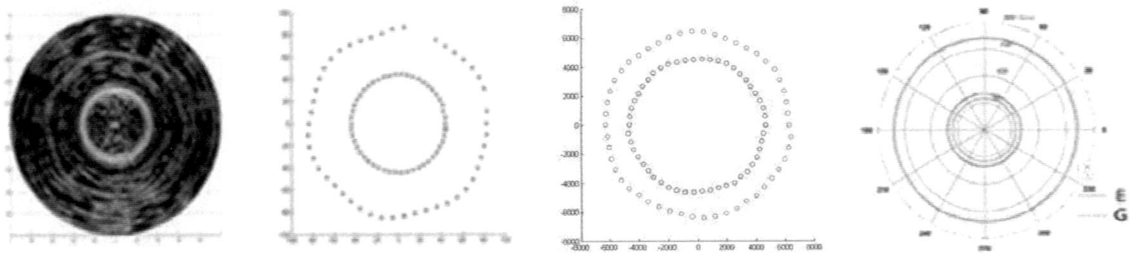

Figure 5: [111] monocrystalline silicon: Angular plot of pulse-echo signal in grey scale, peak positions of pulse-echo signal, speed and slowness curve graph, Young's and shear moduli of monocrystalline silicon.

4. Conclusions

In the present work, the propagation property of LSAW on Silicon wafers have been interrogated by using fabricated PVDF transducer. As a result, it is found that the angular profile of LSAW speeds offers information of crystal structure of silicon.

Acknowledgements

This work was supported by the Research & Education Program at the Korea Science Academy of KAIST with funds from the Ministry of Science, ICT and Future Planning. Authors also thank to Dr. Jeong-ki Lee for valuable assist during fabrication of transducer.

References

1. H. F. Pollard, 'Sound waves in solids.' (Pion Ltd, London, 1977)
2. D. Xiang, N. N. Hsu and G. V. Blessing, 'The design, construction and application of a large aperture lens-less line-focus PVDF transducer,' Ultrasonics, Vol 34, pp 641-647, 1996
3. H. -J. Yoon, K. -L. Ha, M. -J. Kim and J. -R Yoon, 'LSAW Velocity Measurement by Using a PVDF Ultrasonic Transducer,' J. Acoust. Soc. Kor., Vol 20, pp 62-67, 2001

Shortest Ways Finding in Discrete Bodies

Milan CHLADA[1], Zdenek PREVOROVSKY [2]
NDT Laboratory, Institute of Thermomechanics AS CR, v.v.i.,
Dolejskova 1402/5, 182 00, Prague 8, Czech Republic;
Phone: +420 266 053 144, Fax: +420 286 584 695;
e-mail: [1]chlada@it.cas.cz, [2]zp@it.cas.cz

Abstract
Estimation of the shortest wave-path is a crucial task not only in the field of acoustic emission (AE) source location. The choice of most suitable solution method largely depends on the kind and form of the body shape definition. The differential geometry provides useful means in a case of mathematically well-describable surfaces. Further, the graph theory is capable to solve the problem when there exists an interpretation of a body as a system of mass points (nodes) and their interconnections (edges), i.e. a graph. Another recently designed method was inspired by Huygens principle as easy to implement alternative suitable for discretized 2D or 3D 'bitmap' pictures of a body. This approach also enables tracing of elastic waves propagating in structures of complicated shapes without mathematical models. Such approach enables collecting sufficient amount of representative data for AE source location algorithms based on database searching or artificial neural networks. The new method already showed its advantages and potentialities in practical tasks as discussed.

Keywords: Shortest ways finding, discrete bodies, acoustic emission, non-destructive testing.

1. Introduction

1.1 Problem definition

Many problems in real world lead to estimating of shortest way between two given points. Anyhow it seams to be a simple task, there are many complicated cases in practice, as e.g. measuring the distance on a surface of complex bodies. In the field of non-destructive testing (NDT) we meet many situations as illustrated in fig.1 when there is no straight line connecting points A and B. For reliable results of acoustic emission (AE) source location it is needed to trace elastic wave while propagating from initiation point (A) to sensor (B) and estimate the distance between them.

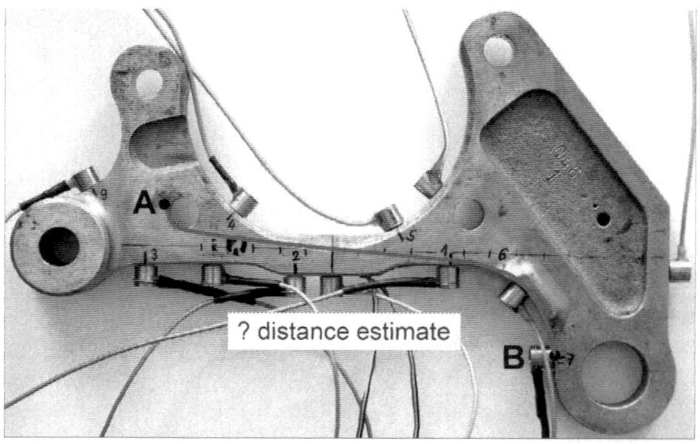

Figure 1. Example of elastic wave tracing on a real aircraft structure part.

The choice of most suitable method for solving the problem depends largely on the kind or form of the body shape definition. It can be given by simple hand drawing on a paper, as a photograph, as 2D or 3D computer sketch, otherwise as a mathematical model etc.

1.2 Differential geometry solution

The theory of differential geometry is suitable and provides necessary apparatus for case of mathematically well-describable surfaces [1]. It supposes that the surfaces are mathematically described as a conjunction of several simple parametric shapes. By solving the system of geodesic equations it finds geodesics (shortest ways) on simple and combined surfaces. The advantage of this approach can be its good precision, however, the method is relatively complicated to implement and there is necessary need to discretize the borders between the two neighbouring model shapes.

1.3 Graph theory solution

The graph theory is capable to solve the problem when there exists an interpretation of a body as a system of mass points (nodes) and their interconnections (edges), i.e. a graph. In graph theory, the finding of shortest path consists in searching a path between two vertices (or nodes) in a graph such that the sum of the weights of its constituent edges is minimized [2]. Not only for graphs representing a body just by rectangular mesh points (see left part of fig.2) there may be more than one different shortest paths, all of the same length. The distance between two points in a grid based on a strictly horizontal and/or vertical path is called "the Manhattan distance" as the simple sum of horizontal and vertical components. Anyway, such approach gives us only very rough estimate of actual distance due to entirely four allowed path directions.

To increase the precision it is needed to suppose more possible directions, i.e. the higher number of graph edges (see right part of fig.2) and consequently very complex graph. There may be also a problem of construction of an adequate graph (with holes in material etc.) in this case similar with shortest path finding itself.

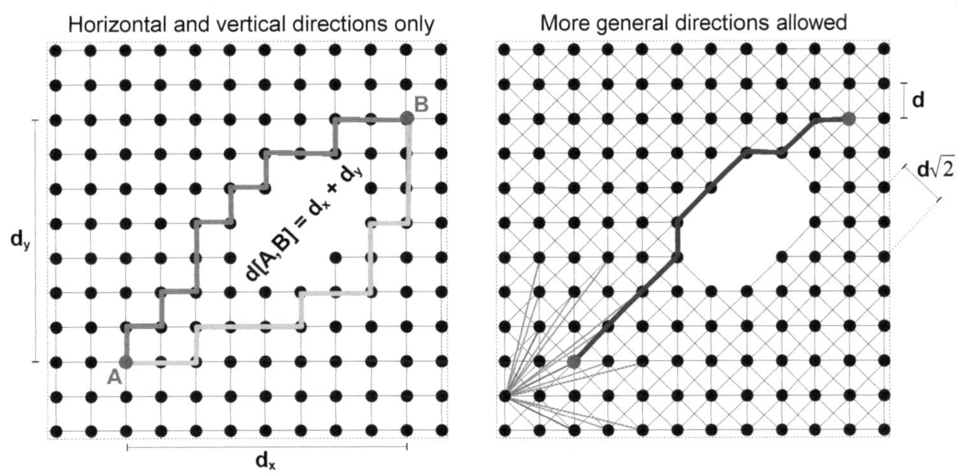

Figure 2. Demonstration of graph designs.

The typical and well known algorithm for shortest way estimation in graphs is Dijkstra's algorithm [3]. The algorithm exists in many variants. Dijkstra's original version finds the shortest path between two nodes, but a more common variant fixes a single node as the "source" node and finds shortest paths from the source to all other nodes in the graph, producing a shortest path tree.

2. Huygens' principle based solution

2.1 Motivation

The new method of shortest ways finding is further generalization of Dijkstra's algorithm and can be considered as a basic model of wave propagation. It can be regarded as a simple to implement alternative suitable for discrete bodies derived from 2D or 3D bitmap pictures. Such approach also enables the tracing of elastic waves propagating through bodies of complicated shape. It is inspired by Huygens' principle: each point on an advancing wave front may be considered to be a new point source generating spherical Huygens' wavelets. For known location of a wave front at one time, the location of the wave front at a slightly later time $t+\Delta t$ can be determined by drawing spherical arcs centred at points along the wave front. Each spherical arc is a Huygens' wavelet. The radius of each arc $(c\Delta t)$ is given by the distance that sound travels in that small amount of time.

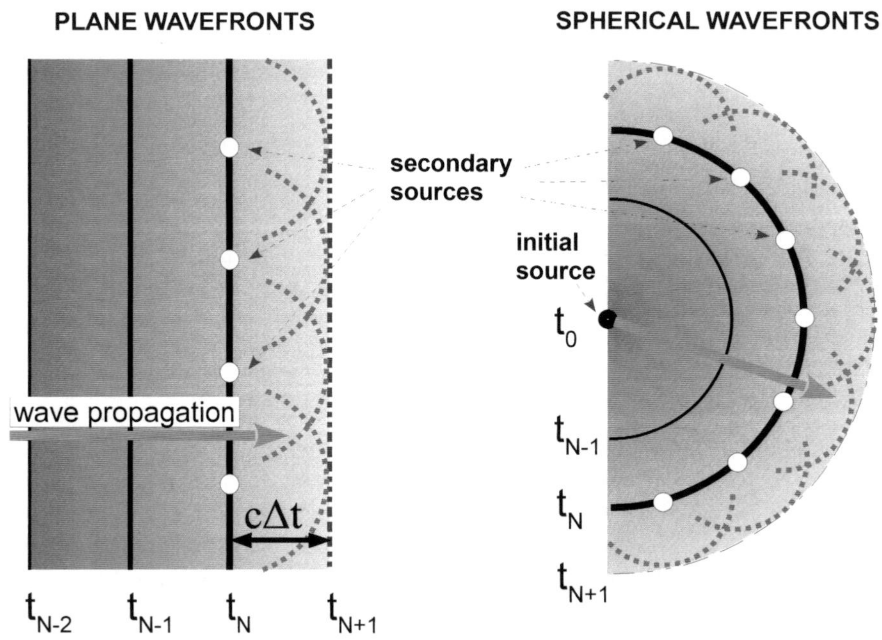

Figure 3. Illustration of Huygens' principle.

2.2 "Flooding" propagation

Let's visualise the principle of new method as a "flooding" propagation in a bitmap. The similarity with the graph theory is simple and clear. Every centre of a pixel can be taken as a node, while the edges can be considered in the meaning of specific body topology. All pixels representing the body can be marked in each step of algorithm as "not reached", "already reached", "actual" and "next" (see fig.4). For every actual pixel, the "next pixels" from the set of body pixels are searched. Those are potentially the points within some fixed diameter around selected actual pixel, representing the pattern of specific level (see fig.4). The "out-of-mass" and "a hole over-jumping problem" pixels are excluded from the pattern as described in following chapters.

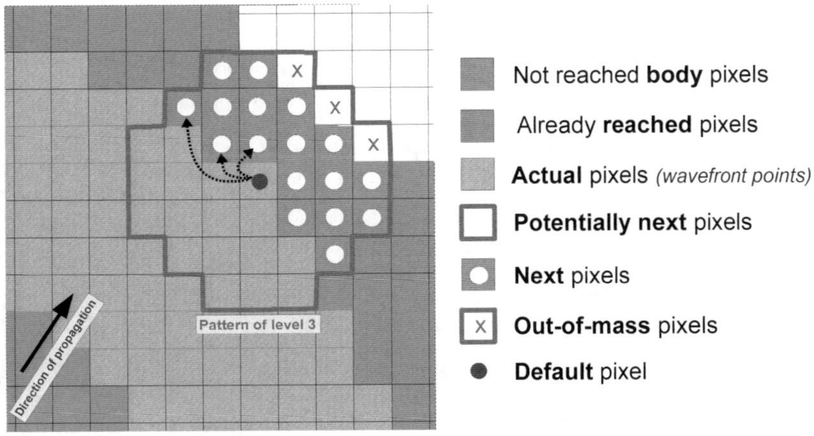

Figure 4. Visualisation of flooding.

2.3 "A hole over-jumping" problem

Similarly as for well identifiable "out-of-mass" points, it is needed to stop propagation in a space angle behind each reached pixel out of mass to avoid over-jumping of a material hole (see fig.5). The width of the space angle should be narrower with the longer distance from particular "out-of-mass" point to default pixel (see fig.6). By such restriction, the propagation continues not in the direct way through material hole, but passes by the problematic area.

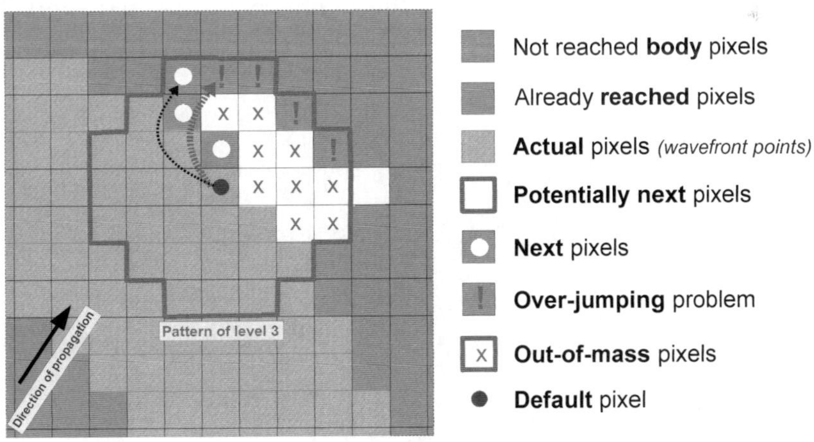

Figure 5. "A hole over-jumping" problem illustration.

Solving the problem which points to exclude from the "potentially next" pixels set is crucial step of the algorithm. The general rule can be based on polar coordinates of "potentially next" points (centres of pixels) in circular pattern (see fig.4-6). The pole of coordinate system is set to "default pixel". Afterwards, it is possible to exclude all points having with some tolerance similar angular coordinate and simultaneously higher radial coordinate than all out-of-mass points. This general rule works well even for 3D data. The level of tolerance is indirectly proportional to the radial coordinate of particular pixel. That means the forbidden section is wider for pixels closer to default point (pole) and vice versa. Some examples of "forbidden pattern pixels" for case of different pattern diameters and "out-of-mass" pixels are shown in fig. 6.

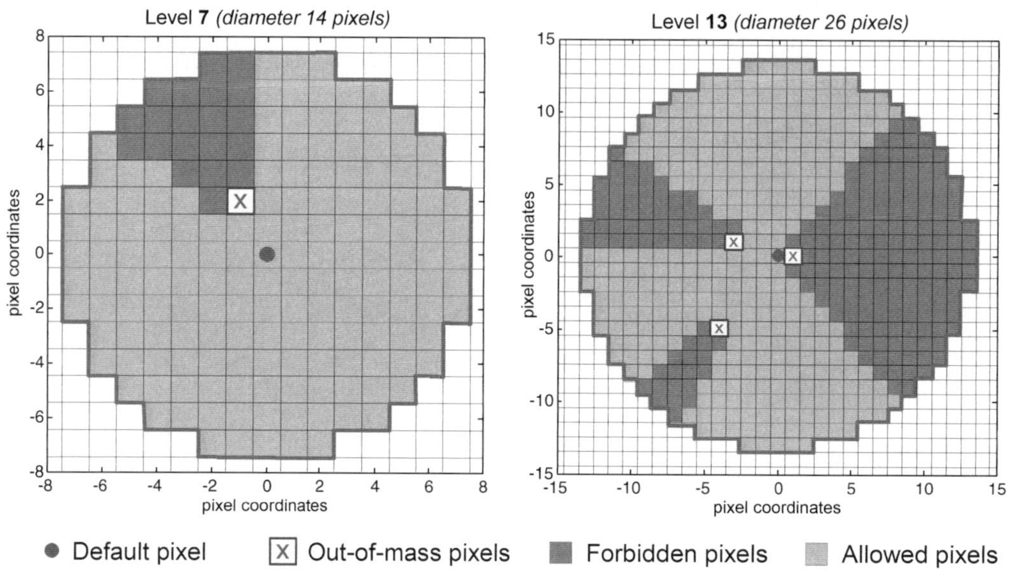

Figure 6. Examples of forbidden pattern sections.

2.4 Discrete distance computation

During each propagation step it is possible to gradually compute the estimate of the distance from initiation to actual pixel. The method is to jump from every actual pixel to all approved next pixels from the pattern (see fig.7).

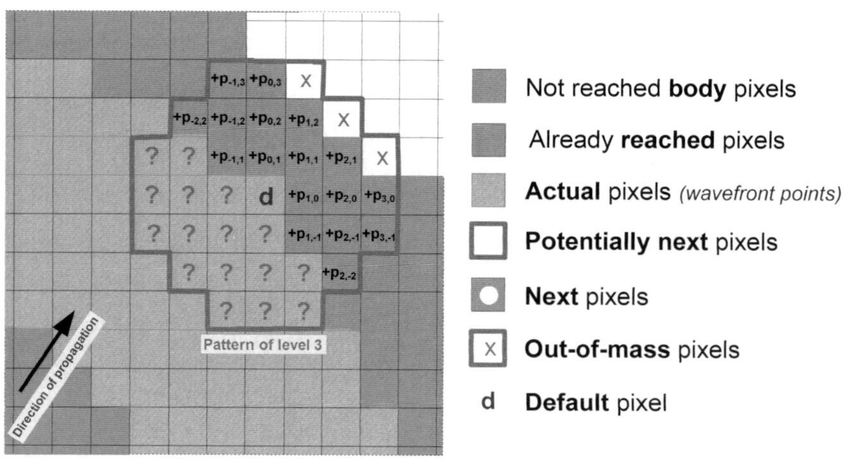

Figure 7. Scheme of actual distance "d" updating.

Let's assume the actual pixel has relative position indexes *[0,0]* and the information about the distance *"d"*. Indexes *[i,j]* represent the relative position of a pixel within the pattern, while number $p_{i,j} = \sqrt{i^2 + j^2}$ is the distance of selected target pixel with coordinates *[i,j]* from actual pixel. In case of a jump onto "not reached pixel", distance $d+p_{i,j}$ is set there. If the target pixel already has an information about previously set distance *d'* that is greater than the number $d+p_{i,j}$, the actual distance *d'* is overwritten by this value $d+p_{i,j}$.

3. Summary

Recently designed method for finding shortest ways was inspired by Huygens' principle as easy to implement alternative suitable for discretized 2D or 3D "bitmap" pictures of a body. This approach also enables tracing of elastic waves propagating in structures of complicated shapes without mathematical models. Such approach enables collecting sufficient amount of representative data for AE source location algorithms based on database searching or artificial neural networks. The new method already showed its advantages and potentialities in practical tasks as monitoring of fatigue crack growth in aircraft wing flange [4] or location of acoustic emission sources in geometrically sparse structures [5].

Above described algorithm can be summarized into following steps:

INITIATION:
- set zero distance to all mass pixels
- set initial propagation pixel(s) as actual (i.e. wave source point)

PROPAGATION:
- for each actual pixels find potential next pixels (laying inside circular pattern)
  → find pixels to forbid *("out-of-mass" or "over-jumping problem" points)*
  → set appropriate new distance number to all allowed pixels of circular pattern
  → set next pixels as actual

REPEAT PROPAGATION until all mass pixels are reached...

Acknowledgements
The work was supported by the institutional support RVO: 61388998 (CR).

References

1. V Kus, M Zavesky and Z Prevorovsky, 'Acoustic Emission Defects Localized by Means of Geodetic Iterative Procedure - Algorithms, Tests, AE Experiment', Proc. of 30th European Conference on Acoustic Emission Testing & 7th International Conference on Acoustic Emission. A Gallego and K Ono (ed.), Granada, University of Granada, 2012, ISBN 978-84-615-9941-7.
2. B V Cherkassky, A V Goldberg and T Radzik, Tomasz, 'Shortest paths algorithms: theory and experimental evaluation', Mathematical Programming. Ser. A 73 (2): 129–174. doi:10.1016/0025-5610(95)00021-6. MR 1392160.
3. E W Dijkstra, 'A note on two problems in connexion with graphs', (PDF). Numerische Mathematik 1: 269–271. doi:10.1007/BF01386390.
4. M Chlada and Z Prevorovsky, 'Remote AE Monitoring of Fatigue Crack Growth in Complex Aircraft Structures', Proc. of 30th European Conference on Acoustic Emission Testing & 7th International Conference on Acoustic Emission. A Gallego and K Ono (ed.), Granada, University of Granada, 2012, ISBN 978-84-615-9941-7.
5. Z Prevorovsky and M Chlada, 'Localization of acoustic emission sources in geometrically sparse structures', Proc. of EWGAE 2014. Berlin, DGfZP, ISBN 978-3-940283-63-4.

Study on Ultrasonic Wave Propagation in Anisotropic Plates Using Large Aperture Line Focusing Transducer

Jaewon OH, Donghyun KIM, Inyoung JUNG, Kyubeom Han, Young H. KIM

Applied Acoustics Laboratory, Korea Science Academy; Busan, Korea
e-mail: curie720@naver.com, ggjjvb@gmail.com, dlsdud1999@naver.com, smarthkb98@gmail.com, yhkim627@kaist.ac.kr

Abstract
Ultrasonic waves are widely used in nondestructive testing and in materials characterization. Guided ultrasonic waves are used to evaluate the properties of a thin layer. Dispersion should be understood in order to proper application of guided waves. In the present work, directional properties of Lamb wave propagation in anisotropic materials were investigated. A large aperture line focusing transducer was fabricated and used to generate and to receive a ultrasonic pulse on a [100], [110] and [111] silicon wafer [6]. The $V(f,z)$ graph was obtained from the waveforms acquired in the experiments and the dispersion curve was derived from calculations. As rotating the specimen, a dispersion curves in the propagation direction were determined. Simulation on Lamb waves on anisotropic plates were also carried out in order to compare with the experimental results.

Keywords: Guided wave, dispersion, anisotropy, large aperture line focusing transducer

1. Introduction

Ultrasonic waves are widely used in nondestructive testing for flaw detection in products and material characterization. Guided waves are preferably used to examine a thin plate or a layered material. In a thin layered material, the two surfaces work as a boundary between two different media and waves are reflected at the boundaries. The reflected and mode-converted waves are combined and formed an elastic wave across the plate; this is called the plate wave. Lamb wave is a kind of plate waves which is formed by addition longitudinal wave and perpendicular transverse waves. Lamb waves have numerous modes. The mode of guided waves is defined by the frequency of the waves, the thickness of the plate and the incident angle. The phase velocity and group velocity differ depending on the mode. Such curves play the most important role in quantitative analysis of guided waves [1].

In order to apply guided waves to real problems, it is required to understand the dispersion of the waves thoroughly. If the wave is dispersive, the waveform changes as the guided waves propagate, so that it is harder to measure the velocity of the wave. For calculating the velocity of a wave without dispersion, one of the most widely used methods is the time-resolved method. Based on the fact that the time gap between the wave that is reflected perpendicularly from the surface and the wave that propagated through the specimen is proportional to z - the distance between the focus of the transducer and the surface of the plate. By controlling z and measuring the time interval, finding the velocity of the curve is only simple calculations. However, for dispersible waves, a more complicated method is required. There have been many preceding research about $V(z)$ and $V(f,z)$ graphs [2-5], but they were mostly limited to isotropic specimens.

In the present work, directional properties of Lamb wave propagation in anisotropic materials were investigated. A large aperture line focusing transducer was fabricated and used to generate and to receive a ultrasonic pulse on a [100], [110] and [111] silicon wafer [6]. The $V(f,z)$ graph was obtained from the waveforms acquired in the experiments and the dispersion curve was derived from calculations. As rotating the specimen, a dispersion curves in the propagation direction were determined. Simulation on Lamb waves on anisotropic plates were also carried out in order to compare with the experimental results.

2. Theory and Experimental setup

2.1 Crystal structures of Silicon

Silicon is one of well-known anisotropic material with different physical properties by direction. The cubic structures of silicon wafer such as [100], [110] and [111] change by the direction of division. Figure 1 shows cubic structure of [100], [110] and [111] silicon crystal.

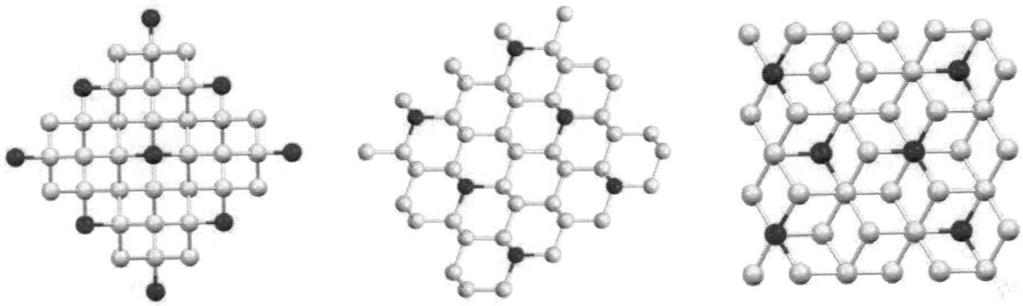

Figure 1 Cubic Structure of [100], [110] and [111].

2.2 large aperture line focusing transducer

Large aperture line focusing transducer (LALFT) was employed to generate and to detect lamb wave. PVDF film of 26 μm thick was used to fabricate LALFT. The time domain waveform and frequency spectrum of pulse-echo signal of fabricated LALFT is shown in Figure 2.

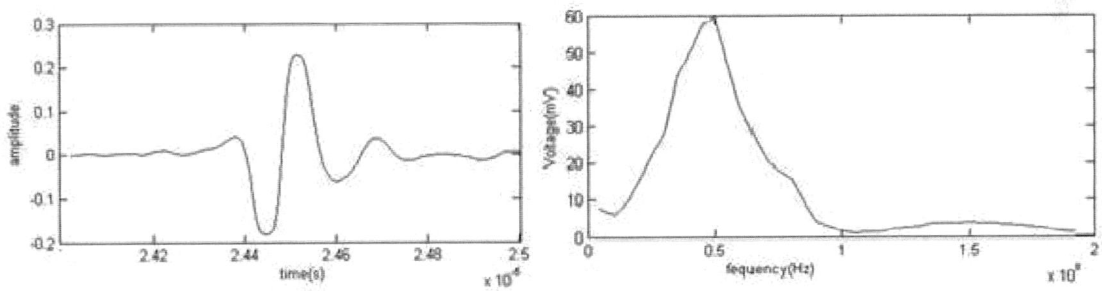

Figure 2 Typical pulse-echo signal of fabricated LALFT. (a) Time domain waveform and (b) its frequency spectrum.

2.3 Experimental setup

Figure 3 shows schematic diagram of measuring system. Two motors are used in this system. Motor 1 moves transducer up and down in order to change defocusing distance, z. Motor 2 rotates the specimen in order to change the propagation direction in the specimen. Three silicon wafers of [100], [110] and [111] with diameter of 100mm, thickness of 0.5 mm were under interrogation in the present work.

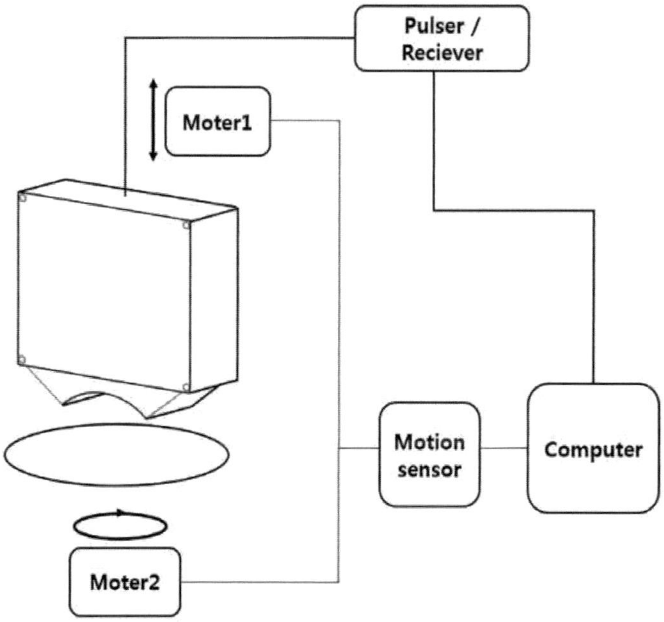

Figure 3 Schematic diagram of experimental setup

2.4 Processing time domain waveform

Since lamb wave is dispersive, complicate waveform processing is essential to estimate its velocity. Lamb wave is generated at certain angle that satisfies condition of phase matching as shown in Figure 4. There are numerous angles satisfying this condition therefore waveform contains numerous frequencies. Thus, it is required to select proper frequencies and calculate phase velocity of lamb wave at the certain frequencies. FFT (Fast Fourier Transform) is used to select proper frequencies.

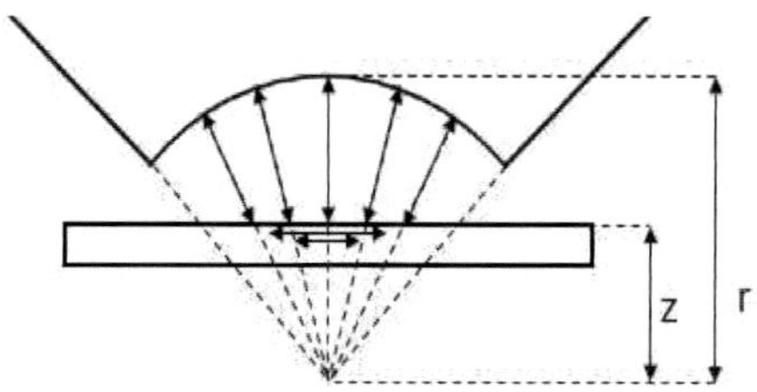

Figure 4 Lamb wave propagation.

$V(f,z)$ graphs were obtained for selected frequencies. $V(f,z)$ graph has a constant period, and phase velocity of Lamb wave for selected frequency can be determined from this period. Some frequencies generate only one mode while other frequencies generate several modes. Each mode has different phase velocity, so $V(f,z)$ graph of this case has several periods. FFT analysis can separate each periods of $V(f,z)$ graph, and phase velocity of each mode can be determined. Finally dispersion curve can be determined from the $V(f,z)$ graphs.

Dispersion curve is dependent on the propagation direction in anisotropic materials. Thus, dispersion curves in each propagation directions should be drawn and 3-dimensional graph can be drawn from these graphs. Nevertheless, it is not easy to analyze, therefore simpler graph is used in this paper. If phase velocity is calculated only at a certain frequency, 2-dimensional graph is drawn finally although each propagation direction is considered. This graph can be compared with the result of computer simulation.

2.5 Simulation

PZFlex which is based on FEM(Finite Element Method) was used in experiment. 20 mm square plates with 0.5mm of thickness were under the simulation. In order to simplify the simulation, one cycle of 10 MHz sine wave is applied center of specimen. Stiffness tensors for materials were used values as follows:

<silicon 100> (GPa) <silicon 110> (GPa) <silicon 111> (GPa)

$$\begin{pmatrix} 166 & 64 & 64 & 0 & 0 & 0 \\ 64 & 166 & 64 & 0 & 0 & 0 \\ 64 & 64 & 166 & 0 & 0 & 0 \\ 0 & 0 & 0 & 80 & 0 & 0 \\ 0 & 0 & 0 & 0 & 80 & 0 \\ 0 & 0 & 0 & 0 & 0 & 80 \end{pmatrix} \begin{pmatrix} 195 & 64 & 35 & 0 & 0 & 0 \\ 64 & 166 & 64 & 0 & 0 & 0 \\ 35 & 64 & 195 & 0 & 0 & 0 \\ 0 & 0 & 0 & 80 & 0 & 0 \\ 0 & 0 & 0 & 0 & 51 & 0 \\ 0 & 0 & 0 & 0 & 0 & 80 \end{pmatrix} \begin{pmatrix} 195 & 54 & 45 & -13.7 & 0 & 0 \\ 54 & 195 & 45 & 13.7 & 0 & 0 \\ 45 & 44.67 & 205 & 0 & 0 & 0 \\ 0 & 0 & 0 & 61 & 0 & 0 \\ 0 & 0 & 0 & 0 & 61 & -13.7 \\ 0 & 0 & 0 & 0 & -13.7 & 70 \end{pmatrix}$$

3. Results

3.1 Simulation in silicon wafer

The frequency of incident wave in simulation is 10MHz. In order to investigate anisotropic propagation, history program saves change of acoustic pressure according to time in the fixed coordinate as shown in Figure 5. Point 1 is the origin of impulse shown in Figure 6, points 2 and 3 are on the horizontal axis and points 4 and 5 are 45° direction from the horizontal axis.

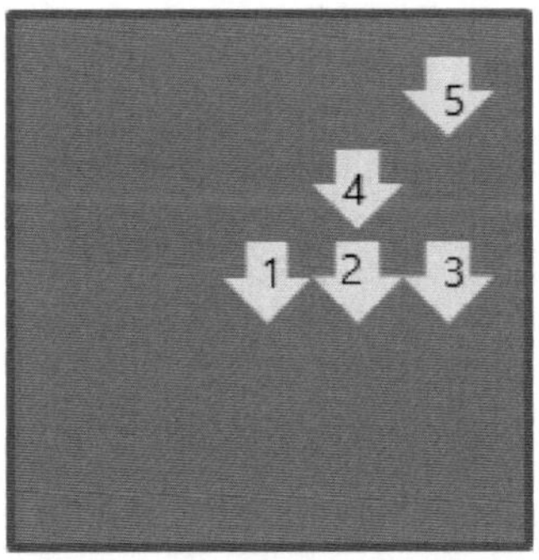

Figure 2 Five points in silicon wafer.

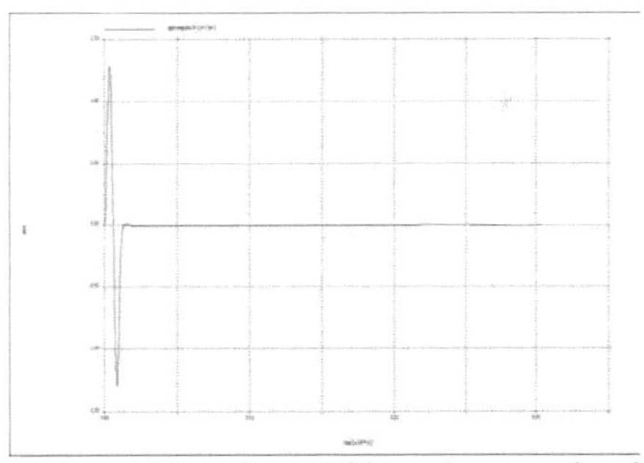

Figure 6 Waveform of impulse at point 1.

Simulation results of [100] silicon plate are shown in Figure 7. Acoustic pressure graphs according to time are also shown in Figure 8.

Figure 7 Propagation of waves in [100] silicon plate.

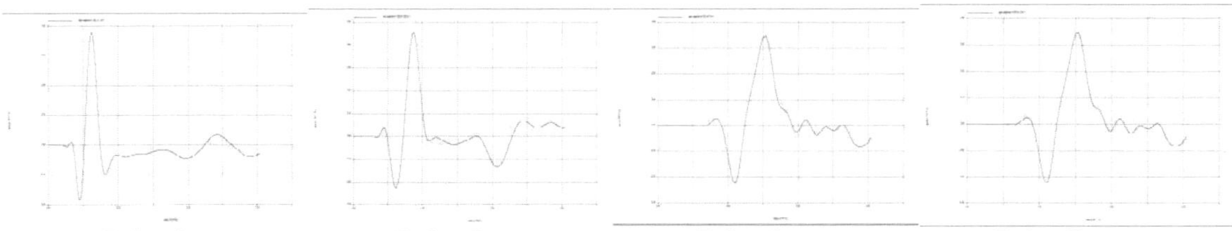

Point 2 Point 3 Point 4 Point 5

Figure 8 Acoustic pressure graphs in silicone [100].

Simulation results of [110] silicon plate are shown in Figure 9. Acoustic pressure graphs according to time are also shown in Figure 10. Propagation of ultrasonic wave in silicon [110] was complicated initially, but it looks like a rectangle with symmetry to x, y-axes.

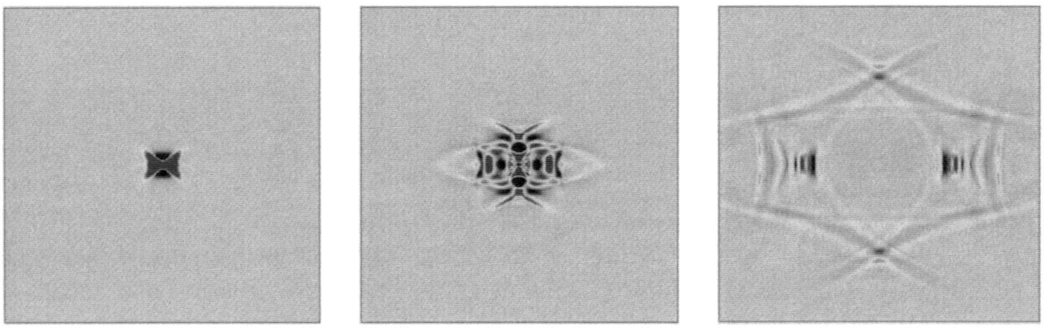

Figure 9 Propagation of waves in [110] silicon plate.

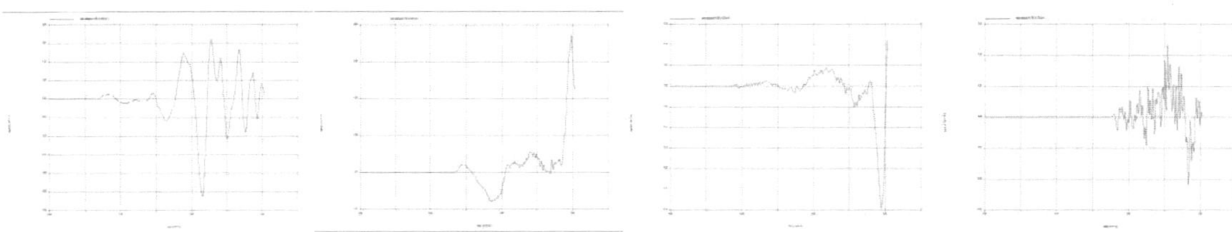

Point 2 Point 3 Point 4 Point 5

Figure 10 Acoustic pressure graphs in silicone [110].

Simulation results of [111] silicon plate are shown in Figure 11. Acoustic pressure graphs according to time are also shown in Figure 12. Propagation of ultrasonic wave in silicon [111] looks like a triangle with symmetry.

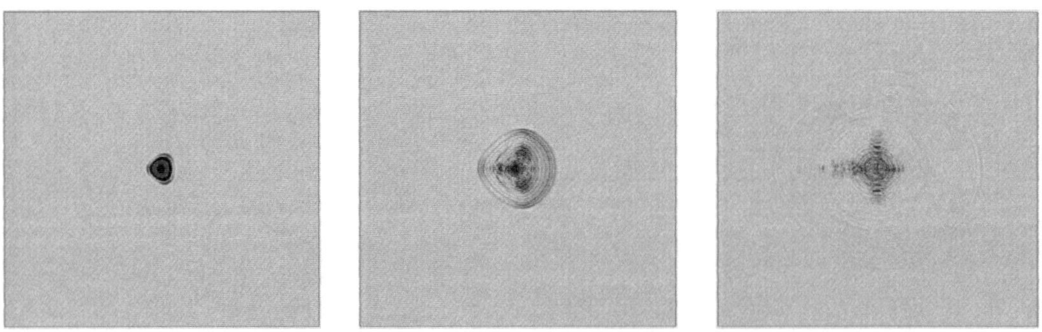

Figure 11 Propagation of waves in [110] silicon plate.

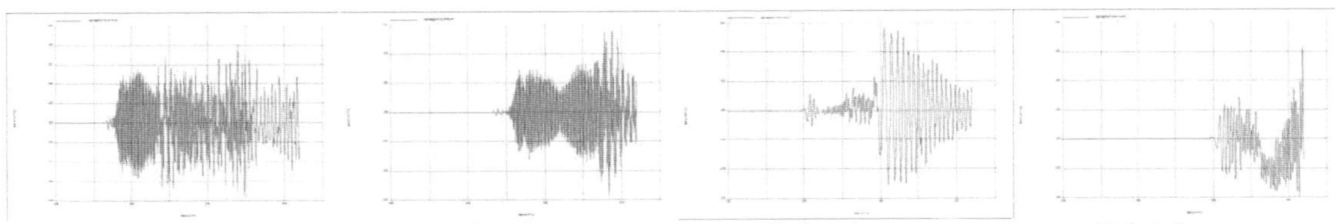

Figure 12 Acoustic pressure graphs in silicone [111].

3.2 Analysis of waveform in silicon wafer

3.2.1 100 Si wafer

Throughout the experiment on Si wafer, the results were given as a figure.

The properties of silicon 100 was obtained by rotating the transducer with 18 degrees. The peak point moved while rotating the silicon wafer.

The moving period was about 90 degrees.

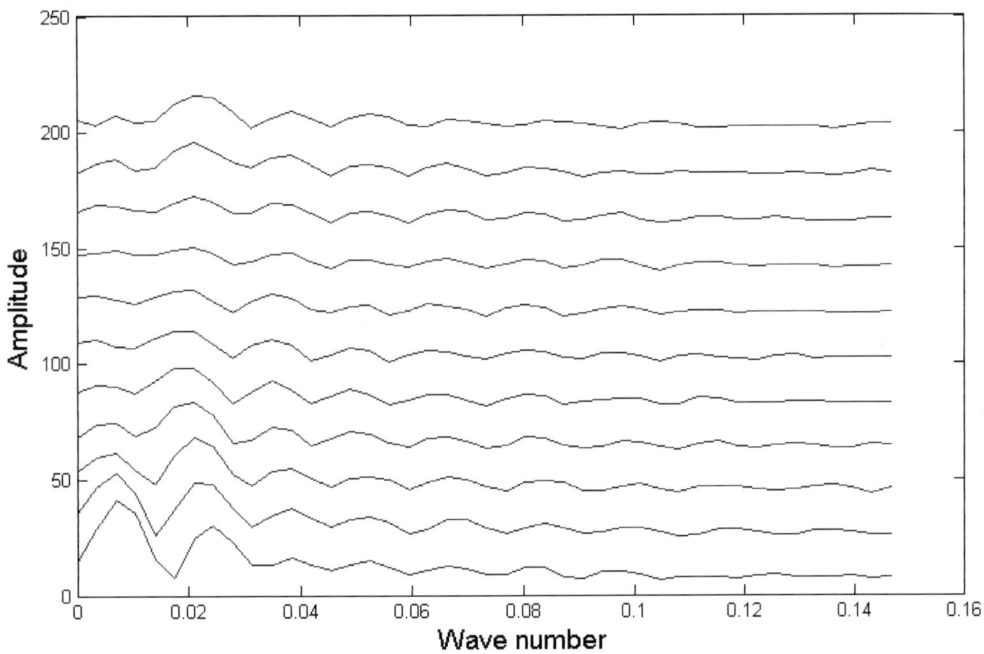

Figure 14 FFT of $V(f,z)$ **for [100] silicon wafer.**

3.4.2 110 Si wafer

The moving period was about 72 degrees.

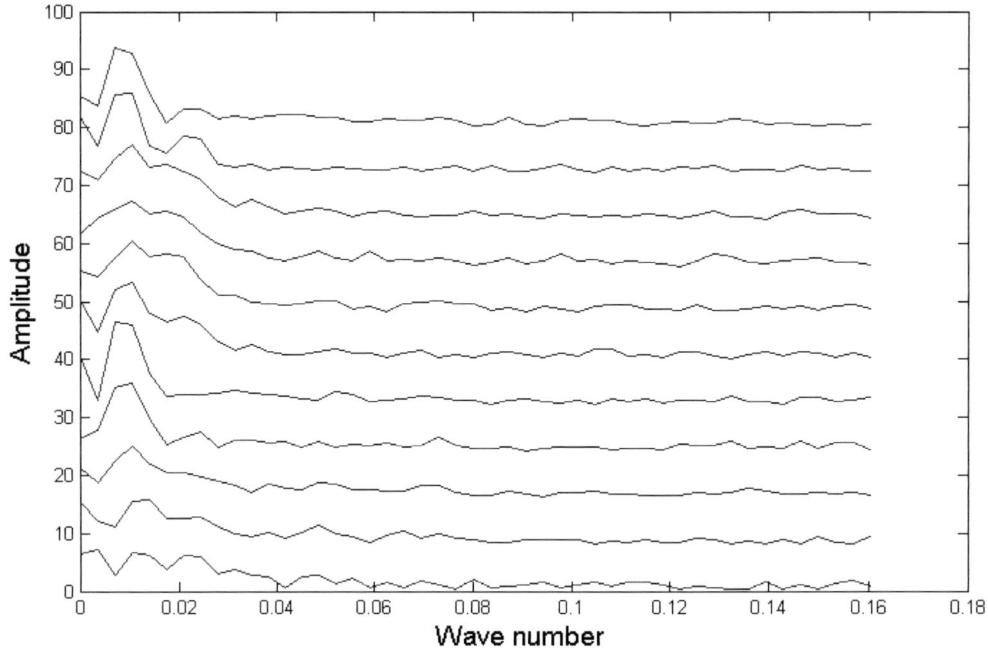

Figure 15 FFT of $V(f,z)$ for [110] silicon wafer.

3.4.3 111 Si wafer

The moving period was about 60 degrees.

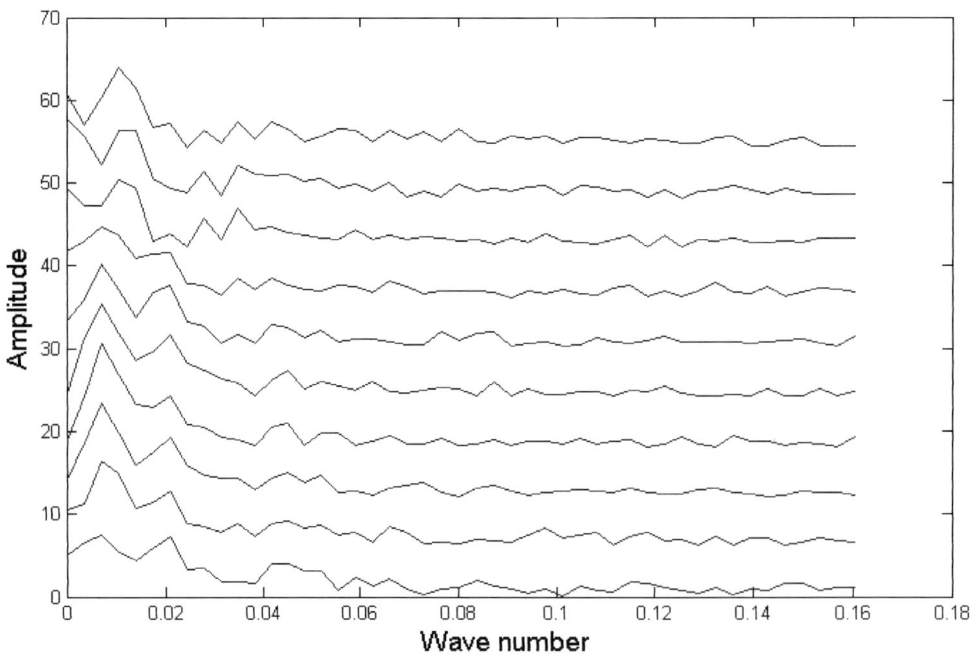

Figure 16 FFT of $V(f,z)$ for [111] silicon wafer.

4. Conclusion

The phase velocity of lamb wave in silicon wafer [100],[110], and [111] are obtained. Using the propagation characteristic of the ultrasonic wave depends on the crystal orientation of silicon crystal will be clear to the anisotropic direction of the plate. Simulation results using a PZFlex was less out yet, but the results are compared with experimental data showing the nail bojineun Lamb waves propagated in the silicon plate a visual appearance success to draw a graph of the sound pressure at a certain point.

Acknowledgement

This work was supported by the Research & Education Program at the Korea Science Academy of KAIST with funds from the Ministry of Science, ICT and Future Planning.

Reference

1. S. J. Song, Y. H. Kimg, S. D.l Kwon, Y. M Cheong, "Determination of Phase Velocity Dispersion Curve and Group Velocity of Lamb Waves Using Backward Radiation," J. Acoust. Soc. Kor., vol. 22, pp. 61-68, 2003.
2. Y. C. Lee, "Measurements of dispersion curves of leaky Lamb waves using a lens-less line-focus transducer," Ultrasonics, vol. 39, pp. 297-306, 2000
3. Y. C. Lee, "Measurements of Multimode Leaky Lamb Waves Propagating in Metal Sheets," Jpn. J. Appl. Phys., vol. 40, pp. 359-363, Jan. 2001.
4. Kenneth K. Liang, Gordon S. Kino, Butrus T. Khuri-Yakub, "Material Chracterization by the Inversion of V(z)," IEEE Transactions on Sonics and Ultrasonics, vol. SU-32, no. 2, Mar. 1985.
5. Jun-Ichi Kushibiki, Noriyoshi Chubachi, "Material Chracterization by Line-FocusBeam Acoustic Microscope," IEEE Transactions on Sonics and Ultrasonics, vol. SU-32, No. 2, Mar. 1985.
6. S. J. Kim, Y. H. Noh, M. C. Son, D. J. Lee, Y. H. Kim, "Study on Propagation of Leaky Surface Acoustic Wave on Monocrystalline Silicon Crystal Using Large Aperture Line-Focusing PVDF Transducer," Joint Conference on Acoustics, Jan. 2015.
7. Miguel A. Torres-Arredondo, Henning Jung, Claus-Peter Fritzen, " A Study of Attenuation and Acoustic Energy Anisotropy of Lamb Waves in Multilayered Anisotropic Media for NDT and SHM applications," 6th NDT in Progress, Oct. 2011.

Influence of Transducer Coupling in Ultrasonic Testing

Ladislav CARBOL [1], Ivo KUSÁK [1], Jan MARTINEK [1], Petra VOJKŮVKOVÁ [2]

[1] Department of Physics, Faculty of Civil Engineering, Brno University of Technology, Veveří 95, 602 00 Brno, Czech Republic; e-mail: carbol.l@fce.vutbr.cz, kusak.i@fce.vutbr.cz, honza@dp.fce.vutbr.cz
[2] Institute of Building Services, Faculty of Civil Engineering, Brno University of Technology, Veveří 95, 602 00 Brno, Czech Republic; e-mail: vojkuvkova.p@fce.vutbr.cz

Abstract
Couplants are necessary in most cases of ultrasonic testing, due to large acoustic mismatch between air and solid material. Signal amplitude as well as wave type are greatly affected by couplant properties. This paper is discussed influence of various coupling methods on received signal in frequency domain. Coupling of both transmitter-specimen and receiver-specimen in a pitch-catch setup are investigated on common building material.

Keywords: Coupling, Transfer function, Pulse compression, Maximum Length Sequence (MLS), Impact-Echo (IE), Young modulus, Impulse response, Linear time-invariant system, time of flight (TOF), Nondestructive testing (NDT)

1. Introduction

Ultrasonic waves introduced into the specimen are changed and reflected as they interact with inhomogeneities. Received signal contains information, which can be interpreted as defects or material properties. The classical defectoscopic approach is to look at reflected echoes or to measure time of flight. However, more sophisticated methods focus on information contained in the frequency spectra, to estimate dynamic modulus of elasticity or to assess severity of micro-cracking. Pure examples of such are Nonlinear Elastic Wave Spectroscopy (NEWS) [1] or Impact-Echo. Regardless of ultrasonic method applied transducer coupling is always problematic. Due to high acoustic mismatch between air and solids, the most convenient air coupling is often out of the question [2]. Immersion testing, on the other hand is more useful, but not in Civil engineering. Cementitious or clay products significantly change their material properties when soaked.

Most defectoscopic method applied for Civil engineering purposes use gel coupling. Unfortunately, the inconvenient soaking effect prevents long-term continuous testing and maintaining the correct gel layer is tricky [3]. If the gap is to wide gel couplant is acting as a separate material and waves are reflected on the boundaries. Also constant force should be applied during measurement.

The answer to those problems might me fixed-coupling, for instance with bee's wax. Significantly higher energy levels are transferred in and out of the specimen, transducers are firmly attached which is useful for long-term measurements and improvement of SNR. These advantages come with a price. The transducer and specimen are bonded together and the resonance frequencies of the sample are altered. Added mass and extended therefore affects resulting longitudinal vibrations [4]. According to the analytical **Eq. 1**, for dynamic modulus of elasticity of longitudinally vibrating 1D element, it is quite clear, that any fixed coupling will result in lowering of fundamental frequency [5,6].

$$E_{dyn} = 4\rho(fl)^2 \qquad (1)$$

It should also be noted that fixed coupling allows propagation of horizontally oriented sheer waves. The effects various coupling has on resonance frequency of the sample are investigated in this article.

2. Experimental setup

Tension/compression properties of all materials as such, can be quantified using Young's modulus. Furthermore Young's modulus can be calculated when the device under test (DUT) fundamental frequency is known. Linear time invariable systems (LTI) can be completely described by a system impulse response (IR). If the stimulus signal fulfils criterions described in [7], IR can be simply calculated using cross-correlation with response signal. Initial delay of calculated IR is principally the same as Time of Flight (TOF) measured directly by classical Ultrasonic pulse testing. Of course, omitting dispersive waves and group velocities resulting in the process.

The great advantage rests in the frequency domain. Spotted peaks correspond to fundamental frequency and higher harmonics. Severity of harmonic distortion can be used to quantify damage of tested sample.

2.1 Specimen

In order to make results mutually comparable, all experiments were conducted on the same cement-mortar specimen with dimensions 40 x 40 x 160 mm, w/c = 0.46. The mortar preparation mixture contained CEM I 42,5R cement and quartz sand in a ratio of 1 to 3. In compliance with Czech standard ČSN 721200, 3 sand grades (0-1, 1-3, 3-4 mm) were used to be blended in weight ratio 1:1:1.

Specimen was aged at temperature of 22°C and a relative humidity of 55 % for 24 hours. Afterwards placed in a water bath for 27 days and then dried at a temperature of 60°C for two days.

2.2 Measurement procedure

Influence of bee's wax coupling and gel coupling is in this article compared with results from classical Impact-Echo (IE). According to results from article [4], frequency measured by IE method, by procedure described below, is in the range of ±1 % from air coupled measurement. Therefore peaks obtained from IE method are a suitable reference. Influence of transmitter (TX) and receiver (RX) coupling was tested by Pseudorandom Binary Sequence (PRBS) of Maximum length sequence (MLS). In order to make both methods comparable the same data acquisition device (DAQ) and the same MIDI receiver was used. Sampling parameters were set close to the highest available limits of NI PCI 6251 card, which is 1 MS/s with 16-bit resolution. NI Labview programming language was used for both signal acquisition and conditioning.

2.2.1 Impact-Echo

Impact-Echo is a well described a commonly used non-destructive testing (NDT) method in civil engineering [6,8]. The variation of IE is sometimes called Hammer resonance method, since it better describes the procedure. Physical principle behind this method is generation of a Dirac delta function and direct recording of a system impulse response (IR). A hammer is used to excite the system. Since no signal processing is required the method is very simple and extremely accurate. However, energy delivered into the system is limited by short contact time. Furthermore, increasing of impact energy leads to undesirable nonlinear effects. As a result, massive elements, soft or fragile substances cannot be tested by this method.

2.2.2 Maximum length sequence

Mathematical properties such as well defined frequency spectrum, repeatability and perfect auto-correlation make MLS a wise choice for excitation signal [9]. There are infinitely many Maximum length sequences, defined by their order and primitive polynomials [10]. For the test is used only one sequence, which is continually repeated [11]. In this way, the signal does not have to be record at the beginning of the sequence, but anywhere as long as the record length is the same as the length of the sequence. This being said, one can deduce the same timing source must be used for both generation and recording. The signal is absolutely repeatable and since it must be synchronized a time averaging (25 times) is applied to improve the Signal to noise ratio (SNR). As long as the sampling rate is an integer multiple of the generation rate, the MLS sequence behaves in a manner that more energy is shifted towards lower frequency bands. Selected divider of the sampling frequency leads to 200 kHz generation frequency.

Previously mentioned PCI card was utilized as D/A convertor of to generate the sequence and A/D convertor to record the signal. Generated electrical sequence of 0 V and 1 V was then amplified to 0 V and 10 V and send to the transmitting transducer [11,12]. Prior to acquisition the signal was amplified and filtered analog bandpass filter (0,3-300 kHz). Analog filtering is crucial to prevent frequency aliasing given by continuous sinc^2 frequency spectra of MLS [13].

2.2.3 Sensors and transducer

Different combinations of wax and gel coupling between DUT and piezoelectric transmitter/receiver were tested. A small (6,5 mm) piezoelectric receiver DAKEL MIDI was used for both IE and MLS testing. Originally three bolt-clamped Langevin transducers were intended as transmitters. However, results from transducer designated as BK03 we inconclusive because of a single dominant resonance at 32 kHz. Other two transducers, designated according to their shape (CONE, CYLINDER) had more convenient frequency characteristics. All Langevin transducers are designed to operate at their resonance frequency in order to create maximum displacement to observe nonlinear behavior. Their use for generation of broadband signals, such as in this article is not common.

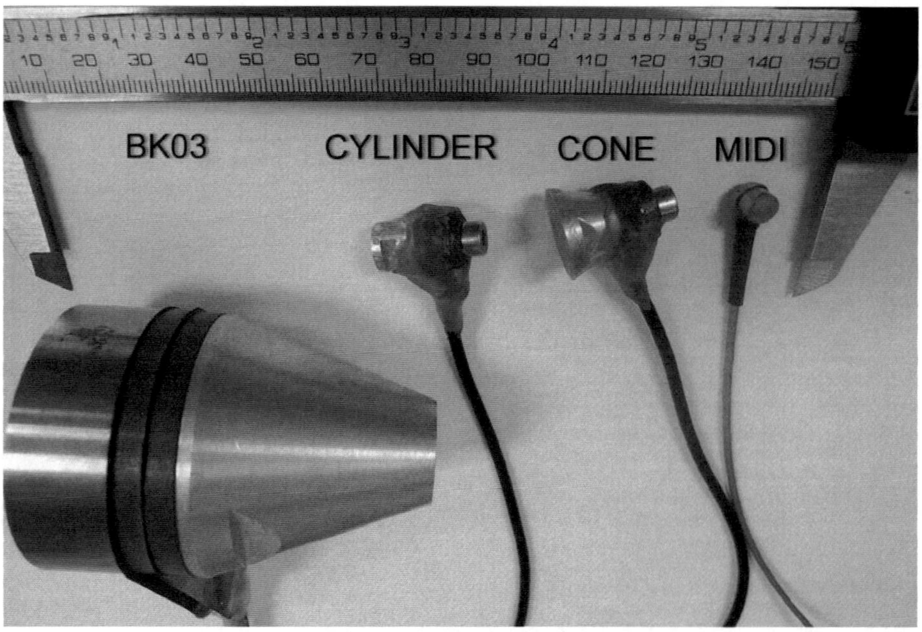

Figure 1. Transducers

3. Results

On **Fig. 1 to 4**, respectively **5 to 7** are displayed frequency spectra obtained with CONE and CYLINDER transducers. Individual plots indicate the coupling between transmitter (TX) and specimen and also between specimen and receiver (RX). Measured spectra are compared to hammer impact resonance measurement (IE) (black plot. Figures clearly show that position of frequency peaks is influenced by coupling. As it was said in the introduction, lower fundamental frequency is to be expected with fixed coupling. Unfortunately all harmonic frequencies seem to be completely erratic, comparing them to both IE and between each TX. The arising of second peak for CONE transducer at first harmonic is clearly a result of fixed coupling between TX and specimen. Although the transfer characteristic of the CONE TX shows no anomaly in the region, TX and specimen fixed together can create such unexpected peaks. This finding makes fixed coupling of TX doubtful at best. On the other hand fixed coupling of small MIDI receiver seems to have no such side effect (green plot).
Best representation of frequency spectra offers Gel-Gel coupling, because the specimen is not influenced by transducers. In fact, measured fundamental frequency was always higher than of fixed-coupled IE, which is perfectly in accordance with the additional mass phenomenon.

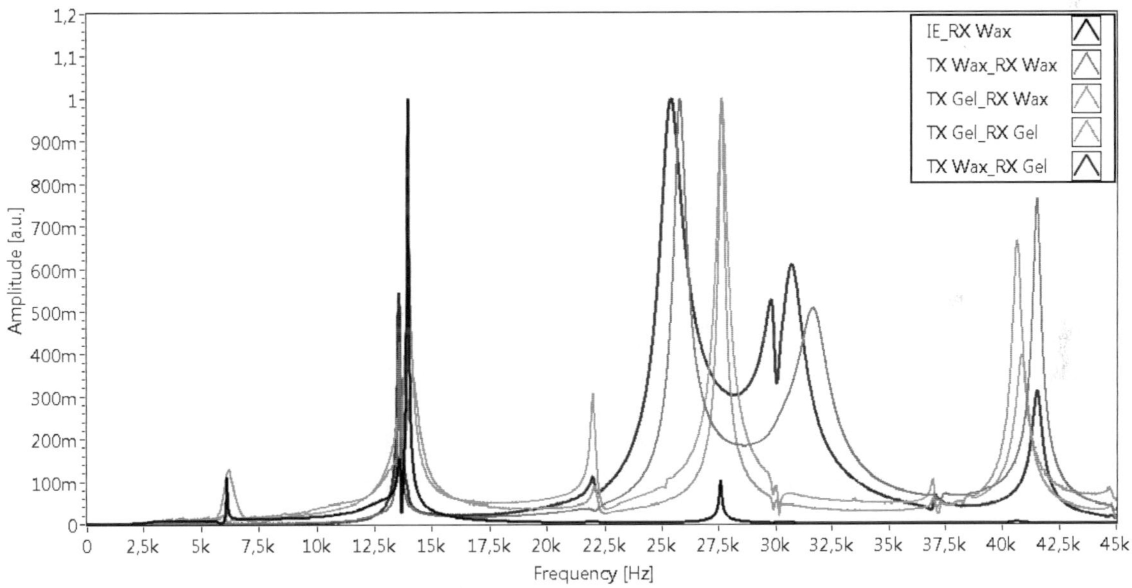

Figure 2. Frequency spectra measured with CONE transducer on mortar sample

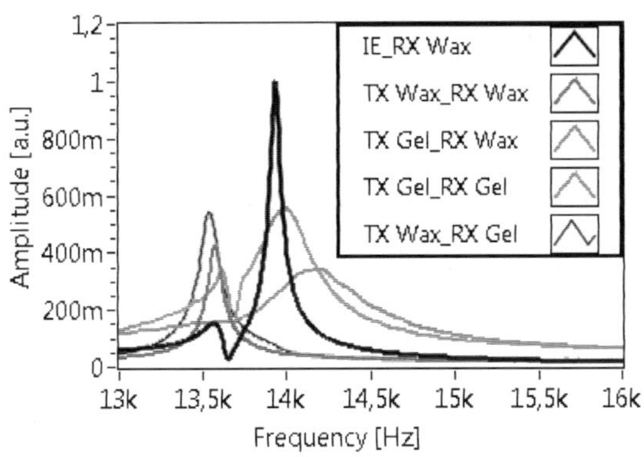

Figure 3. Fundamental frequency CONE transducer

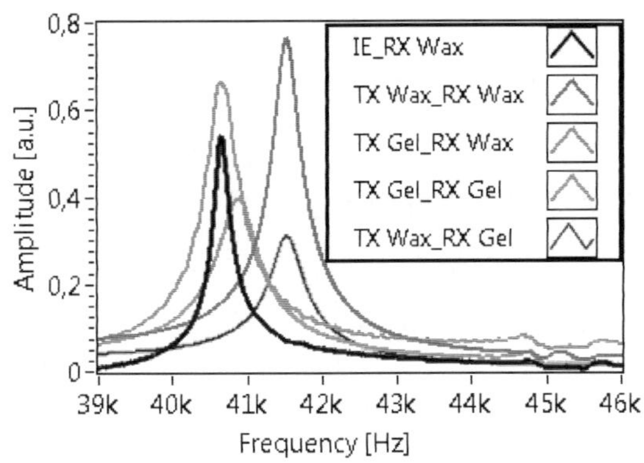

Figure 4. Second harmonic frequency CONE transducer

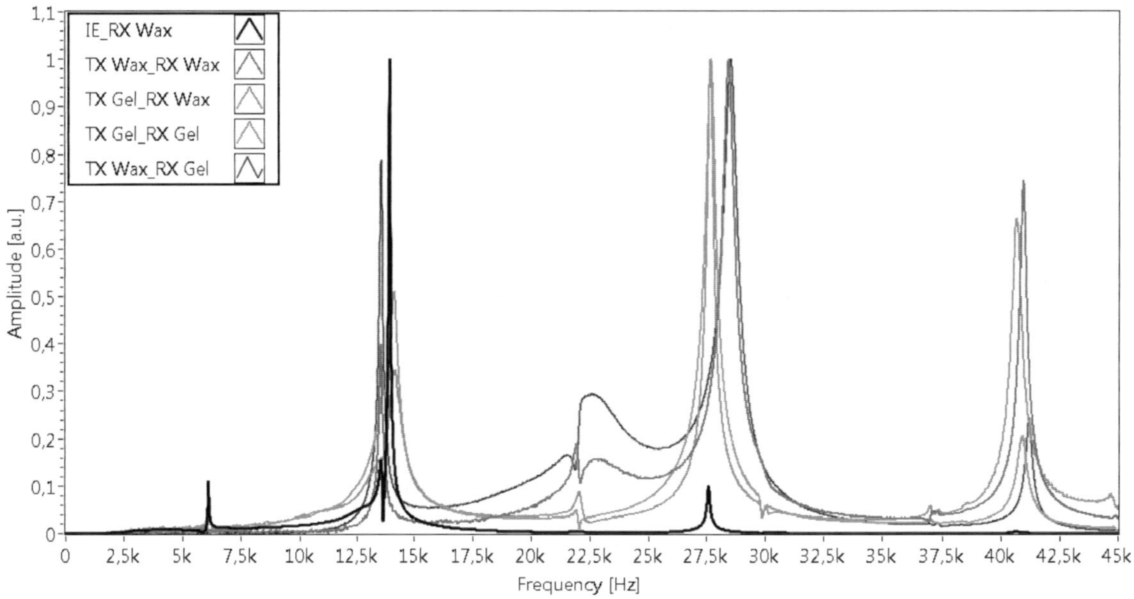

Figure 5. Frequency spectra measured with CYLINDER transducer on mortar sample

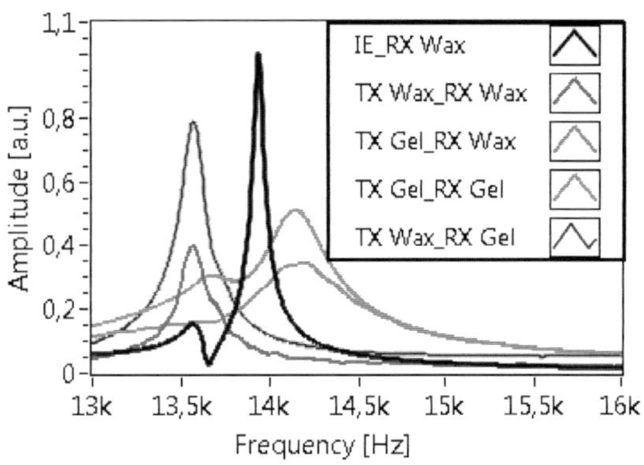

Figure 6. Fundamental frequency CYLINDER transducer

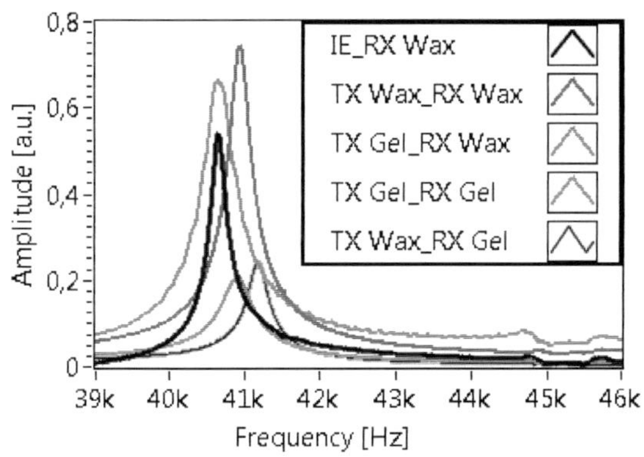

Figure 7. Second harmonic frequency CYLINDER transducer

Table 1. Peak positions

TX name, coupling		1st [Hz]	diff. [%]	2nd [Hz]	diff. [%]	3rd [Hz]	diff. [%]	SNR [dB]
	IE_RX WAX	13936	0,0	27577	0,0	40644	0,0	66
Cone	TX Wax_RX Wax	13580	-2,6	Double	NA	41537	2,2	67
	TX Gel_RX Wax	14198	1,9	27647	0,3	40635	0,0	42
	TX Gel_RX Gel	13994	0,4	27662	0,3	40875	0,6	49
	TX Wax_RX Gel	13543	-2,8	Double	NA	41533	2,2	55
Cylinder	TX Wax_RX Wax	13569	-2,6	28431	3,1	40943	0,7	45
	TX Gel_RX Wax	14198	1,9	27647	0,3	40635	0,0	42
	TX Gel_RX Gel	14143	1,5	27669	0,3	40904	0,6	59
	TX Wax_RX Gel	13570	-2,6	28502	3,4	41179	1,3	54

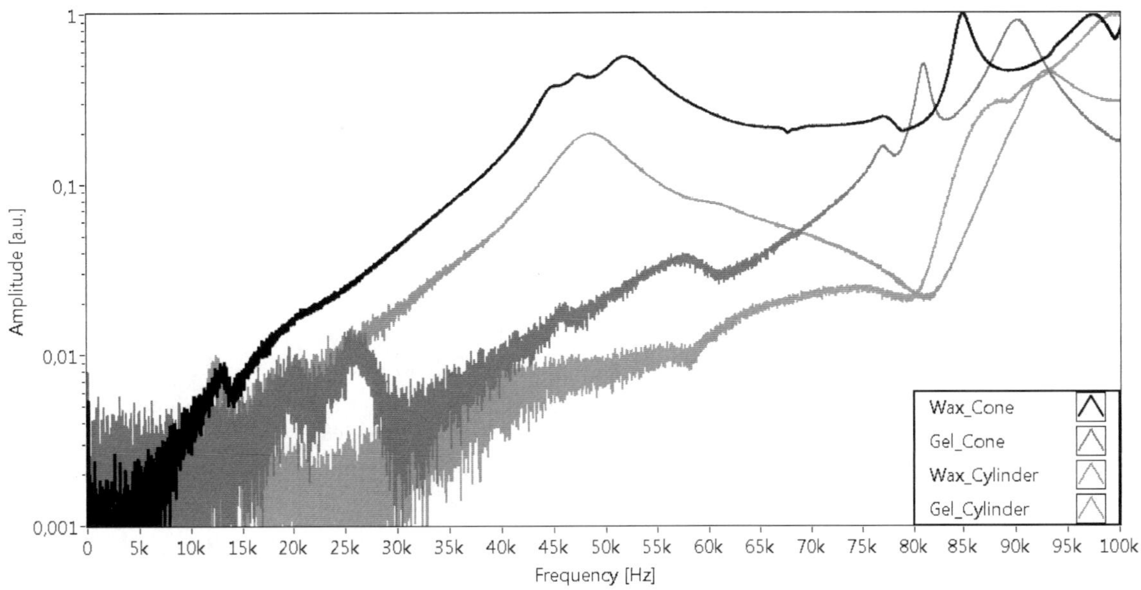

Figure 8. Frequency spectra of directly connected TX and RX.

4. Conclusions

Fixed coupling by bee's wax is very useful when it comes to transferring large energy or for long-term measurement. However, one must keep in mind that new vibration modes can arise from fixed coupling with a transmitter. Fixed coupling of receiver will only slightly shift the frequencies to lower values, which due to the added mass of sensor.
Gel coupling of both transmitter and receiver is certainly the best approach, but it is not suitable for long-term measurement (soaking effect). Also higher amplification is needed. This experiment, frequencies above 50 kHz needed to be filtered out because of noise problems.

Acknowledgements

This paper was elaborated with the financial support of the project No. LO1408 "AdMaS UP - Advanced Materials, Structures and Technologies", supported by Ministry of Education, Youth and Sports under the „National Sustainability Programme I" and under the project of specific research program at Brno University of Technology, project No. FAST-J-15-2650.

References

1. VAN DEN ABEELE, K. E.-A., P. A. JOHNSON, A. SUTIN, Nonlinear ElasticWave Spectroscopy (NEWS) Techniques to Discern Material Damage, Part I: Nonlinear Wave Modulation Spectroscopy (NWMS). Research in Nondestructive Evaluation. 12(1): 17-30. DOI: 10.1007/s001640000002. ISSN 0934-9847.
2. D. HUTCHINS, P. BURRASCANO, L. DAVIS, S. LAURETI, M. RICCI. 2014. Coded waveforms for optimised air-coupled ultrasonic nondestructive evaluation. Ultrasonics. 54(7): 1745-1759. DOI: 10.1016/j.ultras.2014.03.007. ISSN 0041624x.
3. L. CARBOL, MARTINEK, J. KUSÁK, I., Influence of Water Content on Fundamental Frequency of Mortar Sample, příspěvek na konferenci , ISSN 1022-6680, Trans Tech Publications, Switzerland, 2015

4. L. Carbol, I. Kusák, J. Martinek, Measurement of Material Properties Using Deterministic White Noise, NDT net: e-journal of nondestructive testing and ultrasonics (2014) 1-7.
5. J. Wang, T. Chang, Bo-Tsun Chen, and H. Wang. Determination of Poisson's ratio of solid circular rods by impact-echo method, Journal of Sound and Vibration, vol. 331, issue 5, pp. 1059-1067, 2012.
6. X. Lu, Q. Sun, W. Feng, and J. Tian. Evaluation of dynamic modulus of elasticity of concrete using impact-echo method, Construction and Building Materials, vol. 47, pp. 231-239, 2013.
7. L. CARBOL, J. MARTINEK, D. ŠTEFKOVÁ, Correct Choice of Maximum Length Sequence in Nondestructive Testing, ISSN 1022-6680, Trans Tech Publications, Switzerland, 2015
8. M. Tupý, D. Štefková, K. Sotiriadis, L. Krmíček, L. Carbol, V. Petránek, Recycled Poly(Vinyl Butyral) Used as a Barrier to Prevent Mortar Carbonation, Advanced Materials Research vol. 1000 (2014) 28-34.
9. T. Niederdränk. Maximum length sequences in non-destructive material testing: application of piezoelectric transducers and effects of time variances, Ultrasonics, vol. 35, issue 3, pp. 195-203, 1997.
10. A. H. Tan, K. R. Godfrey, A guide to the design and selection of perturbation signals, Proceedings of the 48h IEEE Conference on Decision and Control (CDC) held jointly with 2009 28th Chinese Control Conference (2009) 464-469.
11. L. Carbol, and J. Martinek, Deterministic white noise generator based on microcontroller, in 11th workshop ndt 2013 non-destructive testing in engineering practice, Brno: Akademické nakladatelství cerm, 2013, pp. 72-80.
12. L. Svilainis, A. Chaziachmetovas, and V. Dumbrava. "Efficient high voltage pulser for piezoelectric air coupled transducer," Ultrasonics, vol. 53, issue 1, pp. 225-231, 2013.
13. M. Ricci, L. Senni, P. Burrascano, Exploiting Pseudorandom Sequences to Enhance Noise Immunity for Air-Coupled Ultrasonic Nondestructive Testing, IEEE Transactions on Instrumentation and Measurement vol. 61, issue 11 (2012) 2905-2915.
14. C., and E. Nordlund, Effects of couplants on acoustic transmission. Rock Mechanics and Rock Engineering vol. 26, issue 1 (1993): 63-69.

Acoustic Emission Crack Detection in Injection Molding

Tomaž KEK [1], Dragan KUSIĆ [2], Aleš HANČIČ [2], Janez GRUM *[1]

[1] University of Ljubljana, Faculty of Mechanical Engineering
Aškerčeva 6, 1000 Ljubljana, Slovenia, tomaz.kek@fs.uni-lj.si, janez.grum@fs.uni-lj.si
[2] TECOS Slovenian Tool and Die Development Centre
Kidričeva 25, 3000 Celje, Slovenia, dragan.kusic@tecos.si, ales.hancic@tecos.si

Abstract
This paper presents the experimental results regarding acoustic emission signals measured during the injection molding of those standard test specimens commonly used for examining the shrinkage behaviour of various thermoplastic materials. In daily industrial production of different plastic products we often have to deal with various errors that practically occur on the mold primarily as a result of tool wear and tear, improper storage and improper settings on the injection molding machine. In the testing phase of plastic materials we use many times different inserts that are made from standard tool steels, such as OCR12VM. In case of tool steel inserts after some years of usage a few micro-cracks can occur in the early stage, which can later quickly spread according to the applied loading. With the help of different non-destructive testing methods we know that we can most certainly detect possible formation of cracks on the tool steel inserts. The acoustic emission was measured on an injection mold with the visible sign of a crack on the cavity's surface, using two contact PZT sensors under normal and increased injection pressure loads. In this paper, we focused exclusively on the acoustic emission signal acquisition by using two resonant 150 kHz piezoelectric AE sensors on such tool steel inserts that are already affected by macro-cracks. On such tool steel insert the obtained acoustic emission results were compared with those obtained from a brand new tool steel insert. The final obtained acoustic emission results on the crack defected tool steel insert revealed as expected that the energy and intensity of the captured AE signals is higher compared with the ones that were captured on the brand new engraving insert under same processing conditions.

Keywords: Acoustic emission, injection molding, crack, mold, polypropylene

1. Introduction

Injection moulding is a well-known plastic manufacturing process where heated molten plastic material is forced into a mould cavity under high pressure. The plastic material solidifies into a shape that has conformed to the contour of the mould. Nowadays it is still regarded as the most important and very popular manufacturing process because of simple operation steps. A typical production cycle begins when the mould closes, followed by the injection of the plastic into the mould cavity. Once the cavity is filled, additional pressure compensates the material shrinkage. In the next step, the screw turns, feeding the next shot to the front screw tip. This causes the screw to retract as the next cycle is almost prepared. Once the moulded part is sufficiently cooled, the mould opens and the part is finally ejected.
Chen and Turng [1] defined three categories of variables that describe the hierarchy and dependency of injection-moulding control. First level can be defined as machine variables

that incorporate barrel temperature in several zones, pressure, sequence and motion, ... Second level variables are process dependent variables that incorporate melt temperature in the nozzle, runner, or mould cavity, melt pressure (in the nozzle, cavity), melt-front advancement, maximum shear stress, rate of heat dissipation and cooling. Third level of quality definitions are part weight and part thickness, shrinkage and warpage, ... Most of current work deals with the control tasks at the first two levels. For quality control, this is often achieved indirectly through online dynamic process control or off-line statistics process control (SPC). For monitoring second level variables majority of commercially available in mould sensors are pressure and temperature sensors.

Muller et al. [2] presented a wireless sensor concept for injection moulding which uses structure born sound as transport medium. It is orientated on level two variables. The sound is generated by an acoustic actor which is activated by the passing flow front at certain predetermined positions in the cavity (or cavities).

Lijuan Zhao et al. [3] developed ultrasonic diagnosing technique with a high-temperature ultrasonic transducer to real-time diagnose polymer processing and its morphology changes in injection moulding processing. Ultrasound diagnosis showed that longitudinal wave can realtime characterize the data of the injection process and polymer morphology changes, including melt flow arrival time, the part ejection time, filling and packing stages, polymer solidification process, and the morphology changes during polymer crystallization.

Kusić et al. [4] analyzed measured AE signals and their correlation with the post-moulding shrinkage and warping. Experimnts indicated the possibility of on-line monitoring and the adequacy of the process control by capturing AE signals. Location and advancement of a possible crack on the steel can be detected by the use of acoustic emission technique as already reported by many researchers [5-8]. But the detection of cracks in tools during injection moulding process is a topic that is not sufficiently presented in accessible literature. In this paper the results of AE monitoring to detect cracks in tool steel inserts are presented, that can upgrade the process control using AE signals presented by Kusić et al. [4]

2. Experimental procedure

Acoustic emission signals were captured during the production cycle of standard test specimens that are intended for shrinkage evaluation of various plastic materials. The main aim was to analyse the influence of a possible crack located in tool steel insert on the captured acoustic emission signals. After finishing the first experimental part on a tool steel insert with a macro-crack we repeated the experiment under the same processing conditions also on the brand new tool steel insert. In this way the captured acoustic emission signals could be compared. The captured acoustic emission signals were then correlated with the quality of the produced test specimens. Acoustic emission measurement system AMSY-5 from Vallen-Systeme GmbH was used for capturing and analysing the AE signals. Two piezoelectric AE sensors VS150-M (resonant at 150 kHz) were mounted using silicone grease with two sensor holders on the tool steel insert from both sides as shown on Fig. 1. Both PZT sensors were connected via two preamplifiers AEP4 with a fixed gain of 40 dB on the first and second channel of AMSY-5 measurement system.

Figure 1. Tool steel insert with AE sensors

If we want to produce standard test specimens with good quality, the process parameters must be correctly set. Before the start of the experiment, it was necessary to select and fix the following process parameters: injection pressure was set to 1100 bar in the course of the experiment, holding pressure was set to 500 bar. On the new tool steel insert the melt temperature in cylinder has been set to 230 °C and the injection speed was set to 50 mm/s. On the tool insert with a macro-crack the melt temperature was set to 240 °C and the injection speed was 45 mm/s. The main criterion for the quality of produced test specimens was chosen to be the size of the shrinkage in longitudinal and transverse direction of the melt flow.

3. Experimental results

We carried out the experiments on a brand new tool steel insert and on a tool steel insert with visible sign of macro-cracks. In both cases we used the same polypropylene material from Sirmax manufacturer (H40 C2 FNAT), which is mainly used in the automotive industry. After the test specimens were produced, they were scanned with an optical 3D digitizer ATOS II SO after 24 hours.

Once the test specimens are digitized the measured data can be saved and used later on. Usually we are interested in individual measuring values and sections across the test specimen. Larger deviations and/or dimensional changes compared to the nominal ones are easy to verify and to control. Optical 3D digitizer is based on the principle of capturing images through the camera, which then through an appropriate program prepares a computer model. The accuracy that can be achieved with these digitizers depends largely on the quality of the camera, which records the desired object. Of course, in a very small precision scale the accuracy itself also depends on the wavelength of the light. The biggest advantage of used 3D digitizer lies in extremely rapid procedure of digitizing and excellent precision (precision declared for very small objects are up to 2 microns). Example of produced and later scanned test specimens on a brand new tool steel insert and insert with a visible macro-crack is shown in Fig. 2.

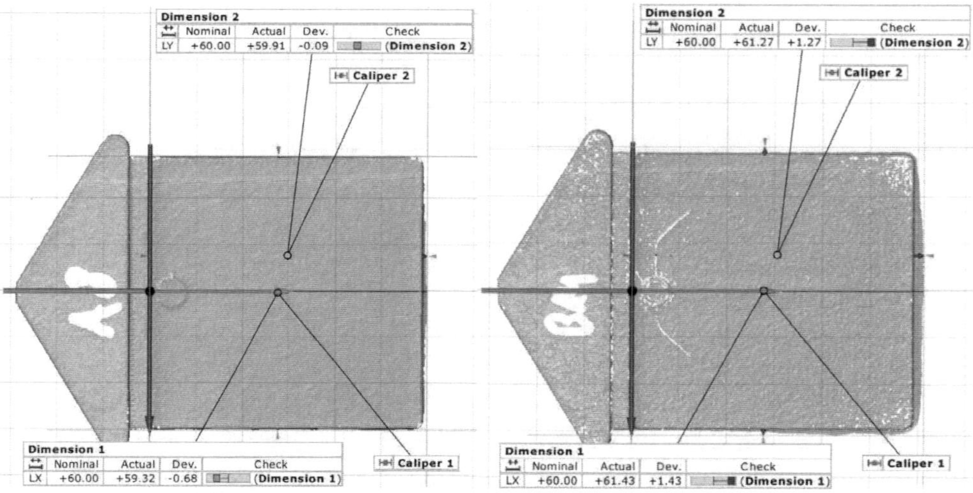

(a) (b)

Figure 2. (a) Scanned test specimens produced on a new tool steel insert and (b) specimen produced on a tool steel insert with a visible sign of macro-cracks

From Fig. 2a we can see that the produced test specimen is within the prescribed tolerances (59.91 mm in length and 59.32 mm in width) and with no visible defects. As can be seen from Fig. 2b the test specimen has a clearly visible macro-crack. Also both dimensions are outside of the nominal 60 mm in the longitudinal (1.27 mm) and transverse (1.43 mm) direction.

An advantage of scanning the test specimens is the fact that quality of surface and its deviation can be compared to the flat (ideal) surface. By comparing scanned test specimens we can notice also significant surface deviations with regard to the ideal surface, which are within the range of 0.17 mm exactly on the place of macro-cracks.

By measuring the simple waveform parameters such as energy, amplitude, hits, counts etc. on acquired AE signals, we can obtain useful information about the AE source intensity and its seriousness. In this way we can determine if the tool steel inserts quality is still good or if it is necessary to be replaced in the forthcoming future.

The peak AE signal amplitudes are normally related to the intensity of one or more AE sources, which are in our case macro-cracks located on the tool steel insert. In the past research works [9, 10] correlations between total counts, count rate and various fracture mechanics parameters (like for example the stress intensity factor) have been established and can be expressed by the following equation

$$N \cong K^{n} \tag{1}$$

where K is the stress intensity factor, N is the total number of counts and n is a constant value between 2 and 10. The fatigue crack propagation rate is defined by

$$\frac{dN}{dc} \cong \frac{da}{dc} \tag{2}$$

where a is the crack size, c is the number of cycles and N is the total number of counts.

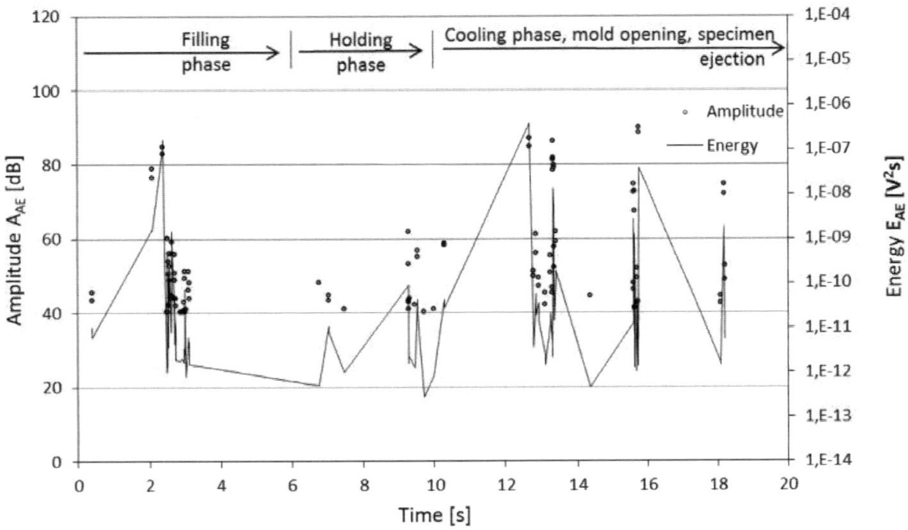

a)

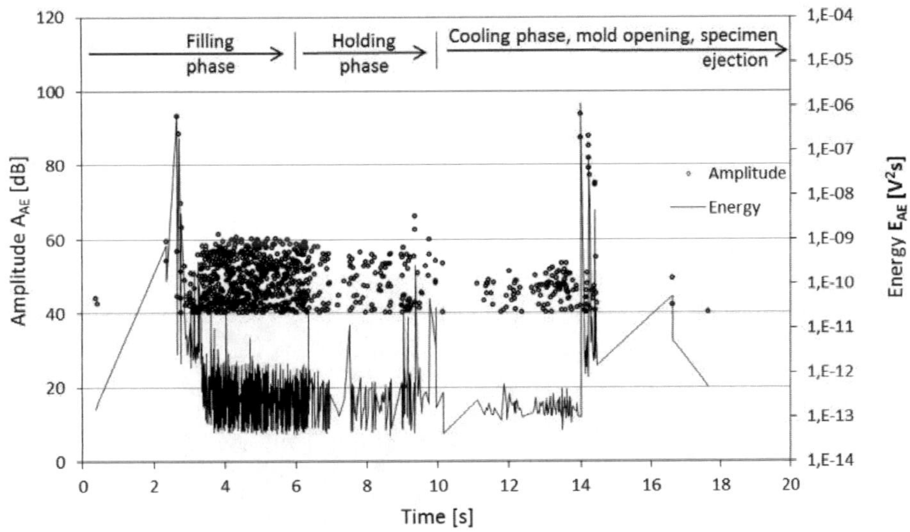

b)

Figure 3. AE signals during production of (a) a test specimen with a new insert, and (b) a test specimen with a damaged insert

Figure 3 shows the amplitudes and AE signal energy during the production of a test specimens with a new tool insert and damaged tool insert. Presence of cracks in the insert can be detected by increase of amplitudes and energies of measured AE signals. On the tool steel insert with macro-crack a significant increase in the number of detected AE signals at the end of filling phase and beginning of holding phase can also be seen. Increase of AE activity is caused by increase of mechanical stresses in the injection moulding tool.

The analysis of the AE signal based on the spectrogram has been utilized to evaluate the behavior of injection molding process. A spectrogram is a visual representation of the spectrum of frequencies in an AE signal as they vary with time. Time–frequency analysis is based on the decomposition of one-dimensional signals into two dimensions, time and frequency, represented through the respective axes. The time–frequency diagram allows observation of the distribution of energy of the acoustic signal. Colours of the time frequency diagram represent the most important acoustic peaks for a given time frame, with red representing the highest energies, then in decreasing order of importance, orange, yellow, green, cyan, blue, and black areas below a threshold decibel level. Fig. 4 shows AE signal waveform measured during filling phase 3 s after the beginning of injection molding of specimen with a new mold. Fig. 5 shows AE signal waveform measured during filling phase 6,5 s after the beginning of injection molding of specimen with a damaged mold and spectrogram of this signal. During the injection molding of a specimen with a new mold we measure predominantly signals with longer duration and not pronounced exponential decay of amplitude. These signals can be characterized as process orientated signals. During the injection molding in a tool with a crack we measured signals with an instant increase of amplitude and pronounced exponential decay of the amplitude. The acoustic emission burst rate during injection molding of specimen with a damaged mold is considerably higher and is a consequence of crack growth in steel tool insert. Pronounced frequency in the measured AE signals is 150 kHz, that is resonant frequency of the sensor.

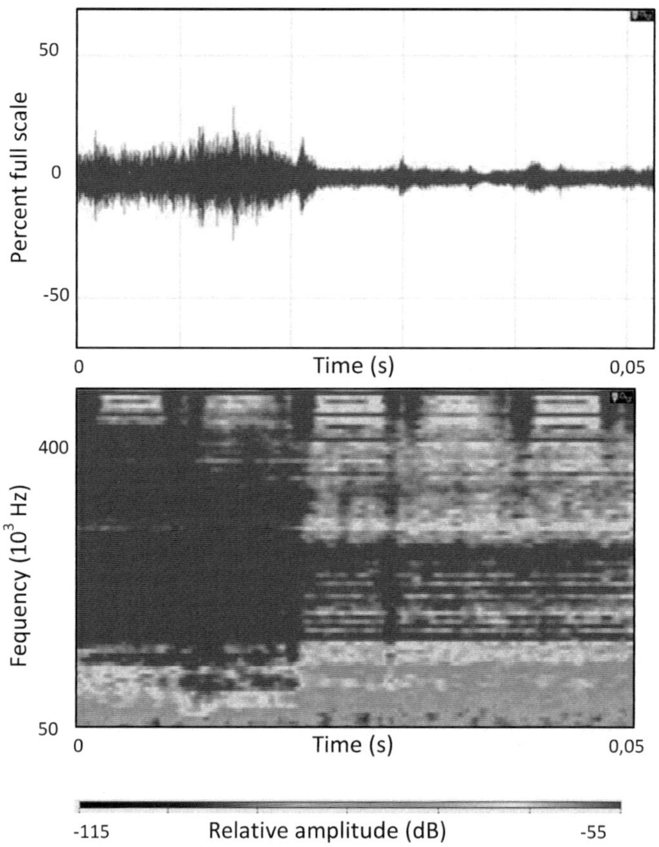

Figure 4. Time frequency diagram (spectrogram) of AE signal during filling phase with a new mold.

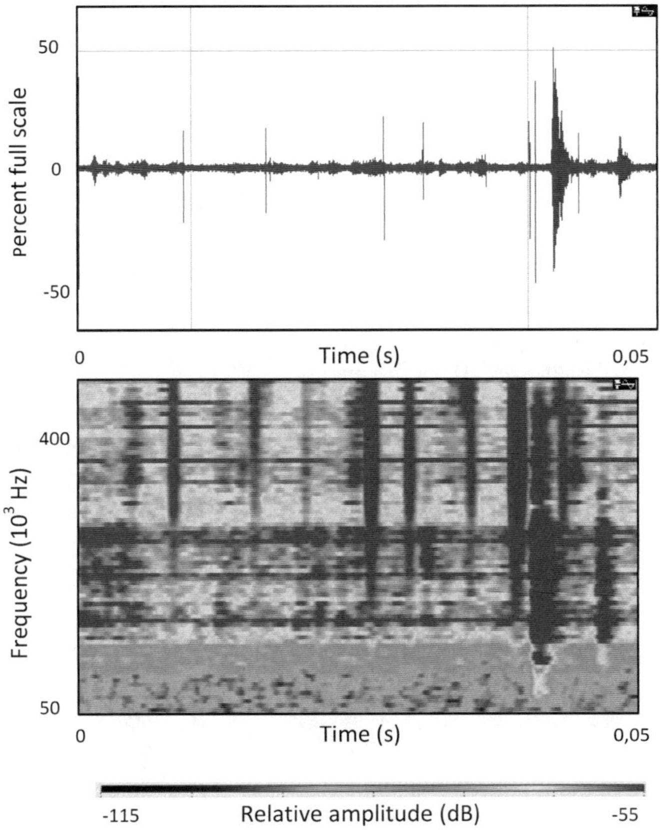

Figure 5. Time frequency diagram (spectrogram) of AE signal during filling phase with a damaged mold.

4. Conclusions

The aim of this research work was to determine to which extent it is possible with AE method to detect the presence of macro-cracks on engraving tool steel inserts by conducting a closer comparison of captured AE signals obtained from new and damaged tool steel insert under the same processing conditions. We have found apparent difference in activity of detected AE signals during the filling phase of commercial polypropylene material in favor of the tool steel insert with visible sign of macro-cracks. The results clearly show also the difference in the maximum amplitudes during filling and holding phase, as well as in energies of captured AE signals. From our experimental results we were able to obtain useful information about the presence of macro-cracks during production of standard test specimens.

Acknowledgement

The research work was partly financed by the European Union, European Social Fund, and by the Slovenian Ministry for Higher Education, Science and Technology.

References

1. Chen Z., Turng L.S, (2005) `A Review of Current Developments in Process and Quality Control for Injection Moulding` , Advances in Polymer Technology, Vol. 24, No. 3, pp.165–182.
2. Muller, F., Rath, G., Lucyshyn, T., Kukla, C., Burgsteiner, M., Holzer, C. (2013) `Presentation of a Novel Sensor Based on Acoustic Emission in Injection Moulding`, Journal of Applied Polymer Science, Vol. 127, pp.4744–4749.
3. Lijuan, Z., Yu, L., Chen, P., Cheng-Kui J., Kuo-Ding W. (2012) `Real-Time Diagnosing Polymer Processing in Injection Moulding Using Ultrasound`, Journal of Applied Polymer Science, Vol. 126, pp.2059–2066.
4. Kusic, D., Kek, T., Slabe, J.M., Svecko, R., Grum, J. (2013) `The impact of process parameters on test specimen deviations and their correlation with AE signals captured during the injection moulding cycle`, Polymer Testing, Vol. 32, pp.583–593.
5. Cao, J., Luo, H., Han, Z. (2012) `Acoustic emission source mechanism analysis and crack length prediction during fatigue crack propagation in 16Mn steel and welds`, Procedia Engineering, Vol. 27, pp. 1524-1537.
6. Biancolini, M.E., Brutti, C., Paparo, G., Zanini, A. (2006) `Fatigue cracks nucleation on steel, acoustic emission and fractal analysis`, International Journal of Fatigue, Vol. 28, pp. 1820-1825.
7. Mukhopadhyay, C.K., Sasikala, G., Jayakumar, T., Raj, B. (2012) `Acoustic emission during fracture toughness tests of SA333 Gr.6 steel`, Engineering Fracture Mechanics, Vol. 96, pp. 294-306.
8. Berkovits, A., Fang, D. (1995) `Study of fatigue crack characteristics by acoustic emission`, Engineering Fracture Mechanics, Vol. 51, pp. 401-416.
9. Miller, R. K., Hill, E.V.K., Moore, P. O. (2005) `Nondestructive Testing Handbook, Third edition, Vol. 6, Acoustic Emission Testing`, American Society for Nondestructive Testing, Inc..
10. Vallen, H. (2006) ´Acoustic Emission Testing: Fundamentals, Equipment, Applications`, Castell Publication Inc., Wuppertal, Vol. 6.

Validation of sensitivity of the UST-Method with a test leak

Dipl.-Ing. (FH) Robert Brockmann
Am TIZ Darmstadt GmbH, Robert-Bosch-Str. 7, 64293 Darmstadt, Germany,
Phone: +49 179 5424359, mail: robert.brockmann@ust-verfahren.de

Abstract
The test sensitivity of classic sniffer test methods is limited by the helium concentration in the air with 5.2 ppm, but by reducing the atmospheric helium concentration of the sniffer method B3 from the DIN EN 1779 the test sensitivity can be significantly improved by the sensitivity factor 500,000. In contrast to conventional vacuum leak test methods, the UST-Method reduces the helium partial pressure in a chamber with a helium free gas. This brings the helium partial pressure down to $1x10^{-08}$ mbar (10 ppt), so the gas tightness of components can be easily and reliably determined with a sensitivity of 10^{-11} mbar*l/sec. During the construction of the fusion experiment Wendelstein 7-X (W7-X), it was firstly shown that leaks at normal pressure (1013 mbar and 20°C) up to 10^{-09} mbar*l/sec are easily detectable. This means that, with the UST-Method, leaks with a theoretical loss of gas of 5 cm^3 in 30 years are well within the detectable range.

Keywords: atmospheric leak detection method, sniffing method, partial vacuum effect, UST-Method

1. Introduction

Fig. 1 magnetic cage of the W7-X with 70 superconducting coils

The fusion experiment W7-X of the Planck Institute for Plasma Physics in Greifswald is the largest stellarator experiment of this kind in the world [1]. The aim of the experiment is the inclusion of a stationary plasma with 3 Tesla in a super-conducting magnetic cage. To run the fusion experiment W7-X, the magnetic cage with its 70 superconducting coils in the cryostat, cooled by liquid helium, must be absolutely free of helium leaks. Before the start of the physical experiment in August 2015, the absolute helium leak security of the cryostat had to be assured. This meant that the absolute helium leakage of all the components of the magnetic cage had to be tested with $1x10^{-09}$ mbar*l/sec. Due to the complex structure of the superconducting coils, it was not possible to localize leaks with classic test methods of the DIN EN 1779 [2]: the vacuum test gas, the overpressure test gas methods. Therfore, the team decided to improve the sensitivity of the overpressure test gas method DIN EN 1779 B3 by leveraging the partial vacuum effect. The overpressure test gas method in combination with the partial vacuum effect, led to an improved sensitivity from 1 ppm to 1 ppt and led to the development of the Ultra-Sniffer-Testgas-Method (UST-Method). Because the extreme increase of sensitivity of the UST-Method by factor 500.000, the method required a special test leak as well as a validation from an independent testing laboratory.

2. The UST-Method

The UST-Method is an overpressure gas test method according to DIN EN 1779. The inherent difference between the classic overpressure test gas method B3 of the DIN EN 1779 and the UST-Method is the partial vacuum effect. This effect allows it, in contrast to the vacuum test gas methods of the DIN 1779, to reduce the helium concentration, not by to evacuating the chamber, but rather by flushing the chamber with a helium free gas. This makes it possible to reduce the partial helium pressure inside in the test chamber with the helium free gas from 5.2 ppm to 10 ppt. To ensure that the test chamber, which is filled with the helium free gas, reserves its helium purity, the test chamber is pressurized with a slight overpressure to ambient pressure [3].

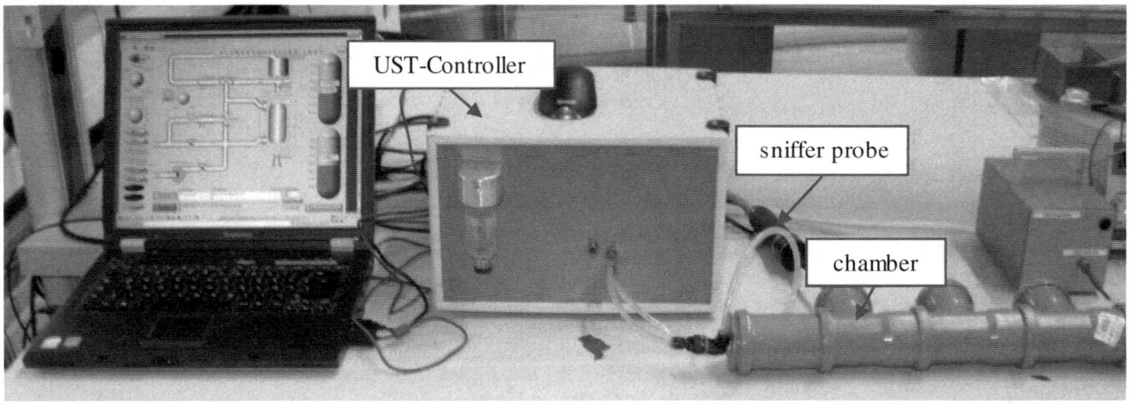

Fig. 2 Components of the UST-Method

Extensive testing with the UST-Method at the Max Planck Institute for Plasma Physics showed that with the UST-Method samples could be tested for the first time at atmospheric pressures with a system sensitivity of $1x10^{-11}$ mbar*l/sec. This means that the UST-Method has a 500.000 times higher measurement sensitivity than the classical sniffing method B3 of DIN EN 1779. Because of the extremely high sensitivity of the UST-Method, the detection of ten helium atoms in one trillion atoms, it is possible to prove the gas loss of a sample with 5 cm^3 in 3.000 years. This corresponds to a leakage rate of $5x10^{-11}$ mbar*l/sec. It is also possible to locate leaks of the size of $1x10^{-09}$ mbar*l/sec precisely. This corresponds with a gas loss of 1 cm^3 in 30 years.

As part of the construction of the fusion experiment W7-X, the UST-Method located several leaks [4], also a leak in a superconducting coil of the W7-X experiment [5][6]. To locate the leak, components of the coil were wrapped in to plastic bag and the superconductor of the coil were pressured with 33 bar helium, so that the leakage could be located in the area of an aluminum weld of a interlayer joint. In addition, measurements with the overpressure test gas method B4 and the vacuum test gas method A2.1 showed that the leakage could not be located be using these leak detection methods.

3. Mathematical Description of the helium concentration rise

To validate the sensitivity of the UST-Method, in the first step of the validation the theoretical rise of the helium concentration in a test chamber was determined. The increase of the helium concentration inside of the chamber follows the function:

$$Q_t = (Q_{leak} - Q_{UST-Gas}) - \left((Q_{leak} - Q_{UST-Gas}) - Q_0 \times EXP\left(\frac{\Delta\, t \times S_{eff}}{V_{chamber}}\right)\right) \quad (1)$$

$Q_t = leakage\ rate\ (mbar * l/s)$
$Q_{leak} =\ leakage\ rate\ of\ the\ leak\ (mbar * l/s)$
$Q_{UST-Gas} = Leakage\ rate\ of\ the\ UST - Gas\ (mbar * l/s)$
$Q_0 = leakage\ rate\ in\ the\ chamber\ (t = 0)(mbar * l/s)$
$\Delta t = testing\ time\ (s)$
$S_{eff} = flow\ rate\ sniffer\ probe\ (l/s)$
$V_{chamber} = volume\ of\ the\ chamber\ (l)$

Measurements showed that the increase of the helium concentration in a metallic chamber could be described very well by the function (1).

4. UST-Method / vacuum test gas method: Comparative Measurement

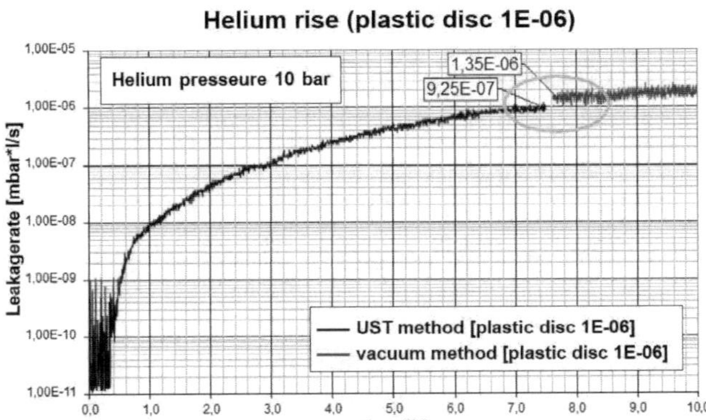

Fig. 3 Determining of the leakage rate with the UST-Method and the vacuum test gas method

To validate the relationship between the UST-Method and the vacuum test gas method comparative measurements were done with plastic slices. For these comparative measurements between the UST-Method and the vacuum test gas method several plastic slices were clamped into a CF flange and the plastic slices was pressured with 10 bar helium. For the determination of the leakage rate, the CF flange with the plastic slices was inserted into a metallic chamber. In the first step of comparative measurements, the helium gas stream was measured by the UST-Method. By raising a certain helium gas concentration, for example $9.25x10^{-07}$ mbar*l/sec, the measurement was terminated with the UST-Method and it was continued with the vacuum test gas method. To illustrate the relationship between the UST-Method and the vacuum test gas method in the range of $4x10^{-09}$ to $4x10^{-06}$ mbar*l/sec, the results were plotted in a graph. It was found out that the measured helium gas stream of the UST-Method and the vacuum test gas method are linear related. It has been shown over the measurement range, that the leakage rate of the vacuum test gas method is 1.2 times higher than the UST-Method. The explanation is that the UST-Method runs with a difference pressure of 10 bar and the vacuum test gas method with a difference pressure of 11 bar. In the assumption that the helium gas flow of a molecular leakage increases linearly with the pressure difference, the relation between the UST-Method and the vacuum test gas method could describe by the function [7]:

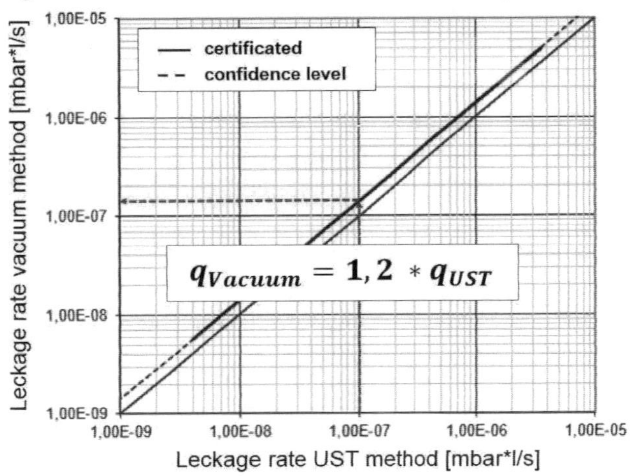

Fig. 4 Relationship between the UST-Method and the vacuum test gas method

$$Q_{Vacuum} = \frac{Q_{UST} \times \Delta p_{Vacuum}}{\Delta p_{UST}} = \frac{Q_{UST} \times 11bar}{10\text{bar}} = 1.1 \times Q_{UST} \qquad (2)$$

$Q_{Vacuum} = leakage\ rate\ vacuum\ test\ gas\ method\ (mbar * l/s)$
$Q_{UST} = leakage\ rate\ UST\ method\ (mbar * l/s)$
$\Delta q_{Vacuum} = pressure\ difference\ vacuum\ test\ gas\ method\ (bar)$
$\Delta q_{UST} = pressure\ difference\ UST\ method\ (bar)$

Based on the measured results with the plastic slices and a pressure difference of 10 bar and 11 bar, it can be assumed, that the leakage rate of the UST-Method and of the vacuum test gas method are linear related by the function (2). This will be accurate also for laminar leakage with a quadratic pressure relationship [7].

5. Development of a test leak for the UST-Method

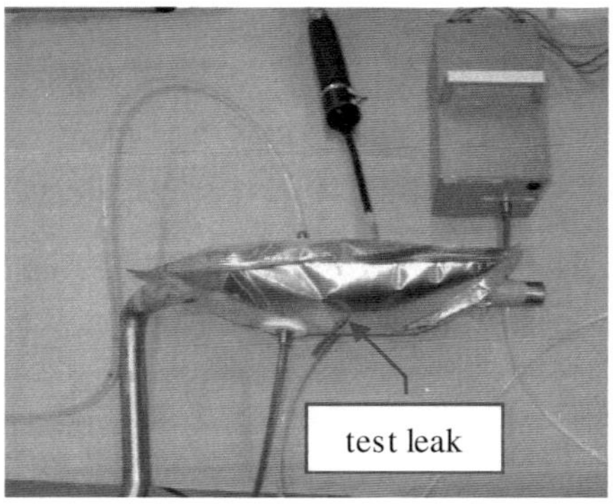

Fig. 5 Proof of set up sensitivity with the test leak for the UST-Method

Test results of the UST-Method could be reproduced for very small leaks, but could not be validated. Therefore, there was a need to develop an own test leak for the UST-Method to validate test results. This test leak allows to inject a steady stream of helium into a chamber. In the first step of the validation of the test leak, the leakage rate of the test leak was determined by a test setup with a metallic chamber. This validation of the leakage rate of the test leak allowed to determine the sensitivity even of a test set up with a flexible plastic bag. To determine of the sensitivity of the set up with a flexible bag, the bag was pressurized with a helium free gas. After reaching a stable leakage rate of less than 1×10^{-09} mbar*l/sec, helium was injected into the flexible bag with the test leak. With a test leak of 1.2×10^{-07} mbar*l/sec, the leakage rate rise in a few minutes from 7×10^{-10} to 9.7×10^{-08} mbar*l/ sec. So, the UST test leak allowed to determine the sensitivity of the UST-Method. For the qualification of these results a further validation of the leakage rate was done by a certificated testing laboratory [8]. The validation through the testing laboratory showed that the leakage rates, which were measured from the testing laboratory, were equal the leakage rate from the earlier measurements.

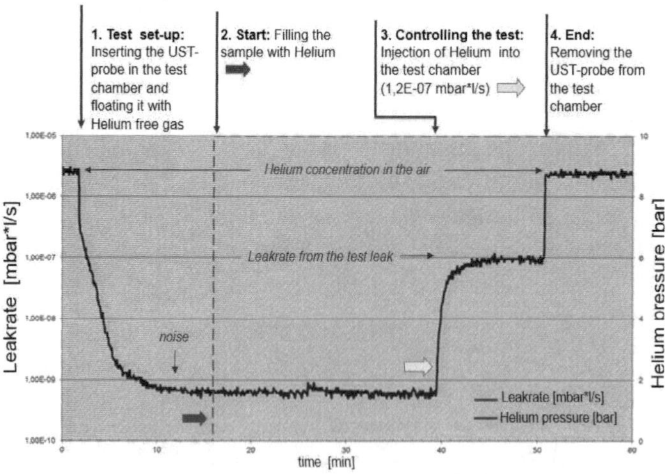

Fig. 6 Proof of the sensitivity of a test set up

6. Conclusions

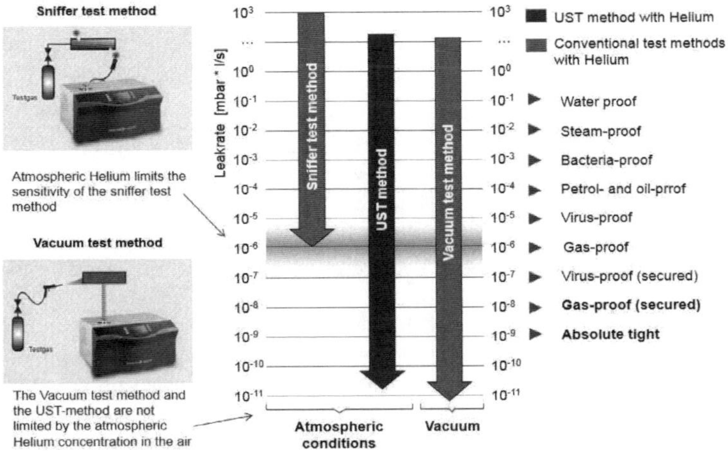

Fig. 7 Sensitivity of the UST-Method

The different steps of the validation of the UST-Method showed that the leakage rate of the vacuum test gas method and the leakage rate of the UST-Method rate correspond. In addition, the leakage rate of the UST-Method has got a proven linear relationship with the leakage rate of the vacuum test gas method from $4x10^{-09}$ to $4x10^{-06}$ mbar*l/sec. Furthermore, it was shown that with the test leak for the UST-Method, even with a test setup with a flexible plastic bag, the sensitivity of the UST-Method can be reliably measured. This means that just like the vacuum test gas method, the integral gas tightness of a sample can be determined with a sensitivity of 1 ppt. In addition, tests during the construction of the W7-X have also shown that the leakage rate of the UST-Method is comparable with the leakage rate from the vacuum test gas method. It was also shown that the expenses for the leak location by using the UST-Method, because of the sensitivity increase of 500,000 of the overpressure test gas method B3, could be significantly reduced.

Acknowledgements

The successful execution of the UST project would not have been possible without the full commitment of Dr. Hans-Stephan Bosch, Johannes Peter Kallmeyer, Mirko Marquardt and Dr. Felix Schauer from the Max Planck Institute for Plasma Physics.

References

[1] Thomas Klinger: 'Stellarators difficult to build? The construction of Wendelstein 7-X', ITER Seminar, Cadarache, France, 14th April, 2012 (internet presentation)

[2] Gerald Schröder: 'Neue Norm zur Auswahl eines geeigneten Verfahrens zur Lecksuche und Dichtheitsprüfung', ZfP-Zeitung 74, April 2001

[3] Robert Brockmann: 'New leak detection method by UST', Physikalische-Technische Bundesanstalt (PTB), Workshop on Vacuum Metrology Industry, Berlin, Germany, June 25-27, 2014

[4] Robert Brockmann: 'Ortung einer Leckage mit dem Ultra-Schnüffler-Testgasverfahren für das Fusionsexperiment Wendelstein 7-X', DGZfP-Jahrestagung 2014, Berlin, Germany, May 26-28, 2014

[5] Robert Brockmann: 'Ultra-Schnüffel-Testgasverfahren auf Grundlage des Partial-Vakuum-Effekts', Vakuum in Forschung und Praxis, Volume 25, Issue 6, pages 19–21, Dezember 2013

[6] Robert Brockmann: 'Verlust der Heliumdichtheit: Verletzung der Passivschicht beim Werkstoff Aluminium', Vakuum in Forschung und Praxis, Volume 26, Issue 2, pages 19–22, April 2014

[7] Dr. Walter Umrath: 'Grundlagen der Vakuumtechnik', pages 110, Köln, March 1997

[8] Prof. Dr. Wolfang Jitschin, Technische Hochschule Mittelhessen, Fachgebiete Physik und Vakuumtechnik, Wiesenstraße 14, 35390 Gießen, Germany

In Situ Calibration of Acoustic Emission Sensors

Jan KOBER[1,2], Zdenek PREVOROVSKY[1]

[1]Department Impact and Waves in Solids, Institute of Thermomechanics, Academy of Sciences of the Czech Republic, Dolejskova 5, 18200 Prague 8, Czech Republic, Phone: +420266053832, e-mail: kober@it.cas.cz, zp@it.cas.cz

[2]Department of Materials, Faculty of Nuclear Sciences and Physical Engineering, Czech Technical University in Prague, Brehova 7, 115 19 Prague 1, Czech Republic

Abstract

Calibration of acoustic emission (AE) sensors is recommended when comparing results of separate experiments or when studying frequency content of AE events. However conventional methods of AE sensor calibration require a reference sensor and/or special test specimen, equipment which is not readily available for most. Furthermore the calibration results are only approximately transferrable to practice, because they depend on bonding conditions and mechanical properties of the medium. A partial solution of this problem is proposed in this contribution: a reference-free in situ method of sensors frequency responses determination. It is based on standardized reciprocal calibration and time reversal method. It will be shown that it is universally applicable for reversible sensors and when certain conditions are met even to irreversible sensors. The theoretical development of the method is accompanied by experimental validation – assortment of five sensors was tested on an airplane wing flange specimen. Analysis revealed that three sensors can be considered reversible and two shown a resonance frequency shift in lower frequency region between reception and transmission.

Keywords: standardisation, Acoustic Emission (AE), time reversal, acoustic emission, calibration, frequency response

1. Introduction

Acoustic emission (AE) sensors are designed for maximum sensitivity to incident elastic waves. Therefore the majority of sensors uses undampened piezoelectric elements working in relatively narrow band around dominant resonances. As a consequence AE sensors exhibit complicated impulse responses which significantly influence the voltage output signal. Although basic application of AE methods can utilize such raw signals, in cases when results of separate experiments need to be comparable, an elemental sensor calibration is necessary. Furthermore it is advantageous when physical interpretation of AE events is attempted and it can improve AE events localization.

In this paper a reference-free, in-situ method of frequency calibration is proposed. The method is based on reciprocal calibration [1] and time reversed acoustics (TRA) [2]. The method is theoretically developed in section 2. The practical implementation is described in section 3. Signal processing and calibration results are presented in section 4.

2. Time reversal for sensor calibration

The introduction of TRA in sensor calibration was motivated by elimination of the need for special test specimens. TRA provides an effective way how to remove the influence of Green's function, which is a significant part of the signal chain from a source to a measured sensor output (1), from the measurement. The wave propagation between sensor *i* and *j* can be written in frequency domain as

$$U_{ij}(\omega) = R_j(\omega)G_{ij}(\omega)\tilde{R}_i(\omega)S(\omega)\,, \tag{1}$$

where R_i, resp. $\tilde{R}_i$ is the frequency response of *i*th sensor for reception, resp. transmission, G_{ij} is the Green function and *S* the transmitted source function. Near the focal time can be the result of classical TR experiment written as

$$W_{ji}^{C}(\omega) \approx R_i(\omega)\tilde{R}_i^*(\omega)R_j^*(\omega)\tilde{R}_j(\omega)S^*(\omega), \tag{2a}$$

where asterisk marks complex conjugation. The result of reciprocal TR experiment on the other hand reads

$$W_{ji}^{R}(\omega) \approx \tilde{R}_i^2(\omega)R_j^2(\omega)S^*(\omega), \tag{2b}$$

i.e. sensor i acts only as a transmitter and sensor j only as a receiver. It is useful to introduce frequency transfer function defined as

$$T_{ji}^{C/R}(\omega) = W_{ji}^{C/R}(\omega)/S^*(\omega), \tag{3}$$

where the source function is deconvoluted from (2). The assumption of transducer reversibility will be taken at this point, i.e. $\tilde{R}_i(\omega) = R_i(\omega)$. Now the number of unknowns in (2) reduces to two. When a third transducer is introduced, the set of equations (3) can be solved, e.g. for transducer 1

$$\frac{T_{12}T_{31}}{T_{23}} = |R_1|^4. \tag{4}$$

The relation (4) forms the fundamental principle of the time reversal calibration method proposed in this paper.

3. Experimental setup and procedure

3.1. Sample and sensors

The experiment was conducted on a section of an airplane wing flange (fig. 1). The sample was machined from Al alloy 2024 T3 equivalent and previously used for a fatigue test. Sample dimensions in wide (narrow) end were 365 x 145(117) x 55(48) mm. This sample was chosen for its complex geometry and low attenuation which on one hand suits the single-channel TR experiment [3] and on the other hand provides a challenging conditions for in situ sensor calibration. An assortment of sensors of various types and sizes (see tab. 1) was attached to the sample with cyanoacrylate glue. Sensors were placed over thick and thin cross-sections.

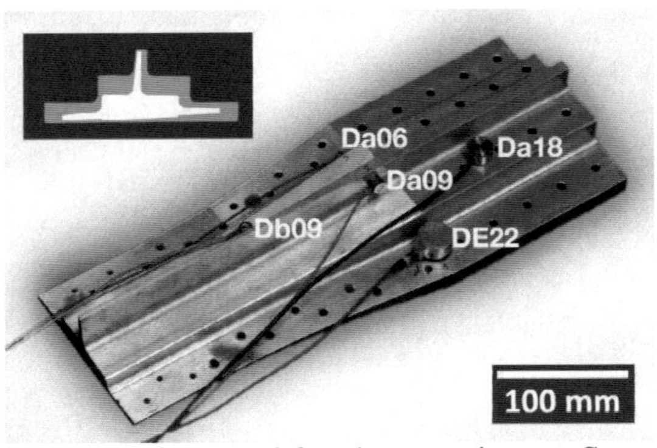

Fig. 1: Airplane wing flange section used for the experiment. Sensor placement is shown.

Table 1: Overview of tested AE sensors

Manufacturer	Dakel	Dakel	Dakel	Dakel	Dunegan/Endevco
Type	MIDI	IDK-09	IDK-09	custom	S9204
Size	6 mm	9 mm	9 mm	18 mm	22 mm
Abbrev.	Da06	Da09	Db09	Da18	DE22

3.2. *Source functions, experimental setup and measurement procedure*

A source function must balance its frequency bandwidth and duration for optimal results of TR reconstruction. A symmetrical chirp function was selected as a base function:

$$s_0(t) = \begin{cases} \cos\left(2\pi\left(f_0 t - \beta t^2\right)\right) & for\; t \in \langle -T/2, 0) \\ \cos\left(2\pi\left(f_0 t + \beta t^2\right)\right) & for\; t \in \langle 0, T/2 \rangle \end{cases}$$

$$\beta = (f_1 - f_0)/(T/2), \tag{5}$$

which was then applied for 10 partially overlapping frequency intervals covering range from 70 kHz to 1,2 MHz as was the expected bandwidth of used sensors. The duration of source function decreased from 200 us for low frequency intervals to 100 us for high frequency intervals.

Using this set of 10 source functions a classical and reciprocal TR experiment [2] was performed. The experiment was controlled through Matlab environment. At first forward propagation signals were recorded. The arbitrary wave generator Agilent 33522A was used to transmit a source function with the peak-to-peak amplitude of 10 V and 50 MHz sampling frequency. Responses on remaining four receiving sensors were recorded by TiePie HS3 and HS4 USB oscilloscopes triggered by generator output. Signals were sampled by 25 MHz and were 4 ms long. The measurement was repeated 40 times and the signals were averaged. Then a custom channel multiplexer was used to switch the generator to following sensors and the measurement was repeated until the complete set of forward propagation signals u_{ij} (i, j = 1-5) was obtained. The forward propagation signals were time reversed and TR reconstruction signals were measured.

4. Signal processing and calibration results

Signal processing was performed following the sequence outlined in the theoretical derivation of TRC method in section 2:

1. extraction of source function reconstruction,
2. spectral analysis,
3. deconvolution of time reversed source function,
4. separation of frequency responses by (4).

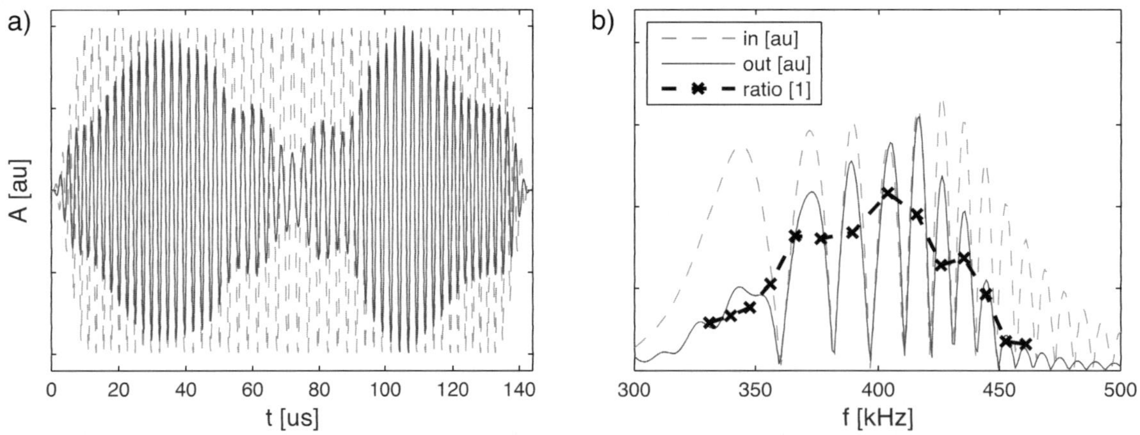

Fig. 2: Results of TR focusing between sensors Da06 and Da09. a) Comparison of input source function and TR reconstruction in time domain. Signals are normalized. b) Extended Fourier spectra of the source function and the TR reconstruction in linear axes. Their ratio in spectral lobes is shown. Data are scaled for presentation purposes.

The TR reconstruction was detected by cross-correlating w_{ij} with the source function. The maximum of Hilbert envelope of this cross-correlation marked the focal time. The TR reconstruction was then cropped from the signal and 10% cosine tapered (see fig. 2a). The spectral analysis was performed on source functions and TR reconstructions by Extended Fourier transform introduced by Liepins [4]. The deconvolution was completed by spectral division in individual spectral lobes in order to ensure its numerical stability (see fig.2b). Consequently; the results obtained for the ten applied source functions were merged into a single set so that nearby data points in overlapping frequency bands were averaged.

The results of the classical TR experiment were used for frequency response evaluation. The responses were separated by (4). Since five sensors were tested, it is possible to form six different groups of three for each sensor. Each group can be regarded as an independent experiment. The frequency responses were normalized by their geometrical mean in order to align the results from all sensor combinations. Frequency responses for tested sensors are given in fig. 3 together with the results of vibrometer measurements. The vibrometer measurements provide a useful reference of sensors resonances and frequency range. The vibrometer measurement of velocity was performed in the center of sensor faces using slowly frequency swept sine as a source (signal duration was 20 minutes). It can be seen from fig. 3, that even though the overall shape of frequency responses is often different for TRC and vibrometer results, the position of most sensor resonances agrees. Note the differing frequency responses of identical sensors Da09 and Db09.

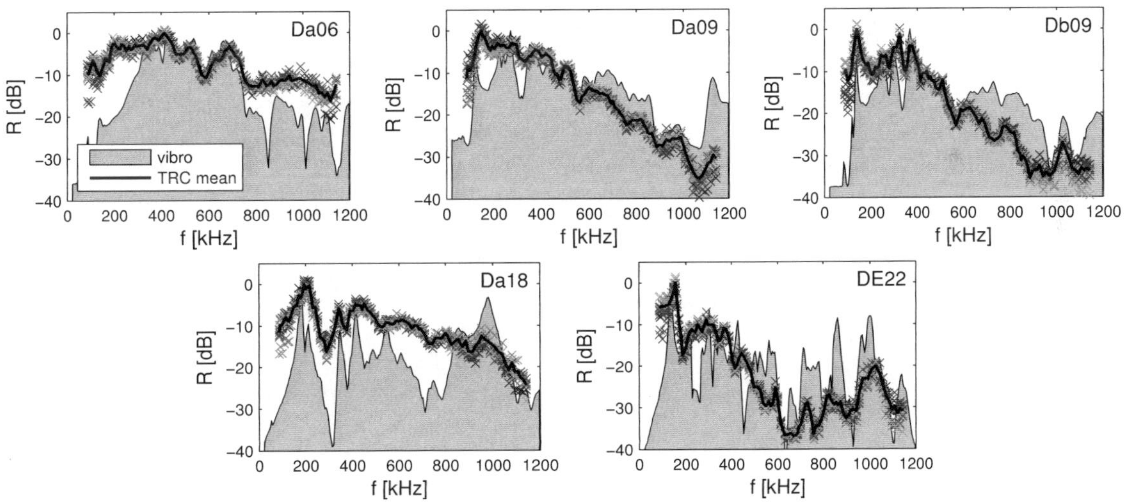

Fig. 3: Frequency responses of tested sensors. Comparison of TRC results and vibrometer velocity measurements. All results are normalized.

So far we have assumed that the sensors are reversible without any verification. However it is possible to test the reversibility of used sensors by evaluating results of reciprocal TR experiment. Whereas the frequency transfer function T_{ij}^{C} is symmetrical with respect to the indexes regardless of transducer reversibility, its reciprocal counterpart T_{ij}^{R} is generally not. The frequency transfer functions from reciprocal TR experiment T_{ij}^{R} can be symmetrical only in case when the transducers are reversible. This serves as a useful reversibility test. From the utilized transducers, three were confirmed reversible (Da06, Da09, Db09) and two irreversible (Da18, DE22).

5. Conclusions

A new method of AE sensor calibration was introduced – the time reversal calibration. It allows reference-free in situ determination of sensor frequency responses. The method is based on time reversed acoustics and reciprocal calibration. The method was introduced with the assumption of reversible sensors and the results were presented for five sensors tested on an airplane wing flange specimen. It proved the validity of utilized model and the robustness of results which exhibit low variation with regards to experimental configuration and sensors used for calibration. In our experiment it was possible to deduce that three sensors were reversible and two irreversible.

Acknowledgement
Institutional support RVO:61388998 is gratefully acknowledged. This conference contribution was granted by the Czech Society for Nondestructive testing.

References

[1] H. Hatano, E. Mori, Acoustic-emission transducer and its absolute calibration, J. Acoust. Soc. Am. 59 (1976) 344–349.
[2] M. Fink, D. Cassereau, A. Derode, C. Prada, P. Roux, M. Tanter, et al., Time-reversed acoustics, Reports Prog. Phys. 63 (2000) 1933–1995.
[3] A.M. Sutin, J.A. TenCate, P.A. Johnson, Single-channel time reversal in elastic solids, J. Acoust. Soc. Am. 116 (2004) 2779–2784.
[4] V. Liepins, Extended Fourier analysis of signals, arXiv Eprint arXiv:1303.2033. (2013) 1–30. http://arxiv.org/abs/1303.2033 (accessed June 09, 2015).
[5] H. Lan, L. Yan, D. Xiao, Z. Jiao, I.-M. Chen, Surface-to-surface calibration of acoustic emission sensors, Sensors Actuators A Phys. 174 (2012) 16–23.

Introduction to the AFF Universal File Format and its Use for NDT Data

Martin DRAB, Michal HANAK

ZD Rpety – DAKEL; Prague, Czech Republic; Phone: +420 241715744, Fax: +420 241717288; e-mail: drab@dakel.cz, hanak@dakel.cz

Abstract

The Acquisition File Format (AFF) is a data container format recently developed by DAKEL within the EU FP7 project ALAMSA with the purpose to have a common, universal and publicly open data container format, which everyone would be able to use for exchange of data produced during various measurements, evaluations, simulations and other data-producing activities with emphasis on streaming data. This text is trying to show you the possibilities and some of its usage.

Keywords: Acoustic Emission (AE), NDT-wide, Data Acquisition, Data Storage, File Format, Acquisition File Format (AFF), Acquisition File Format Library (libaff)

1. Introduction

The Acquisition File Format (AFF) was created as a new open data container format that can primarily be used for storing data from various measurements, evaluations and simulations. It was developed by DAKEL as part of the EU FP7 project ALAMSA [1,2].

1.1 Motivation

The original idea was to have a universal data container format that could be used to exchange data from various experiments, measurements, evaluations and simulations conducted by individual partners within the project, with each partner having a different experiment system, producing various types and formats of data. To have a unified and data format would allow everyone within the project to easily exchange data from their experiments with others and it would also allow the tools produced within the project to have a unified input/output format for data.

Aside from the original idea, the DAKEL company is also intending to apply this file format as a primary file format for storing acquired data measured by their data acquisition devices as well as software data processing tools. The first device to actually use this file format natively is the new DAKEL ZEDO multi-channel synchronnous data acquisition system intended mainly (but by far not exclusively) for acoustic emission, but other DAKEL's devices (such as the DAKEL IPL) are also going to be retro-fitted with this new file format, so that data produced by all these devices could be processed by the same software tools and on the other hand, so that also other people can easily use the data produced by these devices as well.

To make things even easier for the AFF to spread more amongst people, perhaps also to other branches of scientific, industrial and other use, it was decided to let the format be openly available for general public to use, both regarding its specification and its reference implementation library (called *libaff*). For more details and downloads a site referenced at the following address is continuously being built and updated: "*http://aff.dakel.cz/*".

2. The File Format

First thing one needs to realize is that AFF is primarily a data container. It does not necessarily need to be aware of the interpretation of the contents it is carrying, though, in special cases it can be.

Second thing to realize is that AFF is intending to be a universal format in which many variants of data with very different interpretations can be stored. This is a very important fact to take into account when one is trying to work with the file format in its entirety. Loading of general data stored within the AFF into any application with one simple command is next to impossible. One reason being the variability of data types and dimensions (as described later in this chapter) and another reason being the volume of data that can potentially be stored there. The application attempting to load the data may simply not be prepared to handle all variants, types and interpretations of the data. And the application may also not be prepared to handle the volume there can be contained in a single AFF container and in general, one must always count on the contained data volume being larger than the memory available to the specific application.

On the other hand, if the application can assume and is interested only in a specific type of data, loading (and storing) of such AFF containers can be simplified a lot.

2.1 Data Suitable for AFF

The container can store multiple streams of data. The streams can be either time-based, in which case each point of data has time associated with it, or non-time-based (just indexed). It is not recommended to mix time-based and non-time-based data in one container, though.

Primarily, the format is intended for storing time-based data and specifically data organized in sequences of points of identical type, equidistant in time. Where each point consists of a 0- or more-dimensional matrix of defined number of elements of the same type, as you can see on Figure 1.

The elements of the point matrix can be either ordinary numbers, such as integers or floating-points, or complex numbers or a block of data of unknown internal structure, defined only by its total size.

Another type of data, that can be stored within the streams is so called *timed structures*, which is an irregular-time sequence of structures of defined size which each have a time stamp on a specified position within them. And the last type of data is *variable length structures*, where one has to provide a size for each structure along with its data. The last two types can even be combined, but are only supported for 0-dimensional matrix points and there are a few other restrictions when they are used.

Additional data types understood by the format may be added in later versions of the format specification.

2.2 Additional Data Information

In addition to the main streams and main data within them, there is a lot of additional data and meta-data that can be stored within the container as well.

2.2.1 Axis and Data Transformation and Unit

For each spatial axis of the matrix points, the axis points can be defined either by a simple formula derived from the index of each point on the axis or manually by direct mapping and a unit can be assigned to it, so that for instance when the points are displayed somewhere on a graph figure, the axes can be labeled appropriately.

In similar way the actual main data values of the stream can be transformed as well, also using one of the simple predefined mathematical equations, which sometimes can save a lot of space and/or precision. (E.g., for an exponential variable that has a broad range of values, it may be more effective to store just the logarithm of the variable itself as the main stream data values and provide a meta-information saying that the data stored should be interpreted as an exponent of the actual variable they should represent.) And again, a unit can be stored with the data as well.

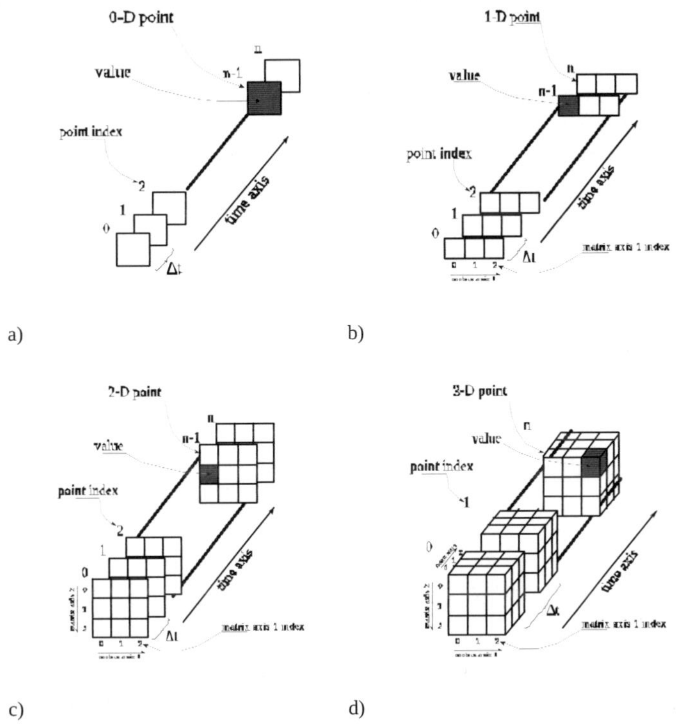

Figure 1: Visualization of an example of time-based equidistant sequences of points of various dimensions: a) 0-D points containing just one element; b) 1-D points of 3 elements; c) 2-D points of 3x3 elements; d) 3-D points of 3x3x3 elements. (Source of the figure: [2].)

2.2.2 *Meta-Information Describing the Stream Data*

Plenty of information can be associated with the data of each stream to further describe them, either for the purposes of the container itself (such as the type of conding and so on) or for the purposes of determining the origin of the data, how they were obtained, especially, if they were measured within an experiment with a physical device and a sensor of some kind.

For such purposes each data stream may have associated one or more (e.g., for the case when one stream of data contains samples from multiple channels) descriptions of a channel. The channel description contains information about the specific device and channel that has been used to obtain the data, about the channel's settings such as bit-resolution, amplifier gain, sampling frequency or voltage range.

The stream may also have associated information about one or more sensors used to acquire the data, where together with the information about the sensor model and number itself, one can also provide information, such as sensor's position or orientation in space, which may then help during modeling or visualizing the situation virtually in the computer.

Database of known and supported hardware shall be updated in the specification as needed or upon request, but there is always the possibility to set the value to *unknown.*

2.2.3 Extra Parameters of the Stream Data

Each stream can also have a number of so called parameters associated with it. These are simple variables, which do not change very often and which somehow provide additional information for the main data of the stream. Every parameter has a name and data type (similar to the main data stream matrix point elements) and optionally it may also have a unit. Special names of the parameters may have special meanings, for instance name "*@gain*" is being used for parameter giving variable value of amplifier gain used to sample the data.

Parameters are coded in a way, which is effective when they change only every now and then, but not too often. When parameter values change very quickly compared to the main data sampling, then the whole stream becomes less effective and it may be worth saving such a variable as a separate stream of data.

2.2.4 Tags Describing the Container and Data

Tags are used to contain strings of variable size describing various parts of the container, such as the entire container contents, individual axes, individual streams, devices and channels for individual streams, creator of the file, etc. with a human-readable text, which can be displayed to the user by the application reading the file.

2.3 AFF features

Among the current features of the AFF, we can mention things such *multi-file* container, which allows storing data from one container in multiple files. It is always better to have multiple not very large files, than to have one huge file from one measurement or evaluation. It always handles better, is easier to manipulate, copy, store on various media and file systems (some of which may have restrictions on size).

Each of the files belonging to one container contain a special *container identification and ordering*, so that even when one looses the filenames (for instance during a disk crash or occasional deletion and renewal), files belonging to the specific container can be found and ordered in the original order.

The container may also contain an *index* of the data, so that one can quickly find specified position on each stream (either position in time or position in points) and start reading from that specific position. There is also a sort of meta-index to quickly find key components of the container (such as the index itself).

More features are intended to be added to the format in the future.

3. AFF Reference Implementation Library

AFF reference implementation library (called *libaff*) is a C library written in C, which implements the specification of the AFF and is publicly available and can be used for your application to access and work with the AFF container files. It can be used to both create AFF containers and write data into them and to read the data from the AFF containers. It tries to implement all the features described in the specification [2]. It also tries to be as effective as possible to handle even large amounts of data as quickly as possible and it is currently implemented for both Linux and Windows operating systems. It is still a work in progress but by the time of the NDT in Progress 2015 conference the first release should already be available for testing and more features and implementations new versions of the format would be added in time.

Since the actual interface of the library encapsulates all the features of the file format, its API interface may seem a little bit complicated for some users (even though we tried to design it as easy as possible, while keeping the functionality), but severe simplifications may be done for specific sub-set of stored data types and even interfaces to other applications (e.g., MatLab) are planned.

Also some tools for basic operations with the AFF containers are currently being worked on, which would hopefully also help and ease the usage of the containers.

So, keep watching the site mentioned in the "*Motivation*" paragraph for updates and progress and other information related to AFF.

4. Conclusions

AFF was designed as a universal but robust file format able to handle large quantities of data for a large variety of possible application uses, but specifically with the NDT data in mind and since all of its specifications and reference implementations are to be openly available for the general public to use, we hope that would help to spread its use in time and be beneficial to everyone in the NDT community and beyond.

Acknowledgement

This work is a part of project ALAMSA which has received funding from the European Union's Seventh Framework Programme for research, technological development and demonstration under grant agreement number 314768 (duration: 11/2012 – 10/2016).

References

1. http://www.bath.ac.uk/ris/about/rpms/projects/alamsa/index.html

2. M Dráb, 'Definition of universal file format for all NIM techniques (Acquisition File Format (AFF) Specification Version 0.1)', EU Seventh Frame Programme ALAMSA, Work Package 5, Deliverable 1; October 2, 2013

Modelling the Influence of Iron Oxide Layers on the Microstructural Damage of Boiler Tubes

J. D. Angelo[1], A. Bennecer[2], S. Kaczmarczyk[2] and P. Picton[2]

[1]Universidade Federal do ABC, Sao Paulo, Brazil.
jonhson.angelo@ufabc.edu.br

[2]University of Northampton, Northampton, U.K.
stefan.Kaczmarczyk@northampton.ac.uk

Keywords: Boilers tubes, heat transfer, oxides layer, creep.

Abstract: Oxide layer growth inside boiler tubes when they are exposed to high temperatures is expected to assume increasing importance as plants are run beyond their design lifetime at maximum load and as steam temperatures and pressures increase with new plant designs. Thus, frequent inspections are required to ensure that the boilers are still safe to use and it is cost effective to keep them running for longer periods of time.

There is a need to determine what measures to be taken to ascertain a safe boiler operation. This need has been recognised and addressed in the past, but the remedial measures devised have often failed to solve the problem. This research explores, through a model developed using a finite element computer simulation platform, the thermal behaviour of slender tubes under a constant working temperature of 728 °K. Our simulation results demonstrate that hematite layers up to 15μm thickness inside the tubes do not act as insulation. They clearly show the process of long term overheating on the outside of boiler tubes which in turn leads to the initiation of flaws.

Understanding how the oxides' growth can damage boiler tubes and lead to the creep process can help to establish inspection programs. These could include oxide thickness measurements, the aim of which would be to prevent failures before the creep process is initiated.

1 INTRODUCTION

Boilers and other heavy industrial equipment operating at high temperatures and pressure are designed for a finite life [1]. The life times of boilers and steam generators are between 20 to 30 years [2]. In recent years heavy industrial structures have aged beyond their design lifetime. Indeed, most boiler tubes have been operating for more than 30 years. Therefore, frequent inspections are required to ensure that the boilers are still safe to use and it is cost effective to keep them running for longer periods of time. Figure 1 shows the areas of one aqua tube wall boiler subject to overheating

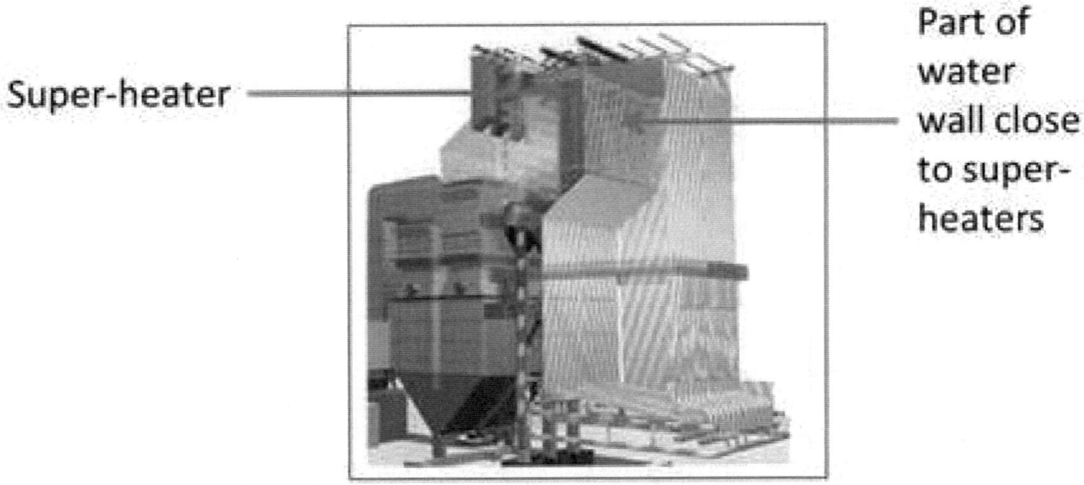

Figure 1: Hot areas shown in red are subject to overheating in a boiler [3].

The temperature inside the water wall tubes can reach 943 ^{o}K. However, super heaters and re-heaters can operate at temperatures of nearly 1163 ^{o}K. At these temperatures, microstructural changes occur in ferritic steel structures but they do not pose a substantial risk of failure [2]. One study conducted by Graham and Waleed [4] at Saudi Aramco on 101 cases of boilers tubes concluded that 24% of tubes such as furnaces fail due long term overheating and 6% of tubes such as super heaters fail due to short term overheating. The distribution of damage mechanisms is depicted in Figure 2.

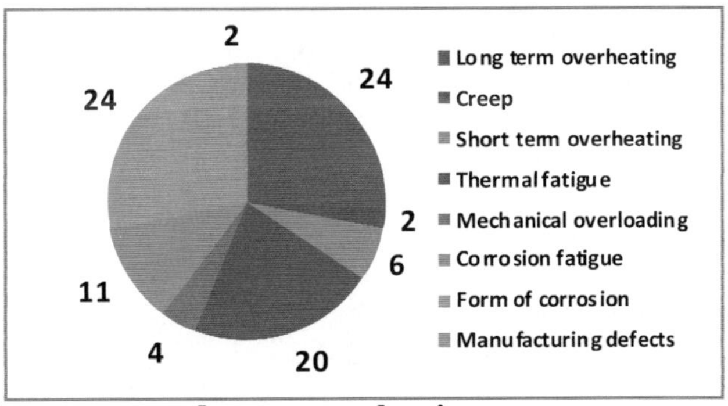

Figure 2: Boiler tubes common damage mechanisms.

Overheating is associated with the formation of different oxide layers with different thicknesses over the internal surface of the tubes where less than 3% Cr is present. At temperatures below 833 °C and high partial pressure of O_2 this film is made of two layers: 1) a magnetite (Fe_2O_3) layer and 2) a hematite (Fe_2O_4) layer. At temperatures higher than 873 °C an additional layer of iron monoxide (FeO) also appears.
The effects of the oxidation layers on the mechanical behaviour of materials manifest themselves in a reduction of the tube cross section which increases the localised stress. The film growth causes an insulation of this section leading to a temperature increase within the cross section and acceleration of material degradation. Also, spalling of oxide scales can be beneficial to reduce the aforementioned effects, but can conversely result in tube overheating if it becomes entrapped in the system, thus reducing the flow rate in this section [5].

The presence of two or more layers of dissimilar materials which present a mismatch between the coefficients of thermal expansion induces stress. As a consequence, a delamination or more likely a crack would occur. In this paper, we study the effects of long term overheating due to the formation of oxide layers inside boiler tubes. We use finite element method in order to model the thermal behaviour of the tube under long term overheating. The study is conducted under stationary conditions to reflect real boiler conditions.

2 METHODOLOGY

The data of operating pressure, design pressure, design temperature and feed water temperature used in our study is sourced from a VU-60 boiler [6] (see Table 1). The structure is made of seamless carbon steel and is 6000 mm long with approximately 60 mm outside diameter and 50 mm inside diameter. It is subject to heating in excess of 450 °C in one longitudinal half. In reality, the tube length, diameter and thickness depend on the work pressure and the quantity of steam height generated.

Table 1: Boiler operational parameters [6]

Model	VU-60
Operating pressure	6.423 [MPa]
Design pressure	6.816 [MPa]
Design temperature	728.15 [^{o}K]
Feed water temperature	463.15 [^{o}K]

One half of the tube is exposed to ambient temperature while the other half is exposed to a combustion chamber. After a certain height, steam starts to evaporate from boiled water inside the tube. It is in this part of the tube where the oxide layer is likely to form and grow causing long term overheating due to the reduction in heat transfer between tube wall and steam. As the diffusion rate of these oxide layers raises exponentially with the temperature new alloys are developed slowing the growth rate. The heat transfer through a steel tube wall with iron oxide layer can be considered unidirectional as shown in Figure 3.

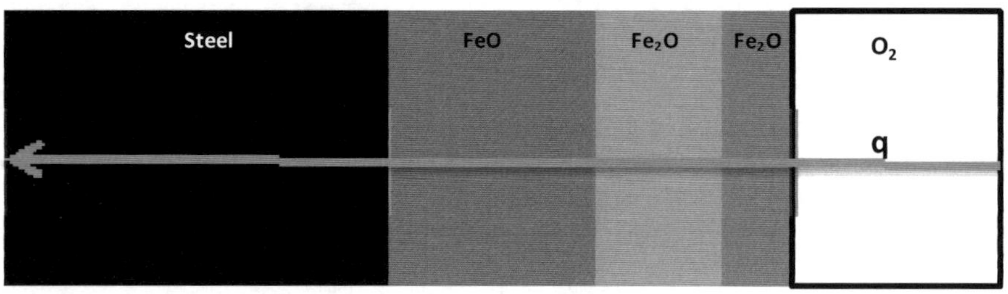

Figure 3: Heat transfer through iron oxide layer [6]

In this model we assume a unidirectional heat flux and there is a perfect adherence between the oxide layers and tube wall and there is no porosity inside the oxides. With this assumption the heat quantity that flows through the oxide layers and tube wall is given by the equation:

$$\frac{q}{A} = k \frac{dT}{dx} \tag{1}$$

Where: q=heat transfer rate,
k=material thermal conductivity,
A=material cross section,
T=temperature in degrees Kelvin, and,
x=material thickness.

In a stationary study equation (1) reduces to:

$$\frac{\mathbf{q}}{\mathbf{A}} = \mathbf{k} \qquad (2)$$

Assuming also that the oxide layer is composed of 90% wustite, 8% magnetite and 2% hematite [3], we impose a maximum oxide layer thickness of 1000 μm. Therefore the wustite layer will have a thickness of 900 μm, the magnetite layer will be 80 μm and the hematite layer will be 20 μm.
We also consider the design temperature (728.15 °K) as the upper limit and the inlet water temperature (463.15 °K) as the steam temperature limit.
We start our study by considering a new tube without any oxide layer and then we gradually add hematite layers of 10 μm, 15 μm and 20 μm to explore the effects on heat distribution on the tube wall. The material properties used in this study are shown in Table 2.

Table 2: Thermo-mechanical properties [8, 9]

Material	K (W/mK)	ρ (Kg/m^3)	C_p (J/KgK)
Carbon steel	44.5	7850	475
FeO	3.2	7750	725
Fe_3O_4	1.5	5600	800
Fe_2O_3	1.2	4900	980

COMSOL Multiphysics®[1] 4.4 is used with heat transfer in solids for modelling these phenomena.

3 RESULTS

The results obtained by isothermal contour analysis shows that the oxide layers act as insulation, making the tube wall facing the combustion chamber hotter than the tube wall inside the refractory wall. We also observe that as the oxide layers grow the internal surface of the tubes facing the combustion chamber becomes colder than the outside surface in the combustion chamber.
The isothermal contour of carbon steel tubes without oxide layers show that the temperature varies from 813 °K on the tube surface in contact with combustion chamber to 693 °K on the steam in contact with tube.
When a hematite layer grows over the internal surface of tube the temperature profile varies from 783 °K to 503 °K for a 10 μm hematite oxide layer. For a 15 μm hematite oxide layer the temperature varies from 783 °K to 635 °K and for a 20 μm hematite oxide layer the temperature varies from 783 °K to 582 °K.
When a magnetite oxide layer is added to a hematite oxide layer the temperature profile varies from 783 °K to 529 °K for 60 μm and 80 μm of megnetite. This indicates, as observed in the field, that at a certain magnetite thicknes this oxide acts as a protective layer.

[1] Comsol Multiphysics is a trade mark from COMSOL Inc.

When we added a wustite oxide layer to the magnetite oxide layer the temperature profile varies from 783 ^{o}K to 476 ^{o}K for 450 μm and 675 μm wustite and from 783 ^{o}K to 529 ^{o}K for a 900 μm wustite.
Our results, using isothermal contour from COMSOL Multiphysics®, show that there is a critical thickness of oxide layer where the temperature profile changes. These changes in the temperature profile could be associated with coarsening pores and crack growth inside the oxide layer, as shown in Figure 4 [7].

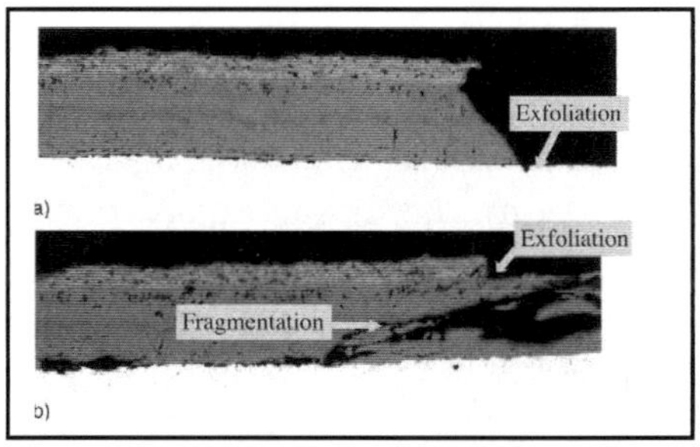

Figure 4: Picture showing coarse pores and exfoliation in hematite oxide layer [7].

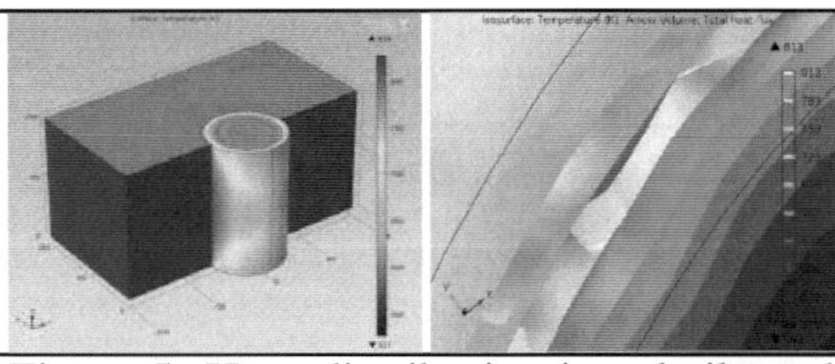

Figure 5: Heat distribution in a boiler tube without any oxide layers and a close-up of the heat distribution.

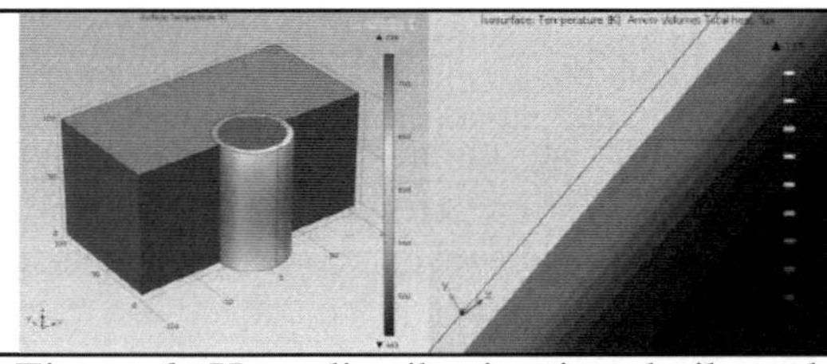

Figure 6: Heat distribution in a boiler tube with a 10 μm hematite layer and a close-up of the heat distribution in the hematite layer.

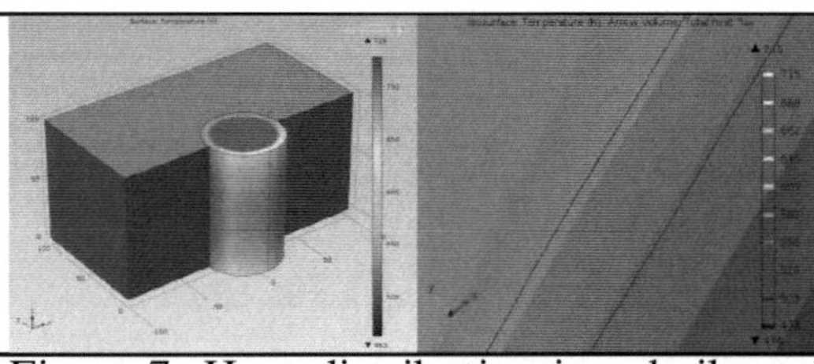

Figure 7: Heat distribution in a boiler tube with a 15 μm hematite layer and a close-up of the heat distribution in the hematite layer.

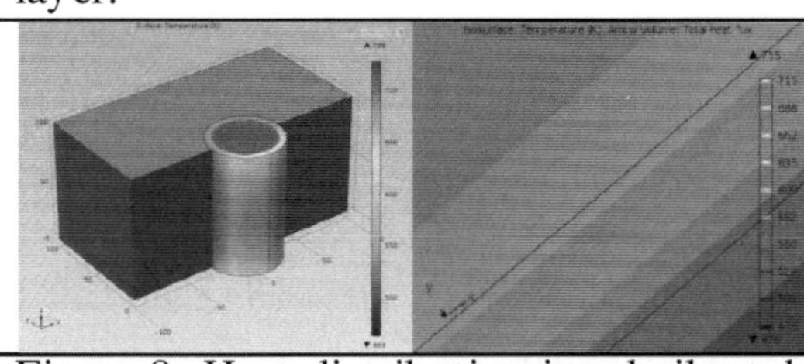

Figure 8: Heat distribution in a boiler tube with a 20 μm hematite layer and a close-up of the heat distribution in the hematite layer.

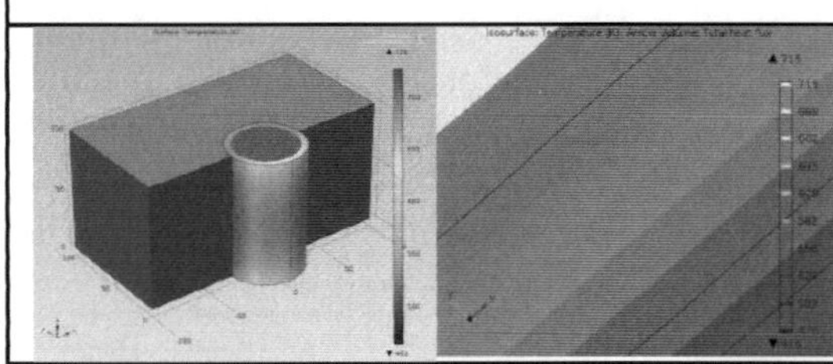

Figure 9: Heat distribution in a boiler tube with a 40 μm magnetite and 10 μm hematite layer and a of heat distribution in the layers.

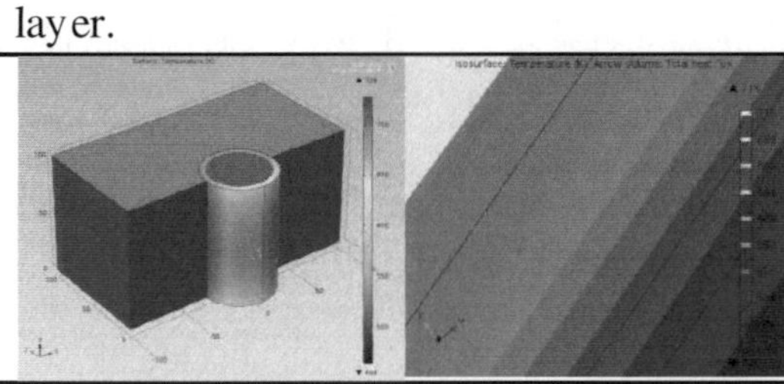

Figure 10: Heat distribution in a boiler tube with a 60 μm magnetite and a 15 μm of hematite layer and a close-up of the heat distribution in the layers.

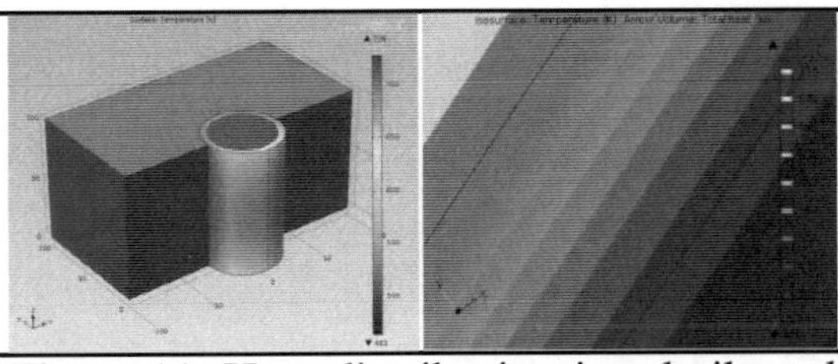

Figure 11: Heat distribution in a boiler tube with an 80 µm magnetite and a 20 µm hematite layer and a close-up of heat distribution in the layers.

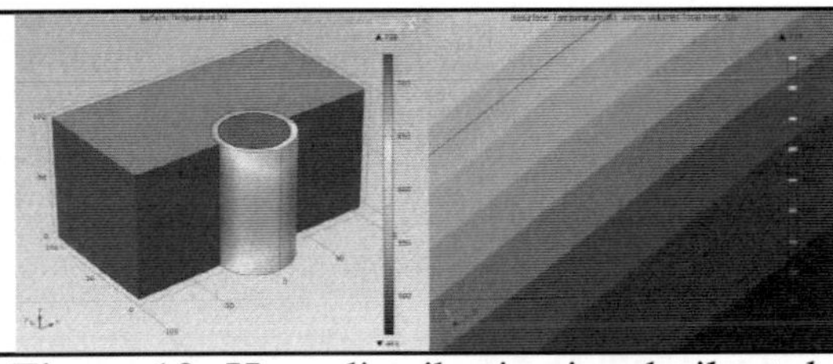

Figure 12: Heat distribution in a boiler tube with a 450 µm wustite, magnetite and a 10 µm hematite layer and a close-up of the heat distribution in the layers.

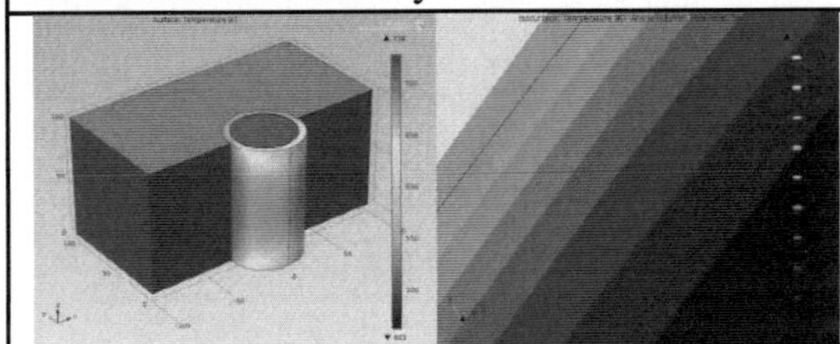

Figure 13: Heat distribution in a boiler tube with a 675 µm wustite, 60 µm magnetite and a 15 µm of hematite layer and a close-up of the heat distribution in the layers.

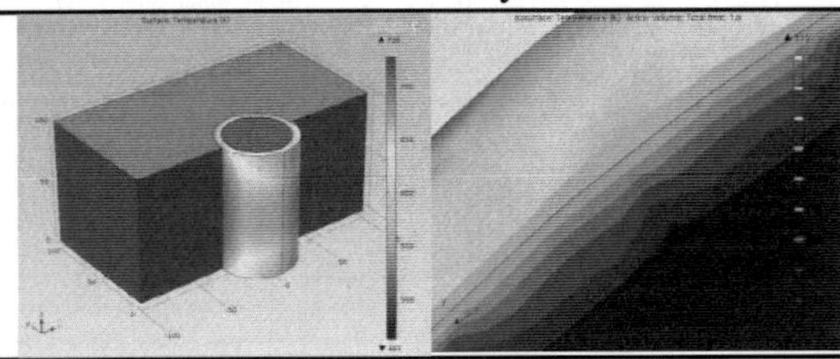

Figure 17: Heat distribution in a boiler tube with a 900 µm wustite, an 80 µm magnetite and a 20 µm hematite layer and a close-up of heat distribution in the layers.

Figures 17 to 20 shows the refratory effects of each iron oxide layer as it grows

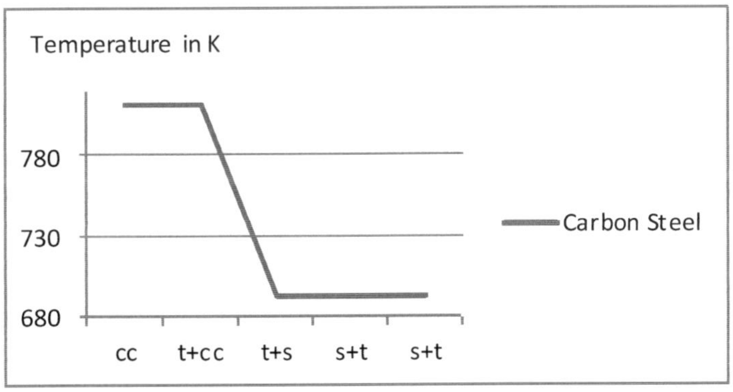

Figure 17: Graphic showing a carbon steel without oxides.

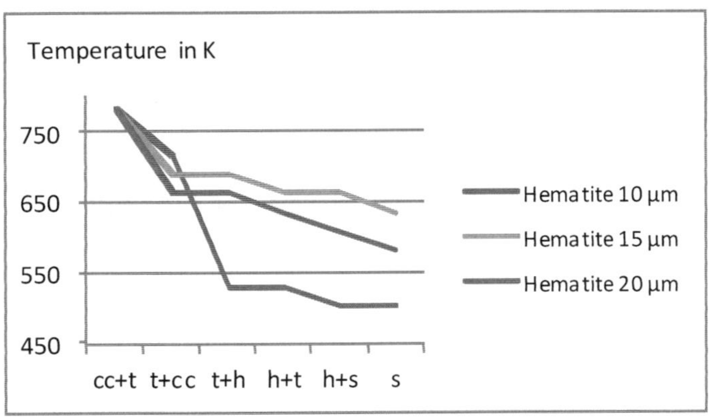

Figure 18: Graphic shows refractory effect of hematite layer growing

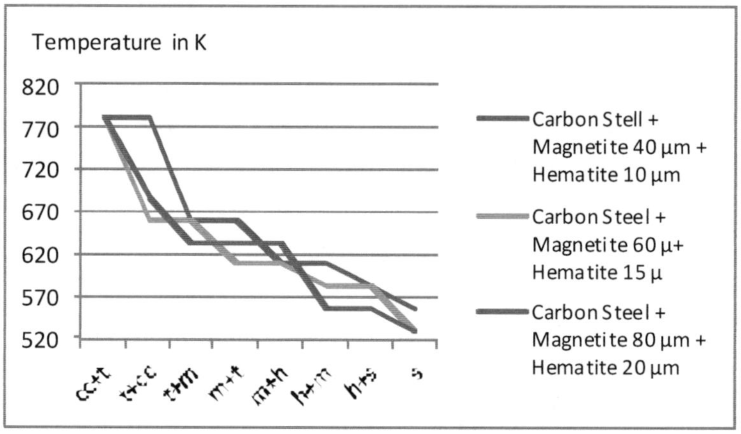

Figure 19: Graphic show refractory effect of hematite and magnetite layer growing.

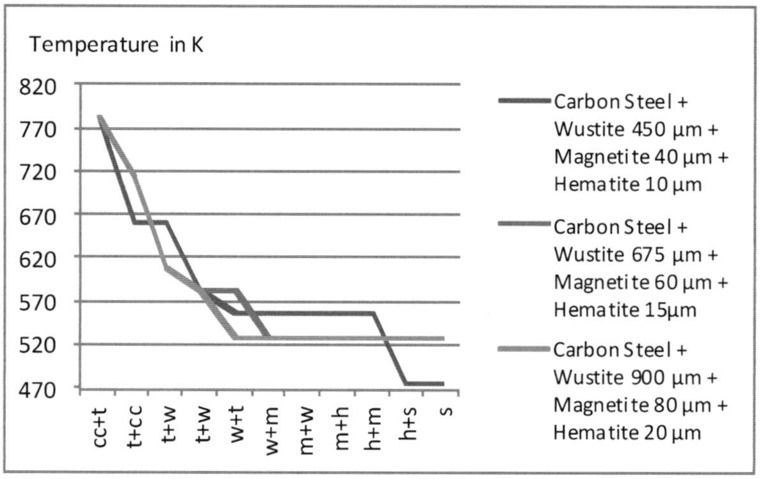

Figure 20: Graphic shows refractory effect of wustite, magneite and hematite layer growing.

3 CONCLUSIONS

The simulating of oxides growing indicates that:
1 – A thinner layer of hematite up to 10 µm is less refractory than a thicker hematite layer over 10 µm. This seems be due to the fact that a thin layer of hematite is more compact than a thick layer which in turn permits oxygen to penetrate the hematite oxide to form magnetite.
2 – Magnetite and hematite layers are less refractory than hematite,
3 – A wustite, magnetite and hematite layer is the worst composition of oxides layers and more refractory. This composition affects the heat distribution causing microstructural changes that leads to tube failure. The refractory behaviour of this composition of oxide layers could be associated with the fact that wustite is an iron and chromium oxide. This composition, due to the chromium content, is more refractory.

The overall conclusion is that an effective control of oxygen content in the feed water, the operating temperature and the use of NDT techniques which can measure hematite thickness could prevent tubes failure.

Acknowledgments

J. D. Angelo would like acknowledge to CNPq – Brazil, for the research grant which funded part of his research.

References

1 Singer, J.G, Combustion, Fossil Power. ABB Combustion Engineering, Windsor, Connecticut, 1991.

2 French, D. N., Microstructural Degradation. The National Board of Boiler and Pressure Vessel Inspector web site, Classic Articles, available at www.nationalboard.org/classics.

3 Catalogue Energia, Sermatec-Zanini, available in http://media.sermatec.com.br/uploads/produtos_arquivo/2_catalogo_sermateczanini_energia_2.pdf.

4 Graham R. Lobley and Waleed I. Al-Otaibi, Diagnosing Tube Failures Related to Overheating, Advanced Materials Research, Vols 41-42, pp 175-181, 2008.

5 Bruno Reis Cardoso et alli, J. of Mater. Res. Tecnol., 2012; 1(2): 109-116.

6 Martin Torres and Rafael Colas, A model for heat conduction through the oxide layer of steel during hot rolling, Journal of Materials Processing Technology, 105 (200), 258-263.

7 Program on Technology Innovation: Oxide Growth and Exfoliation on Alloys Exposed to Steam. EPRI, Palo Alto, CA:2007. 1013666.

Institute of Thermomechanics AS CR, v. v. i.
Academy of Sciences of the Czech Republic
NDT Department

PROGRESS IN NDT METHODS AND TOPICS DURING THE LAST 15 YEARS

Zdenek PREVOROVSKY

Institute of Thermomechanics
Czech Academy of Sciences, Prague, Czech Republic
www.it.cas.cz, E-mail: zp@it.cas.cz

8th Workshop NDT in Progress 2015
Prague, 12-14 October 2015

1

Motivation : **11th ECNDT in Prague**
– report [NDTWB 2014, SI No. 3-4]

Registered Participants & Exhibitors	1461
Short Stay Registration (two days registration)	143
Invited Representatives of NDT Societies	51
ISO meetings (6.10. - 11.10.2014)	45
Defektoskopie 2014 (8.10. - 10.10.2014)	58
Total	**1758**

8th Workshop NDT in Progress 2015
Prague, 12-14 October 2015

2

11 ECNDT 2014, Prague

38 ORAL SESSIONS	No.
01. NDT General	11
02. Training, Certification, Education, Accreditation, Standardization	12
03. Signal and image processing in NDT/NDE	22
04. Modeling and Simulations in NDT	22
05. Transducers and Sensors	6
06. Ultrasonic Testing / TOF (UT)	22
07. Guided / Lamb waves (UT)	17
08. Phased Array (UT)	22
09. Acoustic Emission	16
10. Visual / Optical (VT) and Penetrant and Leak Testing	7
11. Infrared Testing, Thermography, Thermal NDT	22
12. Eddy Current and Electromagnetic Testing	16
13. Magnetic Testing	12
14. Magnetoresistance Methods (MR) Magnetic Sensors	11
15. Radiology and Radiographic Testing, Radiation Protection	22
16. Computed Tomography	27
17. Terahertz Waves Testing	6
18. Structural Health and Vibration Monitoring	6
19. NDT Methods Based on Nonlinear and Time Reversal	11
20. Novel and Non-traditional NDT Techniques	11
21. Material Characterization - Metals	11
22. Material Characterization - Composites	22
23. Material Characterization - Polymers, Wood and Ceramics	5
24. Structural Health Monitoring, Condition Monitoring and Maintenance	11
25. Process Monitoring and Control	6
26. NDT in Pressure Equipment, Pipelines and Welding	17
27. NDT in Transportation (automotive, marine,..)	6
28. NDT in Aerospace	11
29. In-service Inspection (chemical / petrochemical industry)	5
30. NDT in Power Generation	11
31. NDT in Civil Engineering and Concrete Structures	22
32. NDT in Conservation, Restoration, Heritage, Art and Architecture	7
33. Public Security and Safety and Noninvasive Methods in Medicine	5
34. NDT in Railways	22
35. Wave Modelling in Solids (mini-symposium)	11
36. SIMPOSIUM (FP7 project - simulations in NDT)	11
37. EUROPEAN RESEARCH DAY (ACADEMIA NDT Int. & ICNDT)	18
38. DEFEKKTOSKOPIE (CNDT Czech National session)	36
Oral presentations - total	**538**

10 POSTER SESSIONS	
P01. General	7
P02. Modeling and Simulations	10
P03: Ultrasonic testing	33
P04: Acoustic Emission	11
P05: Infrared Testing	3
P06: Eddy Current Testing	12
P07: Magnetic Testing	20
P08: Radiography and Computed Tomography	11
P09: Industrial Applications and Material Testing	24
P10: NDT in Civil Engineering and Concrete Structures	9
POSTERS TOTAL	**141**
TOTAL	**679**

8th Workshop NDT in Progress 2015
Prague, 12-14 October 2015

3

NDT CONFERENCES WCNDT / ECNDT

TOWN	YEAR	E/W CNDT	PAPERS	sessions (oral)	presentations	posters	abstracts
Rome (Italy)	2000	WCNDT	438	50	?		666
Barcelona (Spain)	2002	ECNDT	419	19	?		490
Montreal (Canada)	2004	WCNDT	480	27	321	159	501
Berlin (Germany)	2006	ECNDT	706	40	477	229	783
Shanghai (China)	2008	WCNDT	490	18	441	63	504
Moscow (Russia)	2010	ECNDT	696	24	553	143	696
Durban (South Africa)	2012	WCNDT	471	34	304	68	678
Prague (Czech Rep.)	2014	ECNDT	603	36	465	138	603
Total			4303				4921

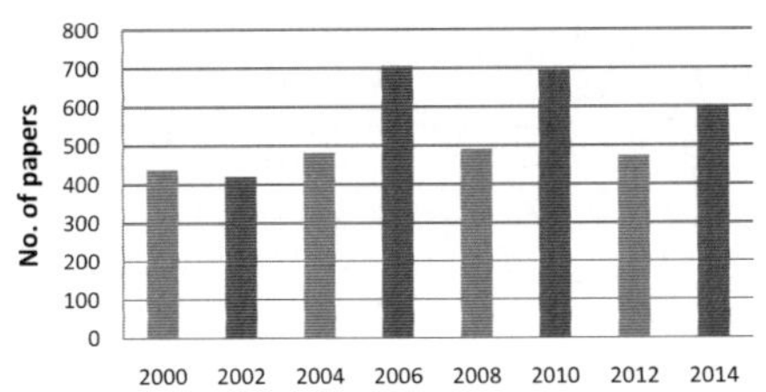

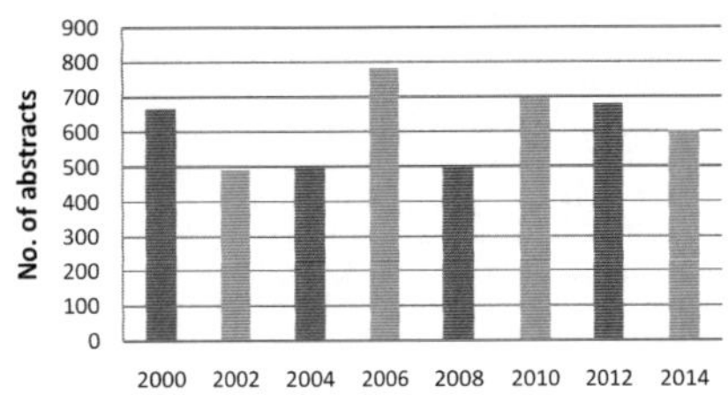

8th Workshop NDT in Progress 2015
Prague, 12-14 October 2015

4

Prague 2014 - 36 topics	
Acoustic Emission	36
Aerospace	9
Civil Engineering and Concrete Structures	44
Computed Tomography	24
Conservation, Restoration, Heritage, Art and Architecture	9
Eddy Current and Electromagnetic Testing	38
General	19
Guided / Lamb waves (UT)	15
Infrared Testing, Thermography, Thermal NDT	18
In-service Inspection (offshores, chemical / petrochemical industry)	5
Magnetic Testing	45
Magnetoresistance Methods (MR) and Magnetic Sensors	9
Material Characterization - Composites	18
Material Characterization - Metals	13
Material Characterization - Polymers, Wood and Ceramics	5
Modeling and Simulations in NDT	37
NDT Methods Based on Nonlinear and Time Reversal Techniques	9
Novel and Non-traditional NDT Techniques and Applications	7
Phased Array (UT)	21
Power Generation	9
Pressure Equipment, Pipelines and Welding	16
Process Monitoring and Control	6
Public Security and Safety and Noninvasive Methods in Medicine	4
Radiology and Radiographic Testing, Radiation Protection	18
Railways	22
Signal and image processing in NDT/NDE	20
SIMPOSIUM (FP7 project - simulations in NDT)	7
Structural Health and Vibration Monitoring	3
Structural Health Monitoring, Condition Monitoring and Maintenance	4
Terahertz Waves Testing	6
Training, Certification, Education, Accreditation, Standardization	11
Transducers and Sensors	6
Transportation (automotive, marine,..)	4
Ultrasonic Testing / TOFD (UT)	71
Visual / Optical (VT) and Penetrant and Leak Testing (PT and LT)	10
Wave Modelling in Solids (mini-symposium)	5

DURBAN 2012 - 34 topics	
Acoustic Emission	19
Certification, Qualification, Education, Training of Personnel, Standardization	10
Computed Tomography	13
Conditional Monitoring, SHM	12
Electro-magnetic Methods	14
General	8
Guided Waves	9
Magnetic Methods	16
Materials Characterization	40
Materials: Composite,Polymers, Ceramics	9
Metals; Microstructure	5
Modeling, Simulations, Signal & Image Processing	28
NDT aerospace	7
NDT in Civil Engineering, Art and Architecture	6
NDT in Power Industry, Power Generation and Nuclear	9
NDT Instrumentation & Techniques	7
NDT of Pipelines	8
NDT of Transportation (Automotive,..)	2
NDT of Wind Turbines, Energy	5
NDT railroad, railway	20
NDT Security, Anti-Terrorist Diagnostics	3
New and Novel Application	7
Optical, Infrared and Microwave Testing	20
Penetrant and Leak Testing	2
Pressure Equipment	6
Radiation Protection	5
Radiology and Radiography Methods, Neutron Radiography	35
Reliability in NDT	8
Review of NDT	0
Steel Industry	5
Thermography	22
Time of Flight Diffraction	5
Ultrasonic Methods (Nonlinear, Laser, Phased Array...)	51
Welds	5
	421

Moscow 2010 -24 topics	
Acoustic Emission	26
Anti-Terrorist Diagnostics	21
Certification, Qualification, Training of Personnel	26
Computed Tomography	33
Ecological Diagnostics	23
Electro-magnetic Methods	29
Magnetic Methods	38
Materials Characterization	37
Modeling and Signal Processing, Image Reconstruction	34
Nanotechnologies	18
NDT Civil Engineering	31
NDT in Metallurgy and Chemical Industry	16
NDT in Metrology	14
NDT in Power Industry	29
NDT of Pipelines	24
NDT of Transport Means	45
Optical, Infrared and Microwave Testing	32
Penetrant and Leak Testing	12
Radiology and Radiography Methods	42
Residual Life Evaluation	26
Standardization	6
Transducers and Sensors	13
Ultrasonic Methods	91
Vibration Analysis	30
	696

WCNDT 2008 Shanghai - 18 topics	
Acoustic Emission (15)	15
Comprehensive NDT methods	6
Comprehensive studies and applications in NDT	49
Computed Tomography & Digital Radiology	33
Electromagnetic Methods	65
Infrared Thermography	21
Magnetic Memory	20
Magnetic Testing	15
NDT techniques for diverse pipelines and structures	6
Novel applications of NDT techniques	7
Optical Techniques	5
Penetrant Testing	4
Radiographic Testing & Computed Radiologic Testing	32
Review of NDT	9
Ultrasonic Testing	109
Ultrasonic Testing - Guided Wave	30
Ultrasonic Testing - Phased Array	30
Ultrasonic Testing - Signal Processing	27
Ultrasonic Testing - TOFD	7
	490

Berlin 2006 - 40 topics	
Acoustic Emission	17
Aerospace	49
Aerospace (Health)	3
Art and Architecture	28
Automotive	15
Certification, Qualification, Training, Personnel	15
Chemical and Petrochemical	10
Computed Tomography	33
Digital Radiology	15
Digital Radiology / Radiography	9
Dimensional Measurement	5
Humaniterian Safety and Public Security	10
Infrared and Optical	21
Leak Detection	4
Lifetime Management	11
Magnetic / Electromagnetic	39
Maritim - Pipelines (offshore)	8
Materials Characterization	57
Metals	15
Modelling and Signal Processing	52
Non linear	12
Penetrant Testing	4
Pipeline In Service Inspection	20
Polymer and Composites	15
Power Generation - Fossile	13
Power Generation - Nuclear	27
Power Generation - Regenerative Energy	1
Process Monitoring	9
Public Security and Safety Technology	14
Qualification, Validation, Reliability of Systems	18
Radiation Protection	9
Radiography	5
Railway	18
Standardization	8
TeraHertz	14
Transducers and Sensors	20
Ultrasonic	55
Ultrasonic - Phased Array	16
Ultrasound in layered and anisotropic media	8
Vibration Analysis	4
	706

Montreal 2004 - 27 topics	
Aerospace	35
Array Transducers	13
Automotive	14
Civil Structures	23
Eddy Current	13
Frontiers of Research in NDT	11
General NDT/NDE	18
Guided Waves	19
High Resolution Ultrasonics	6
En-Process NDT / NDE	24
Inspection of Micro-Systems	6
Laser Ultrasonics	13
Magnetic Techniques	10
Materials Characterization	47
New Techniques	13
Non-Contact Ultrasonics	22
Nuclear	22
Optical Techniques	19
Other Topics	5
Petrochemical Industry	19
Power Generation	12
Radiography	43
Railroad Inspection	3
Reliability	9
Signal Processing	33
Steel	4
Thermography Thermal Techniques	24
	480

Barcelona 2002 - 19 topics	
Aerospace	49
Art	11
Automation	9
Chemical & Petrochemical	30
Civil Enginering	10
Computer Processing , modeling	37
Material Characterisation	80
Method & Instrumentation	30
Mines Detection	14
Non Metalic Materials	8
Nuclear	34
Offshore Industry	7
Pipelines	28
Power Stations	7
Pressure Vessels	6
Probability Detection & validation	10
Railroads	13
Steel Industry	10
Training & Certification	26
	419

WCNDT Roma 2000 - 49 topics	
Acoustic Emission (4)	4
Activities of NDT Organizations (4)	4
Aeronautics and Aerospace (29)	29
US-Europ. W-shop on Nondestruct. Inspection Reliability	9
Civil Engineering (28)	28
Computer Processing and Simulation (8)	8
Conservation and Restoration in Art and Architecture (24)	24
Digital Radiography (4)	4
Eddy Current (6)	6
Education, Training and Certification (10)	10
Electromagnetic Methods (6)	6
General (21)	21
General (Welding) (1)	1
Guided Waves (4)	4
Industrial Plants and Structures (40)	40
Inverse Problems (3)	3
Landmine Detection Equipment (3)	3
Materials Characterization and testing (1)	1
Methods and Instrumentation (55)	55
Methods and Instrumentation for Welding 1	1
MT and Applications (3)	3
MT and Low Frequency Current (4)	4
NDT for material properties (3)	3
Neutron workshop (4)	4
Nuclear Industry (29)	29
Pattern recognition (5)	5

Penetrant and Eddy Current (2)	2
Plenary (8)	8
Radiography (2)	2
Radiography: Applications (1)	1
Radiography: Film and other Detectors (5)	5
Reliability and Validation (14)	14
Signal Processing (3)	3
Simulation of EM NDE (4)	4
Simulation of US NDE (3)	3
Simulation of X-ray NDE (5)	5
Special NDT Methods (8)	8
Standardization (4)	4
Terrestrial and Marine Transportation (12)	12
Ultrasonic Signal Processing and Defect characterisation (6)	6
Ultrasonic Testing (5)	5
Ultrasonic: Applications (4)	4
Ultrasonic: Crack detection and TOFD (3)	3
Ultrasonic: Equipment (11)	11
Ultrasonic: Equipment and Probe parameters (4)	4
Ultrasonic: Phased Array Application (4)	4
Ultrasonic: Probes and coupling (6)	6
Visual testing (5)	5
X-ray and US Tomography (5)	5
X-Ray Tomography and Detectors (5)	5
	438

TOPICAL COMPARISON

33 Thematical Sessions - number of contributions	Prague 2014	Durban 2012	Moscow 2010	Shanghai 2008	Berlin 2006	Montreal 2004	Barcelona 2002	Rome 2000	SUM
Acoustic Emission	36	19	26	15	17			4	117
Certification, Education, Qualification, Accreditation, Standardi.	11	10	32		23		26	18	120
Civil Engineering, Conservation, Restoration, Art & Architecture	53	6	31		28	23	21	52	214
ET- Eddy Current and Electromagnetic Testing	38	14	29	65	39	13		6	204
IT- Infrared Testing, Thermography, Thermal NDT	18	22		21	21	24			106
Material Characterization - Polymers, Composites. Ceramics, Wood	23	9			19		8		59
Materials Characterization - Metals (properties, microstructure.)	13	45	37		72	53	80		300
Modeling & Simulations, Signal & Image Processing	69	44	34	27	52	33	37	1	297
MT- Magnetic Testing	54	8	38	35		10		31	176
Nanotechnologies, Inspection of Micro-systems			18			6		7	31
NDT General, Overview, Comprehensive methods, Other topics	19	44	23	69		34			189
NDT in Metrology			14		5			76	95
NDT in Security, Safety, Anti-Terrorist and Medical Diagnostics	4	3	21		24		14	25	91
NDT Instrumentation & Techniques, Transducers & Sensors	6	7	13		20		39	3	88
NDT Qualification & Reliability		8			18	9	10		45
Nonlinear and Time Reversal Techniques	9	7			12			23	51
Novel and Non-traditional NDT Techniques and Applications	7	7		7		13			34
Power Industry, Energy, Nuclear	9	14	29		41	34	41	8	176
Pressure Vessels, Pipelines, Welding	16	19	24	6	8	4	34	69	180
Process Monitoring and Control, Metalurgy, Petro & Chemical Indust	11	5	16		19	19	71	1	142
RT (UT)- Computed Tomography	24	22	33		33			10	112
RT- Radiography, Radiation Protection, Neutron Radiography	18	36	42	32	38	43		10	219
SHM, Condition Monitoring, Maintenance, Residual Life Evaluation	4	12	26		31			14	87
Terahertz Waves	6	3			14				23
Transport general (automotive, marine,..)	4	8	45		15	14			86
Transport- Aerospace	9	7			52	35	49		152
Transport- Railways, Railroads	22	20			18	3	13	29	105
UT- Guided waves	15	9		30		19		12	85
UT- Phased Array	21	12		30	16	13		4	96
UT- Ultrasonic Testing (general) & Methods (Laser UT, TOFD,..)	71	39	91	116	55	43		4	419
Vibration Aanalysis	3		30		4			18	55
VT, PT, LT- Optical, Penetrant & Leak Testing	10	22	44	9	8	19	19		131
Total	603	481	696	462	702	445	462	425	4285

Topics evolution - % of contributions	2014	2012	2010	2008	2006	2004	2002	2000	SUM
UT- Ultrasonic Testing (general) & Methods (Laser UT, TOFD,..)	11,8	8,1	13,1	25,1	7,8	9,7		0,9	9,89
Materials Characterization - Metals (properties, microstructure. testing)	2,2	9,4	5,3		10,3	11,9	17,3		7,13
Modeling & Simulations, Signal & Image Processing	11,4	9,1	4,9	5,8	7,4	7,4	8,0	0,2	7,02
RT- Radiology, Radiography, Radiation Protection, Neutron Radiography	3,0	7,5	6,0	6,9	5,4	9,7		2,4	5,15
Civil Engineering, Concrete, Conservation, Restoration, Art & Architecture	8,8	1,2	4,5		4,0	5,2	4,5	12,2	4,84
ET- Eddy Current and Electromagnetic Testing	6,3	2,9	4,2	14,1	5,6	2,9		1,4	4,80
NDT General, Overview, Comprehensive methods, Other topics	3,2	9,1	3,3	14,9		7,6			4,54
Power Industry, Energy, Nuclear	1,5	2,9	4,2		5,8	7,6	8,9	1,9	4,14
MT- Magnetic Testing	9,0	1,7	5,5	7,6		2,2		7,3	4,04
Pressure Vessels, Pipelines, Welding	2,7	4,0	3,4	1,3	1,1	0,9	7,4	16,2	4,02
Transport- Aerospace	1,5	1,5			7,4	7,9	10,6		3,62
Process Monitoring and Control, Metalurgy, Chemical &Petrochemical Industry	1,8	1,0	2,3		2,7	4,3	15,4	0,2	3,39
VT, PT, LT- Optical, Penetrant & Leak Testing	1,7	4,6	6,3	1,9	1,1	4,3	4,1		3,10
RT (UT)- Computed Tomography	4,0	4,6	4,7		4,7			2,4	2,80
Certification, Education, Qualification, Accreditation, Standardization	1,8	2,1	4,6		3,3		5,6	4,2	2,75
Acoustic Emission	6,0	4,0	3,7	3,2	2,4			0,9	2,73
IT- Infrared Testing, Thermography, Thermal NDT	3,0	4,6		4,5	3,0	5,4			2,53
Transport- Railways, Railroads	3,6	4,2			2,6	0,7	2,8	6,8	2,38
UT- Phased Array	3,5	2,5		6,5	2,3	2,9		0,9	2,27
NDT Instrumentation & Techniques, Transducers & Sensors	1,0	1,5	1,9		2,8		8,4	0,7	2,07
NDT in Security, Safety, Anti-Terrorist and Medical Diagnostics	0,7	0,6	3,0		3,4		3,0	5,9	2,04
UT- Guided waves	2,5	1,9		6,5		4,3		2,8	2,00
Transport general (automotive, marine,..)	0,7	1,7	6,5		2,1	3,1			2,00
SHM, Condition Monitoring, Maintenance, Residual Life Evaluation	0,7	2,5	3,7		4,4			3,3	1,97
NDT in Metrology			2,0		0,7			17,9	1,95
Material Characterization - Polymers, Composites. Ceramics, Wood	3,8	1,9			2,7		1,7		1,38
Vibration Aanalysis	0,5		4,3		0,6			4,2	1,21
Nonlinear and Time Reversal Techniques	1,5	1,5			1,7			5,4	1,11
NDT Qualification & Reliability		1,7			2,6	2,0	2,2		1,07
Novel and Non-traditional NDT Techniques and Applications	1,2	1,5		1,5		2,9			0,82
Nanotechnologies, Inspection of Micro-systems			2,6			1,3		1,6	0,70
Terahertz Waves	1,0	0,6			2,0				0,53

Topics ordered by number of papers	No.	%
UT- Ultrasonic Testing (general) & Methods (Laser UT, TOFD,..)	419	9,78
Materials Characterization - Metals (properties, microstructure. testing)	300	7,00
Modeling & Simulations, Signal & Image Processing	297	6,93
RT- Radiology, Radiography, Radiation Protection, Neutron Radiography	219	5,11
Civil Engineering, Concrete, Conservation, Restoration, Art & Architecture	214	4,99
ET- Eddy Current and Electromagnetic Testing	204	4,76
NDT General, Overview, Comprehensive methods, Other topics	189	4,41
Pressure Vessels, Pipelines, Welding	180	4,20
Power Industry, Energy, Nuclear	176	4,11
MT- Magnetic Testing	176	4,11
Transport- Aerospace	152	3,55
Process Monitoring and Control, Metalurgy, Chemical &Petrochemical Industry	142	3,31
VT, PT, LT- Optical, Penetrant & Leak Testing	131	3,06
Certification, Education, Qualification, Accreditation, Standardization	120	2,80
Acoustic Emission	117	2,73
RT (UT)- Computed Tomography	112	2,61
IT- Infrared Testing, Thermography, Thermal NDT	106	2,47
Transport- Railways, Railroads	105	2,45
UT- Phased Array	96	2,24
NDT in Metrology	95	2,22
NDT in Security, Safety, Anti-Terrorist and Medical Diagnostics	91	2,12
NDT Instrumentation & Techniques, Transducers & Sensors	88	2,05
SHM, Condition Monitoring, Maintenance, Residual Life Evaluation	87	2,03
Transport general (automotive, marine,..)	86	2,01
UT- Guided waves	85	1,98
Material Characterization - Polymers, Composites. Ceramics, Wood	59	1,38
Vibration Aanalysis	55	1,28
Nonlinear and Time Reversal Techniques	51	1,19
NDT Qualification & Reliability	45	1,05
Novel and Non-traditional NDT Techniques and Applications	34	0,79
Nanotechnologies, Inspection of Micro-systems	31	0,72
Terahertz Waves	23	0,54

8th Workshop NDT in Progress 2015
Prague, 12-14 October 2015

11

COMBINED NDT TOPICS

NDT - Common Topics	**925**
NDT Modeling & Simulations, Signal & Image Processing	297
NDT General, Overview, Comprehensive methods, Other topics	189
NDT Reliability, Qualification & Metrology	140
NDT Certification, Education, Qualification, Accreditation, Standardization	120
NDT in Security, Safety, Anti-Terrorist and Medical Diagnostics	91
NDT Instrumentation & Techniques, Transducers & Sensors	88
Methods	**1897**
UT - Ultrasonic Testing general, guided waves, phased array	605
RT- Radiology, Radiography, Radiation Protection, Neutron Radiography	219
ET- Electromagnetic Testing, Eddy Current	204
MT- Magnetic Testing	176
AT -Acoustic Emission	117
RT (UT) - Computed Tomography	112
IT- Infrared Testing, Thermography, Thermal NDT	106
VT, PT, LT- Visual, Penetrant & Leak Testing	105
Novel and Non-traditional NDT Techniques and Applications	91
Nonlinear and Time Reversal Techniques	88
Terahertz Waves	51
Vibration Aanalysis	23
Industrial Applications	**799**
Transportation	343
Civil Engineering, Concrete, Conservation, Restoration, Art & Architecture	214
Pressure Vessels, Pipelines, Welding	180
Power Industry, Energy, Nuclear	176
Transport - Aerospace	152
Process Monitoring and Control, Metalurgy, Chemical &Petrochemical Industry	142
Transport- Railways, Railroads	105
SHM, Condition Monitoring, Maintenance, Residual Life Evaluation	87
Transport general (automotive, marine,..)	86
Materials characterization	**390**
Properties & Testing (all materials)	300
Material Characterization - Polymers, Composites, Ceramics, Wood	59
Inspection of Micro-systems, Nanotechnologies	31

8th Workshop NDT in Progress 2015
Prague, 12-14 October 2015

12

CONCLUSIONS

1. Survey of more than 4500 contributions presented at 8 NDT conferences (4 ECNDT and 4 WCNDT) during the last 15 years shows main research and development interests and activities in NDT field
2. Most published papers are dealing with UT methods (~600), followed by RT (~220) and ET (~200)
3. NDT in transport is referred most often amongst industrial applications, followed with growing applications in civil engineering
4. Between common topics is leading and growing computer modeling and simulations
5. In material characterization and NDT are most of papers oriented on composites

8th Workshop NDT in Progress 2015
Prague, 12-14 October 2015

13

THANK YOU FOR YOUR ATTENTION

hoping to see you again at 9th NDT in Progress 2017

8th Workshop NDT in Progress 2015
Prague, 12-14 October 2015

14

Modal Analysis as a High Sensitive NDT Method of Compressor Blades

Miroslaw WITOS

Air Force Institute of Technology, Warsaw, Poland; Phone: +48 261851353, Fax: +48 261851612;
e-mail: witosm@itwl.pl

Abstract

The article shows the methodology of reliable health diagnostic of compressor blades made of steel and titanium alloys. The methods and diagnostics symptoms for identification of blade cracks have been described. Widely described are theoretical foundations of:

- blade modal analysis;
- operational blade load;
- influence of material crack and fatigue on blade modal characteristics;
- influence of modal characteristics dispersion on fatigue and service conditions.

The topic has been illustrated by practice taken during operation of the SO-3 and TW3-117 engines.

The laser vibration measuring method which is now used at WZL-1 (Military Aviation Works no. 1) has also been described. New experimental possibilities, like early identification of material fatigue and blade crack initiation symptoms have been depicted. These issues has been illustrated by many examples of damaged and proper blades' modal properties measurement (according to LCF and HCF profiles). It has been proven that the shape of the resonance (the level of non-linearity) rather than the natural frequency of the mode better described fatigue or blade crack. The new diagnostic symptoms are resistant (insensitive) to the effect of tuning blades under repair or in operation.

The described symptoms are the basis for non-contact measurement of rotating blades and reliable prognosis of their health. They are also used in algorithms for active control of engine components fatigue.

Keywords: aviation, compressor blades, material fatigue, non-destructive testing (NDT), laser measurement, modal analysis, repair technology, safety

1. Introduction

Rotating fan, compressor and turbine blades are critical elements of the rotary machine. The blades are exposed to accelerated fatigue and operational degradation – Figure 1, resulting from:

- influence of complex real working loads (quasi-static and dynamic), including adverse dynamic phenomena (stall, flutter, surge);
- modal properties of the blade (resulting from its geometry and material parameters), including very low damping of vibration in resonance ranges;
- influence of the environment and operating conditions on the health of the blade surface (corrosion, erosion, foreign objects, icing, salinity);
- errors in operation of the rotary machine, e.g. quality of the power system adjustment;
- hidden defects in construction and repairs.

Blades that broke off (end phase of the material degradation) is the most frequent source of secondary damages to the rotary machine (other blade palisades, rotor bearing, body) and in the case of aircraft engines, also a source of threat to the crew and other people's life.

Non-**D**estructive **T**esting (NDT) and **S**tructural **H**ealth **M**onitoring (SHM) of blades are a challenge for a diagnostician both in operation and overhauls of rotary machines. Using visual methods it is possible to reliably assess only the health of the blade surface in respect to: erosion and corrosion, mechanical damages of the blade surface and detection of big cracks. Ultrasound flaw detection of blades performed within technical service (without disassembly of blades from a palisade) is effective only in a short period of time after stopping the machines. During stops residual compressive stresses occur in blades, which close the crack and decrease the probability that even big cracks (ca. 15 mm) will be discovered after a few days from stopping the engine - in the SO-3 turbojet engine already after 3 days [1].

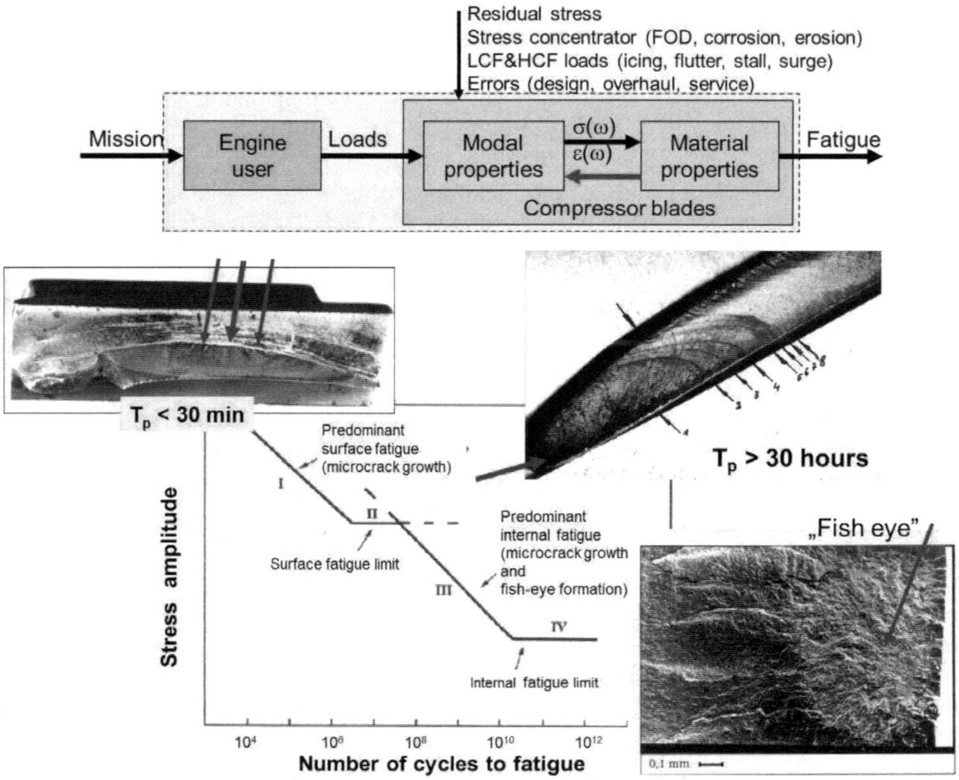

Figure 1. Process of the compressor blades' fatigue - issue of **L**ow **C**ycle **F**atigue (LCF), **H**igh **C**ycle **F**atigue (HCF) and **V**ery **H**igh **C**ycle **F**atigue (VHCF); T_p – crack propagation time from nucleation to being wrenched off the 1st stage compressor blades of SO-3 engines [2-4]

The effect of closing the crack on the NDT results is often omitted in scientific papers (the crack is simulated by a cut), but it is taken into account in repair technology. The blades are warmed directly before conducting the non-destructive testing using penetration (PT), ultrasound (UT) and/or eddy current (ET) methods for the purpose of decreasing residual compressive stresses and opening potential cracks. In such conditions the above mentioned NDT methods ensure the sufficient effectiveness of diagnosis of the current condition of the blades. However, this information is insufficient for providing high quality overhauls and low risk of a repair plant within the guarantee period (connected with bearing of unplanned costs related to removing of faulty blades from operation or results of their braking-offs). In repairs, the ability to detect and classify the early phase of material fatigue is expected, among others detection of cyclically overloaded or weakened blades without symptoms of open cracks detected by typical NDT methods. Only if this additional ability is acquired and implemented in the repair technology of a machine, the following is provided:

- reliable determination of the degradation level of blade material structure and connecting NDT results to the expected material degradation mechanisms;
- reliable evaluation of the real operational durability and horizon of blades' safe work within the assumed load conditions;
- improvement of the quality of overhauls and lowering of the risk connected with unplanned costs.

In the paper the modal analysis method used for objective automated control of the compressor blades' condition which can be applied during production and overhauls of various rotary machine is presented. The described method has been implemented in 2008 in Poland in the repair technology of the TW3-117 turbojets. The obtained test results and verified experimental knowledge concerning modal symptoms of progressing material fatigue degradation have also been used in:

- causal prevention of fatigue problems occurring in the Polish Air Force;
- increasing the functional capabilities of non-contact monitoring of rotating compressor, aircraft engine turbine, industrial fan and steam turbine blades;
- active control of the SO-3 engines' user over the fatigue process of blades' material – Figure 2.

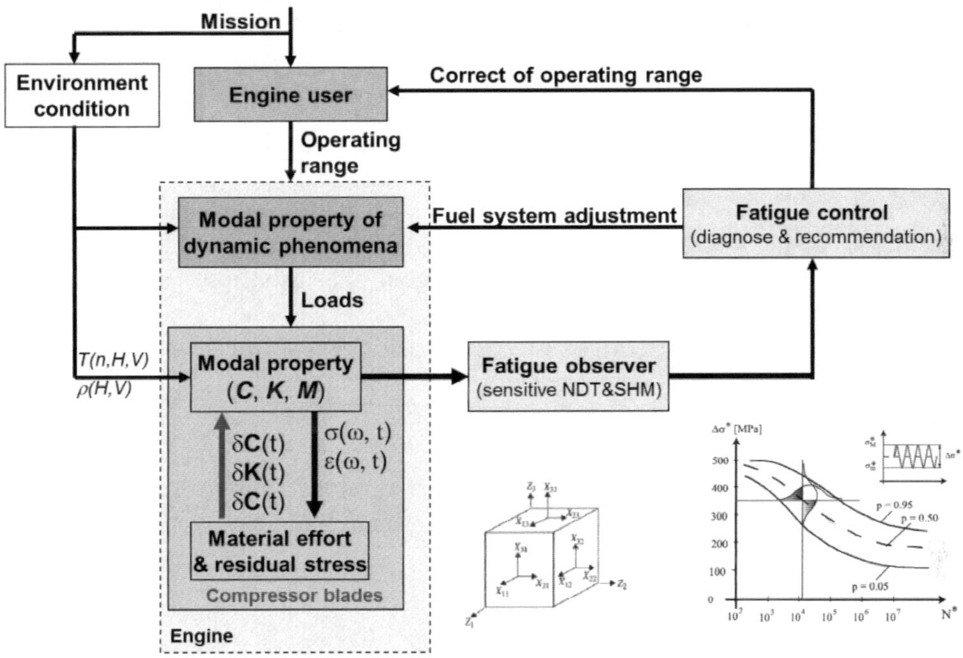

Figure 2. Active control of compressor blades' fatigue [2]

2. Object of the tests

The tested compressor blades are a poorly dumped constant structure working at temperatures from 208 K to 513 K, that is below $0.3 \cdot T_m$ (melting point), within the range of occurring of the cold working or work/strain hardening [5-7]. The blades theoretically have (as a continuous object) infinite number of frequencies and forms of free (resonant) vibration – modes [8]. However, only the lowest N of modes create fatigue conditions, which results from the operational forces' spectrum and limit amplitude of blade vibration (stresses) for particular modes. Example forms of fatigue damages of compressor blades are depicted in Figures 1 and 3. High and low cycle fatigue cracks are initiated in the place of the highest surface or subsurface stresses [4]:

- near a given line of nodes of vibration (blades that were not subjected to the surface hardening used most often in ground rotary machines),
- beyond the line of nodes of vibration (surface hardened blades used in aviation),
- in the focus of erosion or corrosion (influence of a local stress concentrator).

Very High Cycle Fatigue cracks are initiated inside the material (most often on soft inclusions) by low level of the material stress (below the fatigue limit) [3].

Figure 3. Example damages in compressor blades generated by various vibration modes [2]

3. Theoretical premises of blade diagnostic on the basis of modal parameters

Modal analysis is the process of determining the modal parameters of a structure for all modes in the frequency range of interest. The ultimate goal is to use these parameters to construct a modal model of the response [8].
A batch of blades of a given level of rotary machines is made of the same material or materials having similar mechanical parameters in dimensions and tolerances for dimensions determined in the design documentation. This set of new blades shows very high probability of modal properties, among others [2]:

- y_{st} static stiffness - bends and taking apart of the blade under the influence of 3D quasi-static loads: centrifugal force, bending and torsion moments;
- logarithmic decrement

$$\delta = ln\left(\frac{A(t)}{A(t+T)}\right) = \frac{1}{m-n} ln\left(\frac{A_n}{A_m}\right) \tag{1}$$

where A(t) i A(t+T) are dumped vibration amplitudes in moment t and subsequent T or amplitudes A_n and A_m with n and m of free vibration cycle (after the forces disappear);

- free vibration frequencies f_{rez} of subsequent modes;
- width of resonance areas Δf for a given mode;
- amplification factor of vibration in resonance (Quality factor, Q)

$$Q = \frac{f_{rez}}{\Delta f} = 2\pi f_{rez} \frac{Energy\ Stored}{Energy\ dissipaded\ per\ cycle} \tag{2}$$

- symmetry of resonance curve of a given mode (level of nonlinearity of the tested structure - blade material and conditions of its mounting).

On the basis of frequency response of the above mentioned modal parameters the expected reaction of the tested blades on quasi-static loads (occurring by given rotational speed and environmental conditions) and dynamic loads (with the assumed spectrum) may be determined and their technical condition may be objectively assessed.
For the whole batch of blades the modal parameters are described by the multivariate normal distribution. By definition, n-dimensional random variable $X = [x_1, \dots, x_n]^T$ is subject to n-variate normal distribution if a random linear combination $Y = a_1x_1 + \dots + a_nx_n$ of its

components has a normal distribution. The density function of n-dimensional normal distribution of a random vector X with a vector of expected values $\boldsymbol{\mu} = [\mu_1, \dots, \mu_n]^T$ and covariance matrix Σ is described by the following equation:

$$f_{\boldsymbol{\mu},\Sigma}(X) = \frac{1}{(2\pi)^{n/2}|\Sigma|^{0.5}} exp\left(-0.5(X - \boldsymbol{\mu})^T \Sigma^{-1}(X - \boldsymbol{\mu})\right) \quad (3)$$

whose shortened form is:

$$X \sim N(\boldsymbol{\mu}, \Sigma) \quad (4)$$

For the analysed modal parameter PM_i, the values μ_i and Σ_i depend on the expected frequency of vibration of a given blade module f_{ref} (the design characteristics of the blades) and operational term t. The above observation is the basis for using algorithms of the statistical classifier of the condition, e.g. the '2-2-3 sigma' test for the purpose of assessing the fatigue degradation level of a material [2]. Diagnostic criteria are specified with increasing volume of a measurement data base (number of tested blades).

Modal parameters of blades from a given palisade of a rotary machine (from a single object of tests) are most often not described by the normal distribution, which results from intentional human interference in the statistical distribution at the stage of assembling blades in a palisade, e.g.:

- adjusting of the 1st mode's blades' free vibration frequency at the production stage and overhaul of an engine,
- completing blades for a given palisade in accordance with the criterion of the free vibration frequency of adjacent blades and the mass moment of inertia.

The modal parameters determined on the basis of measuring blades of a tested palisade should lie in the diagnostic criteria resulting from the normal distribution determined for blades of a given palisade.

Modal properties of blades change with:

- time of operation - a parameter that is easy to measure by the user;
- progressing material fatigue degradation (increase of the dislocation density and structural anisotropy level, disintegration of the basic phase of a microstructure, precipitation hardening of a microstructure, occurrence of voids in a microstructure, changes in the section area of a cracking blade) - a parameter that is indirectly monitored in the operation by counting working cycles of an engine;
- influence of unknown history of material stress - a parameter that we do not know much about. Only an approximate operation profile of an engine/mission is most often known and the influence of the quality of power system adjustment and real environmental conditions on the blade vibration spectrum is poorly diagnosed. The user most often does not know modal parameters of blades and spectrum of dynamic phenomena.

3.1. Naive method of using modal analysis in blades' diagnostic

In scientific literature and old repair technologies we may find a description of diagnosing blades using chosen single modal parameters. As it results from the author's many years' experience, the control of a single modal parameter e.g. only the frequency of 1st mode identified after pulse excitation or during measurements on an exciter (readjusting with a sinus continuous force in the expected frequency range) is a naive method. The frequency of fractured blade depends on the condition of the crack tip (gap). For example, a frequency of 1st mode for the titanium blade with a 12 mm crack was considered efficient under technical specification conditions – Figure 4.

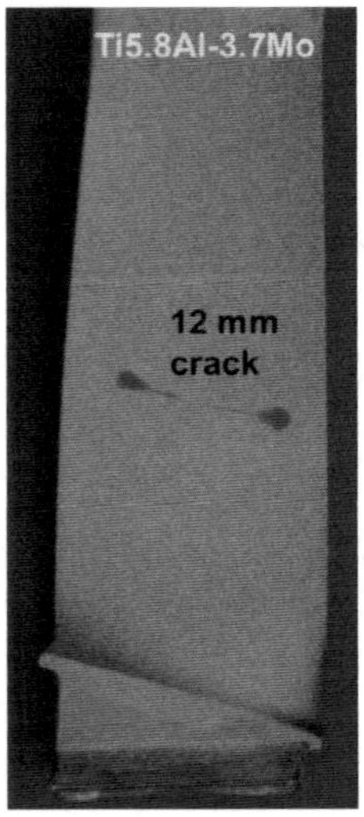

a)

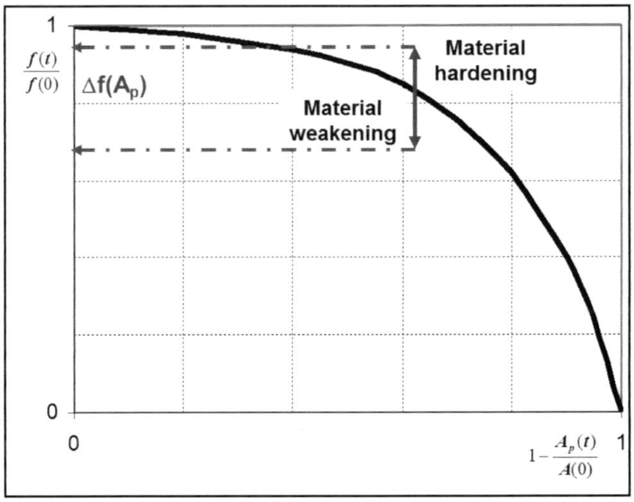

b)

Figure 4. The following has been depicted [2, 9]: a) blade cracked in operation, for which the found frequency of 1st mode f = 635 Hz was consistent with technical conditions (590 – 640 Hz) despite the 12 mm crack - influence of strengthening the material at the crack's tip on the free vibration frequency; b) influence of strengthening/weakening of the material at the crack's tip on the expected frequency of a blade by a given loss in the cross section - **the free vibration frequency does not reliably reflect the crack's size!**

Blades' frequency check does not guarantee safe prognosis, which is expected to ensure high level of the overhaul. So the question arises:

"Can other parameters from the modal analysis be an efficient tool used for blade's material fatigue level diagnosis and to supply quality inspection?"

3.2. Theoretical bases of modal analysis

Following the modal analysis theory [8], the response on diagnosed mechanical system $Y(\omega)$ on the force $F(\omega)$ is conditioned by resonance characteristics $H(\omega)$ of the object (that results from its shape, material, manufacturing technology and health)

$$\mathrm{Y}(\omega) = \mathrm{H}(\omega) \cdot \mathrm{F}(\omega) \tag{5}$$

$H(\omega)$ is called **F**requency **R**esponse **F**unction (FRF). In a complex description, FRF has an amplitude $|H(\omega)|$ and phase $\angle H(\omega) = \phi(\omega)$. The FRF function describes dynamic properties of an object and is independent from force spectrum. The knowledge of the FRF is especially important when solving a reverse problem, e.g. during assessment of health of a tested object on the basis of determined characteristics $H(\omega)$.

Compressor blades may be analysed as a beam or a plate fixed on one side, depending on the proportion between the blade's height, blade's chord and blade's maximum thickness (the parameters changing at subsequent compressor stages). The blade model concerns long and thin blades. Structural response of the blade can be represented in different domains. The modal description relates to descriptions in the spatial, time and frequency domain.

Physical domain – the complex geometrical deflection pattern of the blade can be represented by a set of simpler, independent deflection patterns, or mode shapes.

Time domain – the vibration response of the blade is shown as a time history, which can be represented by a set of decaying sinusoids.

Frequency domain – analysis of the time signal gives us a spectrum containing a series of peaks, shown below as a set of **S**ingle-**D**egree-**O**f **F**reedom (SDOF) response spectra.

Modal domain – we see the response of the blade as a modal model constructed from a set of SDOF models. Since a mode shape is a pattern of movement for all points on the structure at the modal frequency, the single model coordinate q can be used to represent the entire movement contribution of each mode. Each SDOF model is associated with a frequency, clamping and mode shape. An important property of modes is that any forced or free dynamic response of the structure can be reduced to a discrete set of modes. The modal parameters are:

- modal frequency;
- modal damping;
- mode shape.

Modal parameters of all analysed modes (within the frequency range of interest) constitute a complete dynamic description of the blade structure:

- ✍ **blade material;**
- ✍ **blade geometry**;
- ✍ **influence of surface treatment and adding protection coating**;
- ✍ **technical health** (structural heterogeneity, crack and fatigue).

Hence, the modes of vibration represent the inherent dynamic properties of a free structure (a structure on which there are no forces acting).

3.3. Modal model of a healthy blade

A blade without cracks, well fixed (in a technological handle on an exciter or in a disc slot of a given palisade) is a linear object, for which the superposition principle may be used and constant values of the resonance frequencies of particular modes, independently from the blade vibration amplitude, that is $H(\omega)$ may be expected. The expected response of the blade on the constant sinus excitement is periodic vibration with only the excitement frequency ω, which is shifted in phase in respect to the force by the radian φ. The amplitude and phase shift of the blade vibration depend on the forcing frequency and modal properties of the blade. We are to expect pure resonance frequencies for the following modes illustrated by bending or torsion vibration – Figure 5.

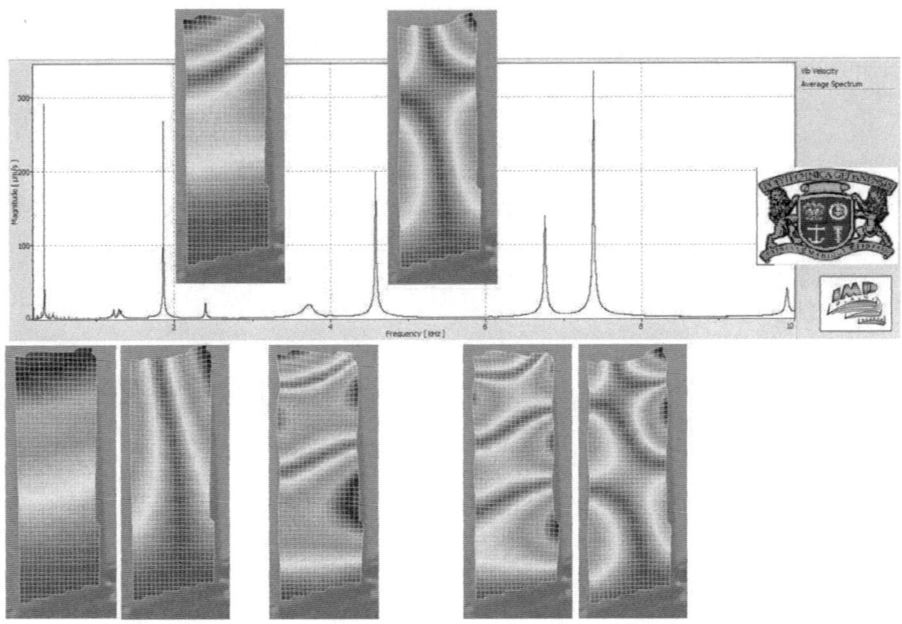

Figure 5. Frequency characteristics and vibration forms of steel blades of the 1st stage compressor of the SO-3 engine determined by the PSV-400 scanning vibrometer [2].

The tested linear system is described by the following equation [2,8]:

$$m\frac{d^2y(t)}{dt^2}+C\frac{dy(t)}{dt}+Ky(t)=F(t)$$
$$F(t)=A(\omega)\sin(\omega t) \tag{6}$$
$$y(t)=B(\omega)\sin[\omega t+\varphi(\omega)]$$

Every mode of the blade can we modelled as a SDOF object with mass **m**, stiffness **K** and damping coefficient **C**. In frequency domain the system can be described as:

$$H(\omega)=\frac{1}{-\omega^2 m+j\omega C+K} \tag{7}$$

In time domain it is a force response function $h(\tau)$:

$$h(\tau)=B_0e^{-\sigma t}\sin(\omega_d t) \tag{8}$$

where: B_0 – complex scale coefficient; ω_d – damped vibration frequency; $\sigma = C/(2m)$ – damping coefficient.
The resultant influence of following modes on blade point movement is the vector superposition. Hence, the movement equation takes the matrix form:

$$[\mathbf{m}]\left[\frac{d^2y(t)}{dt^2}\right]+[\mathbf{C}]\left[\frac{dy(t)}{dt}\right]+[\mathbf{K}][y(t)]=[\mathbf{F(t)}] \tag{9}$$

in which the first part is inertial forces, second – energy dissipation forces, third – resistance forces and the last – external forces.

The main limitation of using equation (9) to data analysis is a lack of knowledge about mass models. Modal mass is not equivalent to the blade mass. However, the measurement data includes necessary information that describes blade in a modal domain. That is why we can translate equation (9) into:

$$[\mathbf{y}]=[\boldsymbol{\varphi}]\,[\mathbf{q}] \tag{10}$$

where: **[y]** – physical coordinate system; **[φ]** – modal matrix; **[q]** – modal coordinate system.

By assumption of elementary modal mass the blade movement equation in the modal coordinate system can be described as:

$$[1]\left[\frac{\mathbf{d^2q(t)}}{\mathbf{dt^2}}\right]+[2\sigma_r]\left[\frac{\mathbf{dq(t)}}{\mathbf{dt}}\right]+[\omega_{0.2}][\mathbf{q(t)}]=[\varphi]^T[\mathbf{F}(t)]=[\Gamma] \tag{11}$$

For each observed mode in chosen blades' point the equation (11) looks like:

$$\frac{d^2q(t)}{dt^2}+2\sigma_r\frac{dq(t)}{dt}+\omega_{0.r}q(t)=[\varphi]_r^T[\mathbf{F(t)}]=[\mathbf{\Gamma}] \tag{12}$$

Characteristics of subsequent modes remain continuous under resonance conditions and exhibit good symmetry around the resonance frequency (within the band-width of 3 dB). The blade displacement at the measuring point can be described with [2]:

- vibration amplitude:

$$B(\omega) = \frac{y_{st}}{\sqrt{\left[1-\left(\frac{\omega}{\omega_0}\right)^2\right]^2+\left(\frac{\delta\ \omega}{\pi\omega_0}\right)^2}} \tag{13}$$

- phase angle of vibration

$$\varphi(\omega) = arctan\left(\left(\frac{\delta}{\pi}\frac{\omega}{\omega_0}\right)/\left(1-\left(\frac{\omega}{\omega_0}\right)^2\right)\right) \tag{14}$$

where: ω_0 – free vibration frequency, δ – logarithmic damping decrement.

3.4. Modal model of a cracked blade

A cracked blade is a nonlinear object, for which significant influence of the blade vibration amplitude on the resonance frequency of given modes, that is $H(\omega, B)$, is expected [10]. The nonlinearity of the blade results from:

- additional phenomena occurring on the crack during its closing and opening;
- local changes in mechanical properties of a material near the crack's tip (in the plasticized area),

which are a source of additional energy dissipation. The expected response of the blade on the constant sinus excitement is periodic vibration with the excitement frequency and its harmonics, which is shifted in phase in respect to the force. The measure of the non-linearity is the energy dissipation ration THD (also called total harmonic distortion) defined as a sum of amplitudes of the following harmonics to the force amplitude ratio:

$$\mathrm{THD} = \frac{\sum_{i=1}^{n} \mathrm{B_i}}{\mathrm{B_1}} \tag{15}$$

The obtained characteristics of the cracked blade cannot be described with a SDOF linear model. The blade crack forms a two-degrees-of-freedom (2DOF) non-linear system for any form of blade vibration. The equivalent linear equation that satisfies the non-linear equation with accuracy ε takes the following form:

$$\frac{d^2y}{dt^2} + 2h_\varepsilon(B)\frac{dy}{dt} + \alpha_\varepsilon^2(B)\cdot y = \varepsilon p sin(\omega t) \tag{16}$$

where: ε - small parameter; p – amplitude of the exciting force; B – steady-state vibration amplitude; $\alpha_\varepsilon(B)$ – equivalent natural (free-vibration) frequency; $h_\varepsilon(B)$ – equivalent elementary damping coefficient.

The measured and analysed parameters of the blade are described with the following relationships [2]:

- vibration amplitude

$$B(\omega) = \frac{\varepsilon\, p}{\sqrt{\left[\alpha_\varepsilon^2(B)-\omega^2\right]^2+4h_\varepsilon^2(B)\omega^2}} \tag{17}$$

- resonance frequency

$$\omega = \sqrt{[\alpha_\varepsilon^2(B) - 2h_\varepsilon^2(B)] \pm \sqrt{4h_\varepsilon^2(B)[h_\varepsilon^2(B) - \alpha_\varepsilon^2(B)] + \left(\frac{\varepsilon p}{B}\right)^2}} \tag{18}$$

- vibration phase angle

$$\varphi(\omega) = arctan\left(-\frac{2h_\epsilon(B)\omega}{\alpha_\varepsilon^2(B)-\omega^2}\right) \tag{19}$$

3.5. Modal model of the progressive degradation of a material

The progressing material degradation (cyclical strengthening, cyclical weakness/softening) is reflected in modal parameters of the blade by symptoms of passing of a given system from the properties of a linear to nonlinear object [1,2].

4. Measuring stand

The identification of early fatigue and cracking of a blade (modified NACA-65 profile) made of 18H2N4WA steel and Ti5.8Al-3.7Mo titanium alloy was made on the B&K electrodynamic exciter 4802T, which is used during overhauls of the SO-3 and TW3-117 engines. The experimental stand included:

- the MTI Instrument laser measurement system MicroTrack™II with the CMOS measurement head LTC-120-40 [11];
- the Vibration Research Corporation VR-8500 controller that includes 24 bit A/D and D/A converters and RISC processor [12];
- the Vibration Research Corporation VibrationView software to control the exciter, data acquisition and analysis [12].

The main idea of the measurement is to illustrate the proportional blade amplitude at the CMOS matrix of a sensor – Figure 6. On the output we get analogue signal whose frequency is limited (be low-pass filter) to 20 kHz. The sensitivity of the measurement system is 100 mV/mm. Such a stand gives the possibility to make precise measurements with the exact test profile and frequency step. The used measurement system gives almost laboratory accuracy (about 2 μm). That is why it enabled [1,2]:

- Precise identification of modal properties of the blade in the measured frequency range;
- Analysis of the influence of metrological factors on the recorded resonance characteristics;
- Analysis of the trends of modal parameters observed during the fatigue tests.

Determination of the modal properties (sine test) was based on the transition function

$$G(\omega) = \frac{Y(\omega)}{X(\omega)} \quad \left[\frac{m}{m}\right] \tag{20}$$

where X(ω) – displacement of the exciter's head (blade root); Y(ω) – displacement of the blade tip. This function gives better resolution on high frequencies than the normally used flexibility function of acceleration.

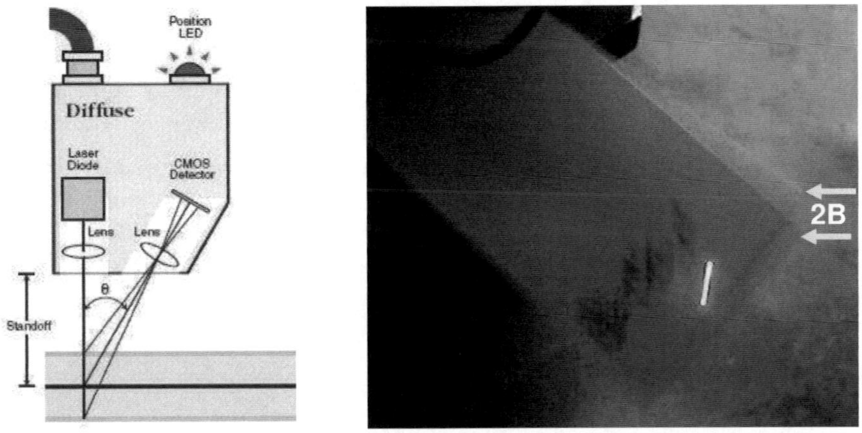

Figure 6. Laser blade vibration measurement [1]

5. Research methodology

The tests were conducted within an active and passive experiment. The active experiments were made in four stages in which [1,2]:
- the measurement method and stand were verified,
- cracking symptoms of the steel and titanium blades were identified in a controlled fracture (using defective blades, including cracked blades in operation),
- early symptoms of material fatigue were identified for the steel and titanium alloy blades (the learning set contains 30 blades with various compressor palisades),
- new diagnostic symptoms were verified for a titanium alloy blade during prevention pursued after finding crack in operation (a set of 2500 blades) .

The passive experiment (objective, automatic control of the blades' health) has been conducted since 2008, after the method had been implemented in the repair technology of the TW3-117 engines.

The control algorithm is used in the tests, which encompasses:
i) fast search for the 1st mode resonance frequency;
ii) one-sided precise measurement of the resonance characteristics of the 1st mode in the frequency range determined in the relation $B(\omega) > 0.05B_{max}$;
iii) automatic analysis of modal parameters and classification of the blades' health.

The activities from i) results from the previous overhaul technology. The labour consumption of the additional control activities from ii) and iii) does not exceed 2 min/blade.

6. RESULTS

On the basis of the active experiment (fatigue tests):
a) reliable diagnostic symptoms (levels and direction of changes in modal parameters' values) of the progressing fatigue degradation of steel and titanium compressor blades, including the cyclic strengthening and cyclic weakening phases of a material preceding an open crack of a blade, have been determined;
b) the jump cycle fatigue (JCF) effect of accelerated crack propagation during vibration of a blade within the range of the resonance characteristics' discontinuity of a given mode has been illustrated;
c) very high effectiveness of the modal analysis for objective diagnostic of the following has been confirmed:

- geometric properties of the blades (resultant influence of the blade's high, profile thickness, changes in the blade's geometry along the height) changing under the influence of the history of loads, erosion and repairs, including adjusting of a given module to technical conditions;
- cyclic strengthening of the material;
- cyclic weakening of the material;
- blade crack.

The above observations have been confirmed in subsequent years of the passive experiment (diagnosing) on over 3,000 blades of the TW3-117 engine 1st stage compressor.

6.1. Modal properties of a healthy blade

Compressor blades can be considered as parallel narrow-phase filters. A blade is forced only around modal frequencies. The health of a blade close to the amplitude resonance frequency f_{rA} is well described by the linear SDOF model characteristics, Figure 7.

6.2. Diagnostic symptoms of a cracked blade

When analysing the resonance curve shape we can observe how different it is from a cracked blade, Figure 7. The blade has all nonlinear properties, which describe the nonlinear 2DOF model. Close to the resonance frequency we can observe two branches of the following characteristics: resonance attractor – S_r, non-resonance attractor - S_n, and jumps between them. The shape of the cracked blade's resonant curve is affected by the blade's material and conditions existing on the edge of the crack gap. The characteristic curve is sloped to the left (towards lower frequencies) for the crack with material weakening. On the other hand, for the gap with material hardening, the curve is sloped to the right (towards higher frequencies) – Figure 8. The knowledge of the resonant curve inclination is essential for the correct interpretation of measurements, including the correct identification of the resonant and non-resonant branches.

There was no asymmetry symptom for a notch on a blade, which was used as a simplified crack model (no friction at a notch hole) – Figure 9. A cut blade is still a linear object differing from a healthy blade only in active cross section of the blade, which is a source of the difference in the blade's stiffness and frequency of particular modes.

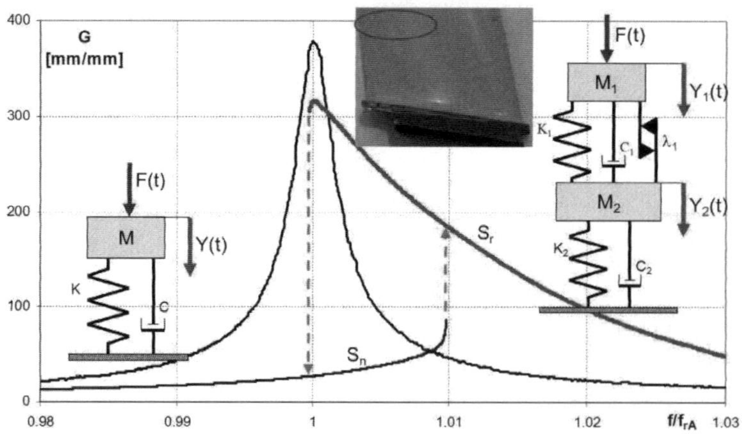

Figure 7. Effect of a crack on the 1st mode characteristics shape (steel blade, a = 19.62 m/s^2)

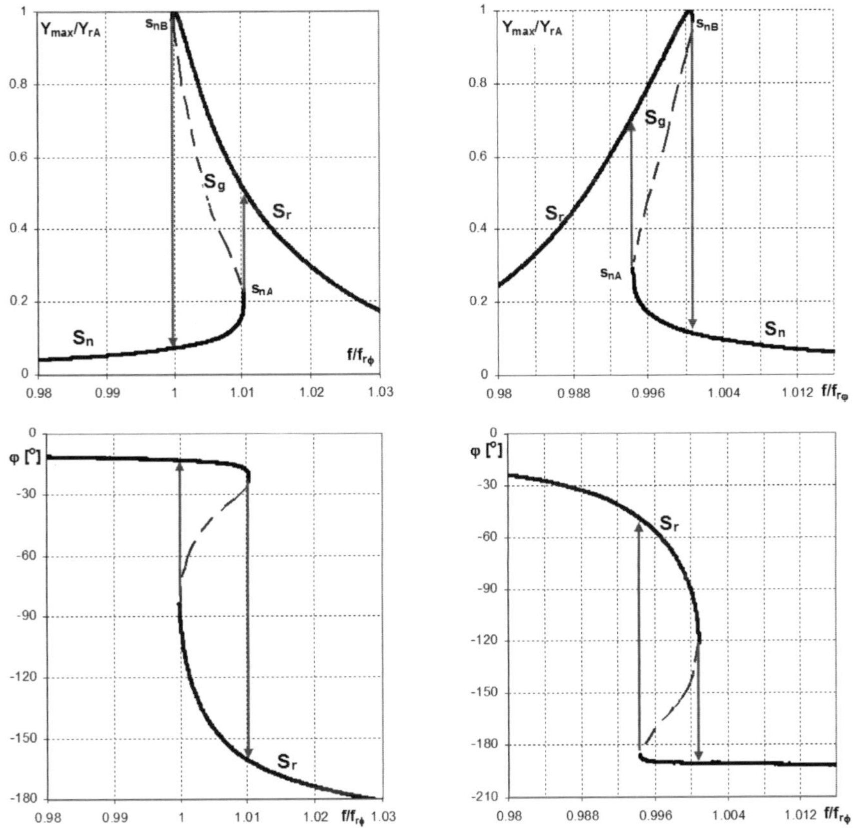

a) b)

Figure 8. Resonant curve of cracked blade [2]: a) gap with material weakening (steel and titanium alloys) b) gap with material hardening (titanium alloys)

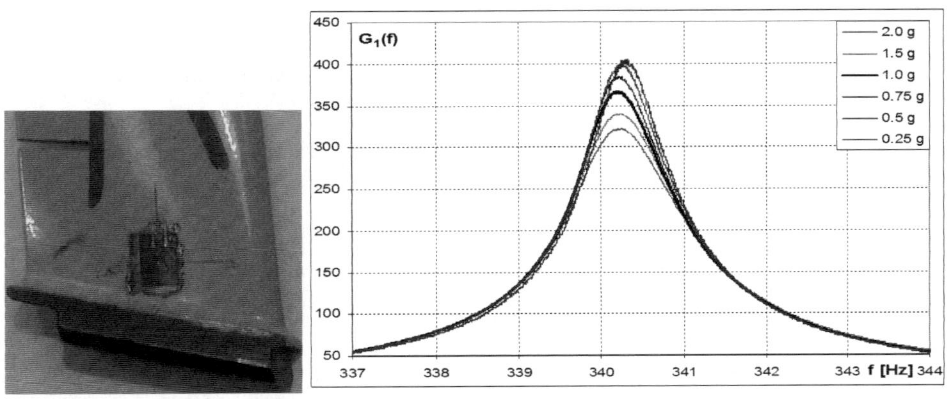

Figure 9. Influence of the blade's cutting on modal properties [1]

The influence of the cracked blade's resonant curve discontinuity on the propagation rate was investigated for blades made of titanium alloy. It was found that in the case of the constant frequency input (HCF tests without fine tuning to current resonant frequency), characteristic curve sloping to right and resonant curve discontinuity helps stopping the crack propagation.

The discontinuity of the resonant curve (blade pulse input discharge and load) is a source of very fast crack propagation during the frequency transient phase – the phenomenon is called **J**ump **C**ycle **F**atigue (JCF) [2,13]. This phenomenon is a reason for the serial material tearing during decrease of the excitation frequency observed in the unstable phase of cracking, Figure 10. Those observations are fundamental to the prognosis of the crack propagation speed and determination of the safe prognosis horizon for blade operation. It also shows that the number of stop lines in the structure fatigue strength may be different than the number of registered force cycles.

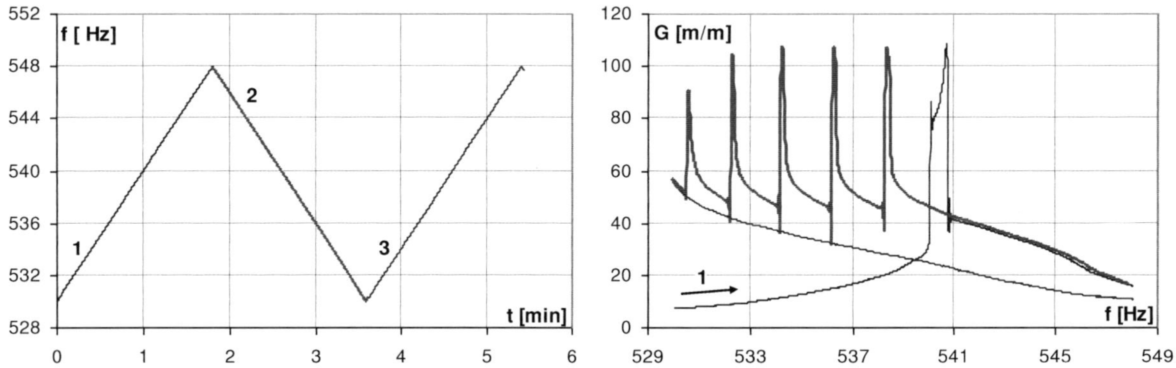

Figure 10. JCF phenomenon for a cracking blade (blade with Ti5.8Al-3.7Mo, sine sweep, a=const) [2,13]

6.3. Early fatigue identification

The LCF and HCF data analysis showed that blade modal properties could be used to observe the material strengthening phase [6,7]. Increase in the 1st mode resonance frequency of approx. 0.4% is a symptom of the initial resonance system quality factor growth – Figure 11. This phase can be described with the linear SDOF model. The orientation indicator of the maximum cyclic material strengthening is $R_m/R_{e0.2}$ (like for the material magnet memory NDT) [14].

The growing asymmetry of the resonance curve was observed only in the final fatigue phase; it preceded the 1st mode frequency decrease – Figure 11. The speed rate of its development was conditioned by the load history of a blade. The asymmetry is a symptom of the material weakening phase [6,7,10]. The speed rate of the resonance curve asymmetry development, from the very first symptom of an open crack, is determined by the blade's history of loads.

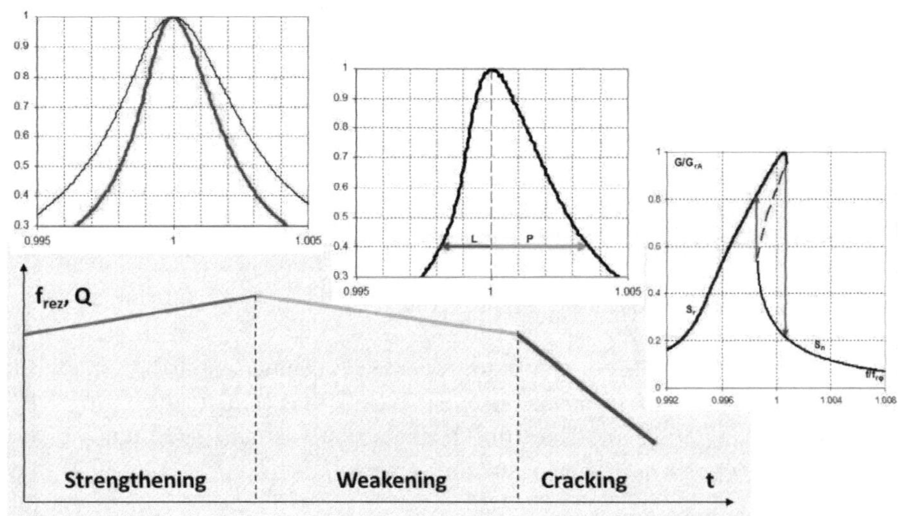

Fig. 11. Symptoms of the material degradation early phase (cyclic strengthening and cyclic weakening phase) [2]

7. CONCLUSION

On the basis of the active experiment and many years' passive experiment, a very high efficiency of the modal analysis method in NDT of compressor blades has been observed.

The modal analysis enables identification and differentiation of reliable symptoms of a blade crack and early phase of material degradation - cyclic strengthening and weakening on condition that the due care of the measurement is preserved, which consists in:

- repeatable way of exciting a blade to vibration,
- control of the strength of fastening of the blade in a handle/slot,
- control of temperature of the blade material,
- selection of the registration method of vibration and measurement range,
- use of the 24-bit resolution of the discretisation of the blade vibration analogue signal.

Predicting the operational durability of blades requires additional knowledge of their real working conditions.

Reference

1. Witos M, Olzak B., 'Modal analysis of compressor cracking blade'. [in] Tip timing and tip clearance problems in turbomachines [ed.] J.F. Brouckaert. VKI Lectures series 2007-03, section 7, pp. 7.1-7.23.
2. Witos M., 'Increasing the durability of turbine engines through active diagnostics and control'. Research Works of Air Force Institute of Technology, Issue 29, 2011, 324 p., doi: 10.13140/RG.2.1. 4341.4560, (pol).
3. Murakami Y, Nomoto T., Ueda T., 'Factors Influencing the Mechanism of Superlong Fatigue Failure in Steels'. Fatigue and Fracture of Engineering Materials and Structures, Vol. 22, pp 581-590, 1999.
4. Shaniavski A.A., 'Tolerance Fatigue of Aircraft Components. Synergetics in engineering applications'. Publishing House of Scientific and Technical Literature "Monography", Ufa 2003, (ros.)
5. Newnham R., 'Properties of materials. Anisotropy, symmetry, structure'. Oxford University Press, Oxford 2005.
6. Socha G. 'Experimental investigations of fatigue cracks nucleation, growth and coalescence in structural steel'. International Journal of Fatigue Vol. **25**(2), pp 139–147, 2003.
7. Almojil M.A., 'Deformation and recrystallization in low carbon steels', PhD Thesis. Manchester Materials Science Centre, 2010.
8. Maia N., Silva J. [ed.], 'Theoretical and experimental modal analysis'. Research Studies Press Ltd, 1997.
9. Witos M. 'On the modal analysis of a cracking compressor blade'. Research Works of Air Force Institute of Technology, Issue 23, pp. 21-36, 2008, doi: 10.2478/v10041-008-0016-0.
10. Awrejcewicz J., 'Dynamika nieliniowa maszyn'. Wydawnictwo Politechniki Łódzkiej, Łodź 2004.
11. http://www.mtiinstruments.com/pdf/products/microtrack2.pdf
12. http://www.vibrationresearch.com/VR8500.html
13. Witos M., 'Analiza modalna pękającej łopatki sprężarki. XI Krajowa Konferencja Naukowo-Szkoleniowa „Mechanika Pękania" 9-12.09.2007 Kielce-Cedzyna; Zeszyty Naukowe Politechniki Świętokrzyskiej, Nr 4, 2007, s. 132-133 + 16 s. CD-ROM.
14. Vlasov V.T., Dubov A.A., 'Physical base of the metal magnetic memory method'. ZAO "Tisso" Publishing House, Moscow 2004.

Non-destructive Testing of Electronic Components Overheating Using Infrared Thermography

Konstantin O. PETROSYANTS [1,2], Igor A. KHARITONOV [1]

[1]Department of Electronic Engineering, National Research University “Higher School of Economics”, Moscow Institute of Electronics and Mathematics , Moscow, Russia
Phone: +007 495 7729590*15214, Fax: +007 495 6287931; e-mail: eande@miem.edu.ru
[2]Institute for Design Problems in Microelectronics, Russian Academy of Sciences, Moscow, Russia

Abstract
Software-hardware IR thermography subsystem for non-destructive testing of electronic components overheating has been developed. IR measuring part included: Flir ThermoVision A40 infrared camera with 17 mkm macro lens, Quantum Focus Instruments Corp InfraScope, precise positioning system. Software part for IR data processing included two standard tools: ThermaCAM Researcher, NEC San-ei Image Processor and several original software tools for extension of the subsystem abilities. The sets of examples are presented to illustrate the efficiency of using IR thermography measuring subsystem for non-destructive testing of electronic components (semiconductor devices, chips, PCBs) to improve thermal stability and reliability of electronic products.

Keywords: Electronic components, chips, PCBs, units, infrared thermography, overheating, hot points, reliability

1. Introduction

Electronic components overheating caused by electrical power is the most important factor reducing their reliability. The increase of 10-15°C in the semiconductor device active region temperature can reduce product lifetime by more than 50%. As a result, the thermal effects are one of the main limiting factors in electronic equipment operation especially for modern components with growing power density and elements temperatures. So the control and monitoring of electronic components thermal behavior is necessary to detect their possible failures and to enhance their reliability. As it is known the most effective way to provide safe thermal regimes of electronic equipment is temperature investigation in the different levels of electronic components: from semiconductor chips to PCBs and units. Thermal imaging with IR cameras [1] provides a nondestructive, nonintrusive, noncontact mapping of temperature in electronic components thus allowing the quick detection and recognition of hot spots and overheating in electronic components.

In this work we describe the infrared thermography system which is used for noncontact mapping of temperature in electronic components. IR images are used then for detection and analysis of electronic components overheating.

2. Infrared thermography system

As it is seen from Fig. 1 the infrared thermography system consists of two parts :

Hardware part is based on Flir ThermoVision A40 infrared camera with 17 mkm macro lens and InfraScope (Quantum Focus Instruments Corp) for IC measuring. The precise 3D positioning system is used providing more opportunities for testing of small-size complex device constructions (semiconductor chips, for example).

<u>Software part </u>for IR data processing includes two standard tools: ThermaCAM Researcher and NEC San-ei Image Processor. Several additional original software tools have been developed to extend the subsystem abilities:

- IRDataProc software tool for generation of emissivity coefficients map of object to improve the accuracy of the object temperature map measurement.
- Tool for IR data conversion between Flir ThermaCAM Reseacher and NEC San-ei Image Processor software tools.

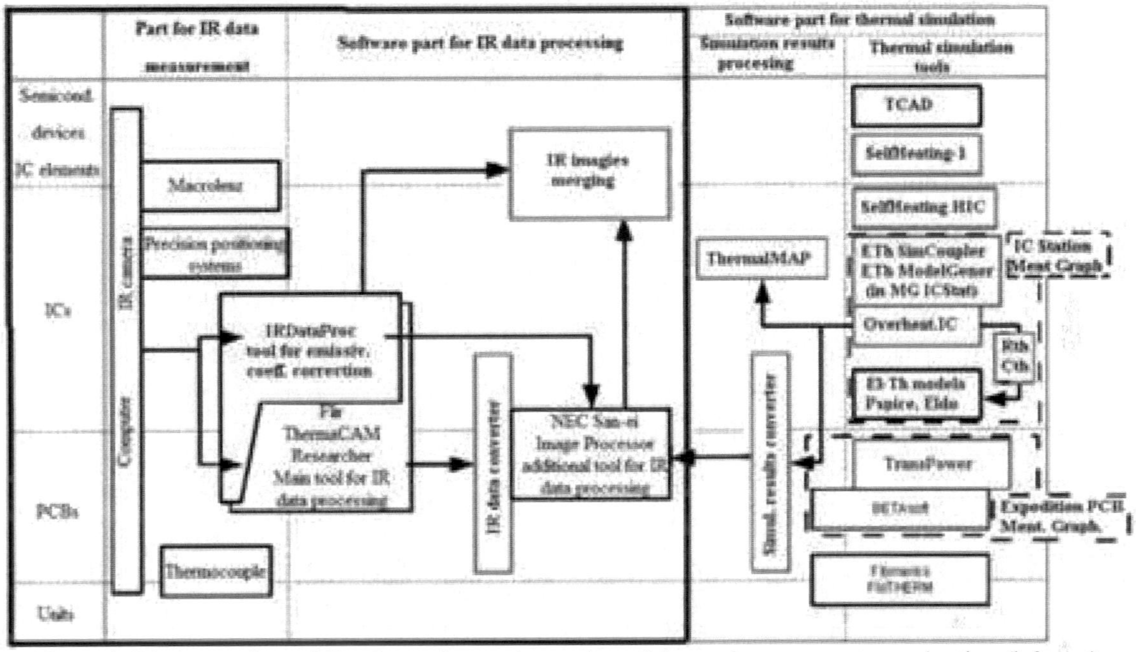

Figure 1. IR system for multilevel non-destructive testing of electronic components overheating (left part) integrated with overheating simulation tools (right part)

3. Non-destructive Testing of Electronic Components Overheating

Using the infrared thermography system the following hierarchy of electronic components was non-destructively tested to detect hot spots and analyze their overheating: integrated circuit (IC) chips, discrete semiconductor devices, printed circuit boards (PCBs), electronic units.

3.1 Integrated circuit chips

The chips for digital ICs with submicron and nanometer elements sizes for low power and high performance; analog middle and high power ICs, high speed and microwave middle and high power ICs were examined.

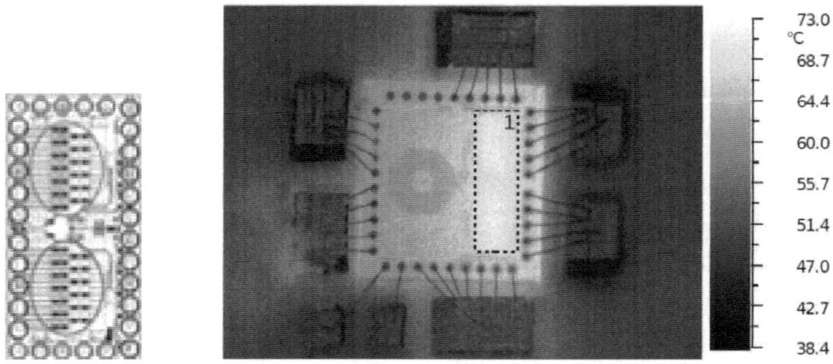

(a) (b)

Figure 2. Layout (a) and IR image (b) of integrated microwave power BJT amplifier (Flir A40 IR camera with 17 mkm macrolenz)

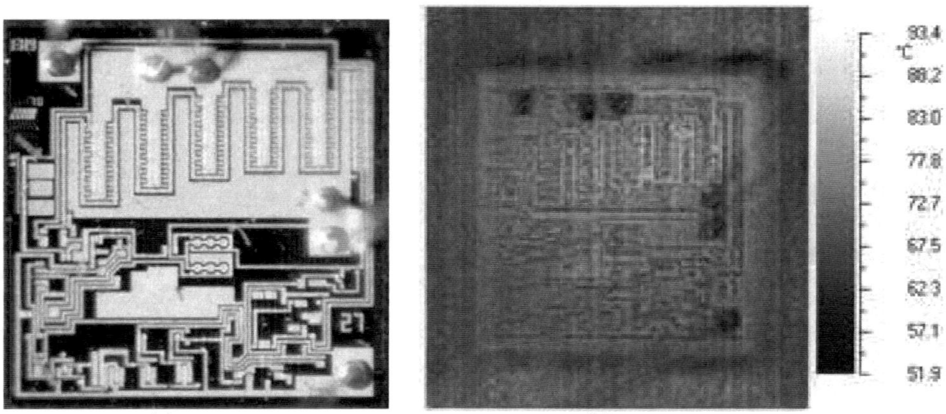

(a) (b)

Figure 3. Photo (a) and IR image (b) of integrated voltage regulator K142EN9, Vin=40V, Vout=27V, Iout=0.27A (Flir A40 IR camera with 17 mkm macro lens)

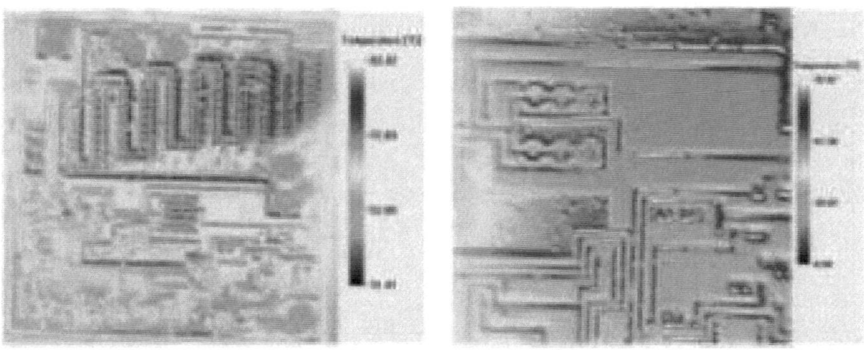

(a) (b)

Figure 4. More detailed IR images of the integrated voltage regulator K142EN9 (see Figure 3) (a) and it fragment (b) (QFI InfraScope).

3.2 Discrete electronic semiconductor components

1) diodes based on p-n and Schottky junctions; 2) low-, middle- and high-power transistors with bipolar, MOS and IGB structures were examined. The devices without case and packaged in case with removed metallic caps were used for surface chip temperature measuring.

For discrete devices and ICs the results of measured data processing were: temperature distributions in device layout, maximal and average temperatures, hot spots and p-n-junction thermal resistances with account for heat sinks and mounting/packaging and cooling conditions .

It is important to note that IR testing of chip temperatures is really the only method of non-destructive testing of whole chip temperature map. The results of IC chips overheating testing were used then to optimize IC layout, case materials, cooling conditions. For discrete semiconductor devices the testing results were used to optimize case construction, cooling conditions, choosing heat sink. Furthermore IR measuring results were used for power MOSFET electro-thermal model parameters determination.

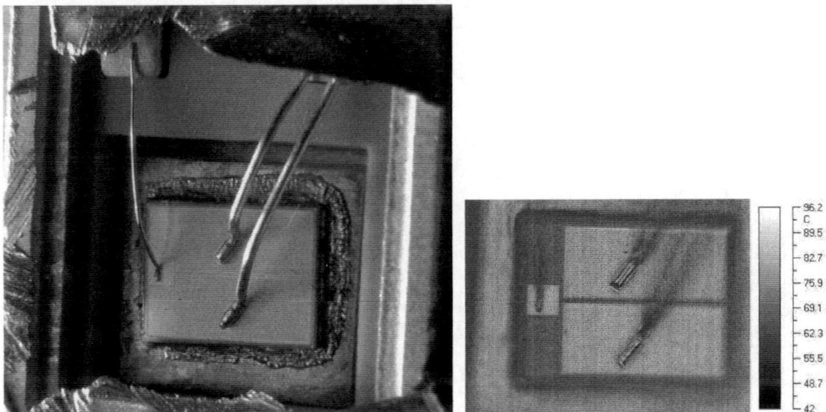

Figure 5. Photo (a) and IR image (b) of power DMOS FET with removed metallic cap placed on heatsink. 10kHz rectangular pulses with 25% duty factor. (Flir A40 IR camera with 17 mkm macrolenz)

3.3 PCBs and electronic units

2D temperature distributions were measured and after that the critical elements with maximal temperatures were recognized. Fig. 6 presents results for power supply PCB and Fig. 7 presents results for AC/DC converter IP 350 electronic unit.

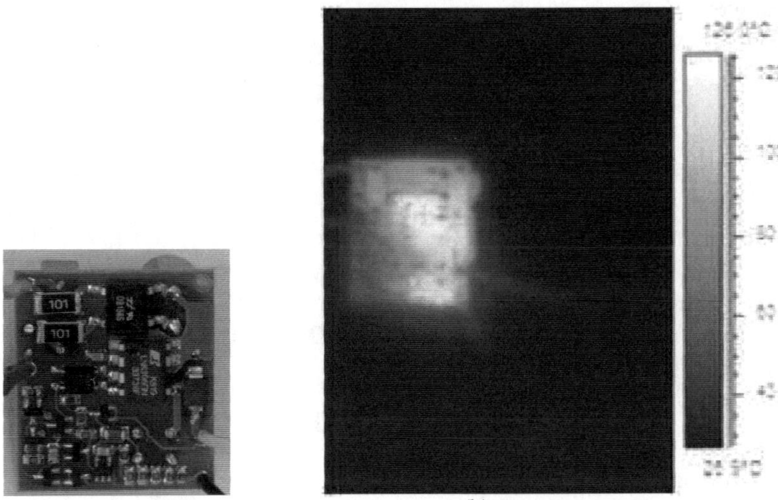

(a) (b)

Figure 6. Photo (a) and IR image (b) of power supply PCB (Vout=12V, Iout=1A) (Flir A40 IR camera)

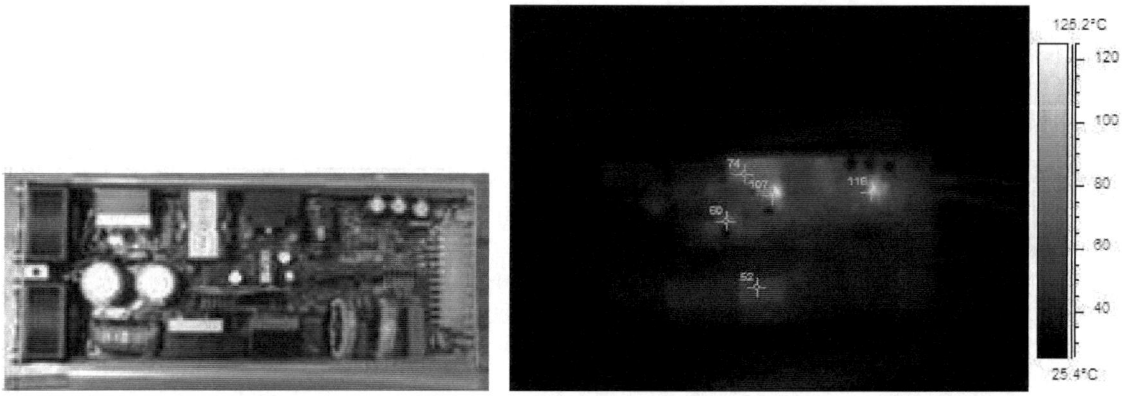

(a) (b)

Figure 7 . Photo (a) and IR image (b) of AC/DC converter IP 350 , P=350W (Flir A40 IR camera)

Critical overheating (temperature up to 126°C) elements have been defined for power supply PCB (Figure 6). Bad location of electrolytic capacitors nearly hot elements (temperature up to 116°C) has been detected. The unit construction was corrected to eliminate capacitors overheating.

4. Conclusions

The described IR measuring subsystem has been included into the general hardware-software system for thermal design, control and management of electronic components [2,3]. Except for hot spots and overheating detection the IR measuring results were used for verification and tuning of mathematical models for thermal simulation of chips, PCBs, units. Furthemore, it is important to note that IR testing of chip temperature is really the only method of non-destructive testing of whole chip temperature map.

This work was carried out within the framework of the National Research University Higher School of Economics' Academic Fund Program under grant 15-01-0165 and the Russian Foundation for Basic Research under grant 14-29-09145.

References

1. Lehtiniemi; Fager; Hynninen; Aapro, 'Applications of infrared thermography in electronics research', Nokia Research Center – 2000.
2. Petrosyants K. O., Kharitonov I. A., etc., 'Multilevel System for Thermal Design, Control and Management of Electronic Components', Intern. Journal of Advancements in Electronics and Electrical Engineering, Vol. 3, No. 2, pp. 22-27, 2014.
3. K. O. Petrosyants, I. A. Kharitonov, etc., 'Thermal Modeling and Experimental Investigation of Electronic Components: Semiconductor Devices, IC Chips, PCBs and Units', Proc. of the 2014 Intern. Conf. on Electrical and Electronic Engineering (EEE2014), April 26-27, 2014, Hong Kong, 2014.

" PowerWave.ndt " : applied to charge wirelessly the batteries of Gas and Petroleum Pipes Crawlers .

Pipelines are regularly inspected from the interior to detect defects, defaults, cracks and corrosion. Such inspections require crawlers that work on battery power and need to be charged often. These charging operations represent a major time loss and cost to pipeline inspection and pipeline construction companies, and therefore to the whole gas and oil industry.
Speed of inspection: the leading companies, such as TecniTes, use robots that are capable of inspecting between 1 kilometer and 2 kilometers per day.

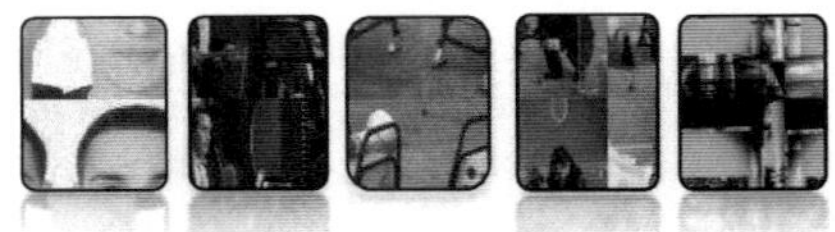

Mohamed Zied CHAARI, PhD Engineer, Tunisia
chaari_zied@yahoo.fr
Patent N°: WO 2012026895 A3

1. Underlying Technology

Pipeline present a big interest for the transportation of big volumes of fluids (gas, chemicals, oil products) that don't pass in transit anymore by the road networks. They present a certain level of risk whose consequences are bound directly to the nature of the elated fluid nevertheless. The first level of security to put consists in place in verifying the state of the pipeline while using a robot «crawler», but this solution presents several problems of localization and visualization.

The wireless power transmission is a technique for the distribution of electric power without using physical media. This technique is intended to be used to supply the places of difficult access. For this reason we invented a system of wireless charging batteries CRAWLER a robot to avoid cutting the line during hydrocarbon depletion of the battery and the blocking of the robot. Our system based on the principle of transfer of microwave energy inside the pipeline. The idea is to use the pipeline as a waveguide to minimize the attenuation of energy and maximize the power recovered to the level of antenna .
Receiving and converting RF energy into electrical energy that makes a patch antenna array that represents a large area, allowing it to have a relatively high effective radiated power and gain acceptable.

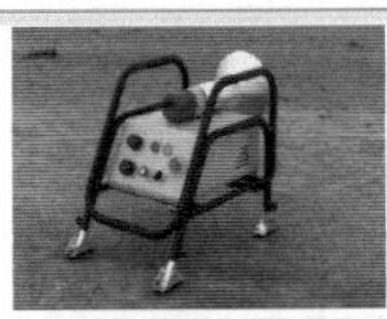

Fig.1 Powerwave NDT

PowerWave's system uses microwave technology, which refers to technology for the transmission of signals at high frequencies. Microwaves are transmitted from an emitter to a receiver through a waveguide. The energy is then converted into direct current electricity thanks to a rectenna.

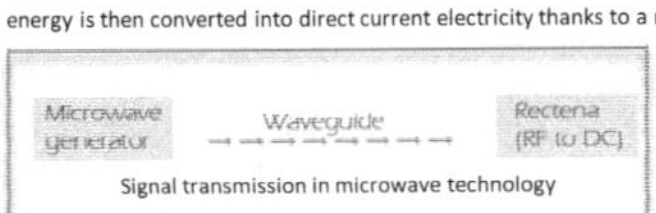

Fig.2 Block diagram of the wireless charging

2. Microwave Power Transmission

- Microwave beam.
- Retro directive beam control capability.
- Power level is well below international safety standard.
- More developed.
- High efficiency up to 85%.
- Attenuation of a wave function of frequency a guidedwave.

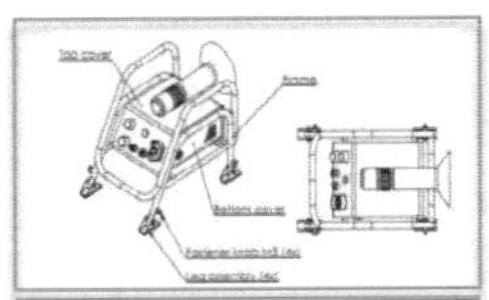

Fig3. Microwave generator

3. Rectenna Design

A rectifying antenna called a rectenna receives the transmitted power and converts the microwave power to direct current (DC) power. The rectifier is a GaAs Schottky barrier diode that is impedance matched to the dipoles by a low pass filter. The rectifying diodes are connected to antenna microstrip.
A rectenna is a receiving device that absorbs incident electromagnetic waves from space and converts it by rectification into a DC power.

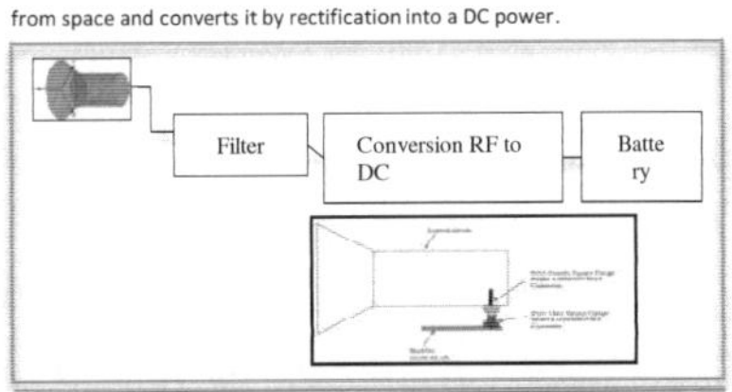

Fig 4. Microwave to DC Converter

4. Pipeline Guidedwave

The idea is to use the pipeline as a waveguide. Waveguides are basically a device for transporting electromagnetic energy from one region to another. They are capable of directing power precisely to where it is needed, can handle large amounts of power and function as a high-pass filter.

The waveguide acts as a high pass filter in that most of the energy above a certain frequency (the cutoff frequency) will pass through the waveguide, where as most of the energy that is below the cutoff frequency will be attenuated by the waveguide. Waveguides are often used at microwave.
frequencies (greater than 300 MHz, with 8 GHz and above being more common). Waveguides are wideband devices, and can carry (or transmit) either power or communication signals. Pipeline circular waveguide is shown in the following figure 3.

For a circular waveguide with diameter a and length d, the mode of propagation with the lowest cut-off frequency is the TE11 mode, as illustrated in Figure 5.

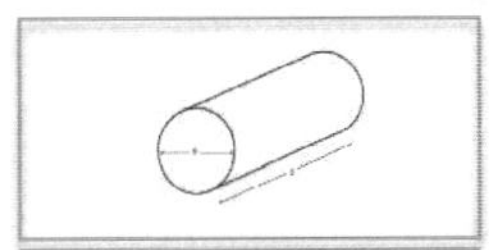

Fig 5. Pipeline circular waveguide geometry

The cutoff wavelength of a circular guide is 1.71 times the diameter of the waveguide. Since the "a" dimension of a pipeline circular waveguide is approximately one half-wavelength at the cutoff frequency, or approximately 1.17 times the "a" dimension of a circular waveguide.

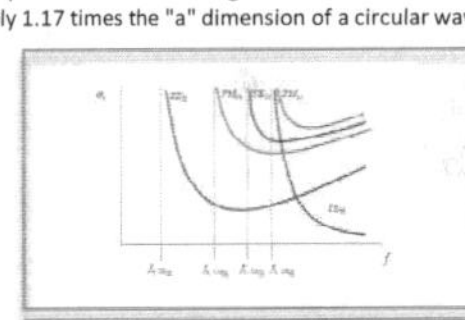

Fig 6. line field of TE11 mode

The TE and TM modes stop growing below a frequency called frequency. To determine if a mode is propagation in a waveguide, calculate its cutoff frequency and compare it to the working frequency

- If f exceeds when that mode is propagated.
- If not there is a mitigation of energy

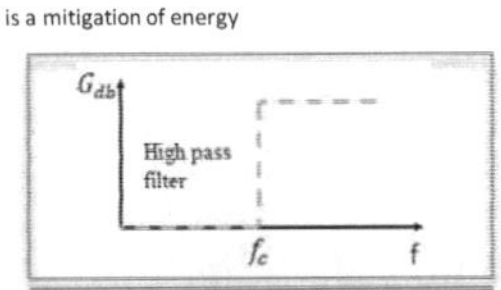

Fig 7. Attenuation of a wave function of frequency a waveguide

The conditions of propagation of a guided wave

$$\lambda_g = \frac{\lambda}{\sqrt{\varepsilon_r}}$$

- λ0 represents the wavelength of an infinite medium having the properties within the guide.
- λg represents the guided wavelength inside the guide.

$$\frac{1}{\lambda_g^2} - \frac{1}{\lambda_0^2} = \frac{1}{\lambda_c^2}$$

- λc represents the cutoff wavelength.

PowerWave's first product is a wireless charging device for pipeline crawlers that allow their batteries to be charged without being brought out of the pipeline. Inspections can then be performed quickly. PowerWave's device is capable of inspecting up to 6 kilometers per day. It also solves the problem of discharged crawlers being blocked inside the pipeline while inspecting since charging can be made wireless.

Damage Detection of Reinforced Concrete Elements Retrofitted with FRP by Using a Wireless Measurement System

Efi A. TZOURA[1], Eleni G. CHOUTOPOULOU[1], Thanasis C. TRIANTAFILLOU[1], Costas P. PROVIDAKIS[2], Aristomenis V. TSANTILIS[1], Corina G. PAPANICOLAOU[1] and Dimitris L. KARABALIS[1]

[1] Department of Civil Engineering, University of Patras
Patras, GR-26500, Greece, +30 2610 997682
Email: eficivil@yahoo.gr, elchou@hotmail.gr, ttriant@upatras.gr, tsantilis.info@gmail.com, kpapanic@upatras.gr, karabali@upatras.gr
[2] Department of Architectural Engineering, Technical University of Crete
GR-73100 Chania, Greece Email: cpprov@mred.tuc.gr

Abstract

In this study lead zirconate titanate (PZT) transducers are employed for damage detection of six reinforced concrete elements (four beams and two columns) retrofitted with carbon fiber reinforced polymer (CFRP) jackets. The PZT transducers were placed externally on the CFRP jackets in an effort to detect the damage underneath the jackets. Measurements of the voltage across the PZT transducers were obtained at various strain values during the experimental procedure. The transducers were placed in positions where the failure of the CFRP jackets was expected to be more intense. The variation of the voltage as the strain increases is the indication of damage. The damage index employed for the quantification of the damage is the root-mean-square-deviation (RMSD) index. The experimental results lead to the conclusion that the measurement system employed in combination with the PZT transducers provide safe evidence that damage detection can be achieved at an early stage.

Keywords: damage detection; FRP jackets; PZT transducers; reinforced concrete; wireless system.

1. Introduction

Recently, a growing interest in retrofitting of existing structures has developed due to their deterioration as a result of ageing but also in order to meet the demands of the contemporary codes. Nowadays one common method for retrofitting of old structures is the application of composite materials and mainly fiber reinforced polymers (FRPs). Although FRPs have been applied in many structures, their application does not provide the ability to detect damage (typically in the form of debonding) underneath. Continuous structural health monitoring of structures retrofitted with FRP is the only way to detect the damage at an early stage by using the appropriate measurement system. One of the most known monitoring techniques is the electromechanical impedance (EMI) or its inverse electromechanical admittance approach (EMA). Recently a new structural health monitoring system was developed by [1]. This system in combination with PZT transducers provides wireless measurements of the voltage across the PZT transducers. Limited research exists combining the damage detection of RC elements retrofitted with FRP sheets, laminates, or jackets, with a wireless measurement system but on small scale specimens. In this study PZT transducers are externally placed on RC beams and columns retrofitted with carbon fiber reinforced polymer (CFRP) sheets or laminates. Each transducer acts both as an actuator and as a sensor. The aim of this research was to detect the debonding of the CFRP sheets or laminates for the case of the beam specimens and also the damage of the RC substrate in the case of column specimens, at an early stage. Measurements of the voltage developed on the transducers for a range of frequencies were obtained. The variation of the voltage as the damage was propagating was the indication of the damage. The damage was quantified by the use of a simple statistical index, the root-mean-square-deviation (RMSD) index.

2. Electromechanical impedance (EMI) method

The system denoted as WiAMS (Wireless Admittance Measurement System) was used in order to obtain the voltage measurements. The PZT transducers were connected to the system as shown in Figure 1.

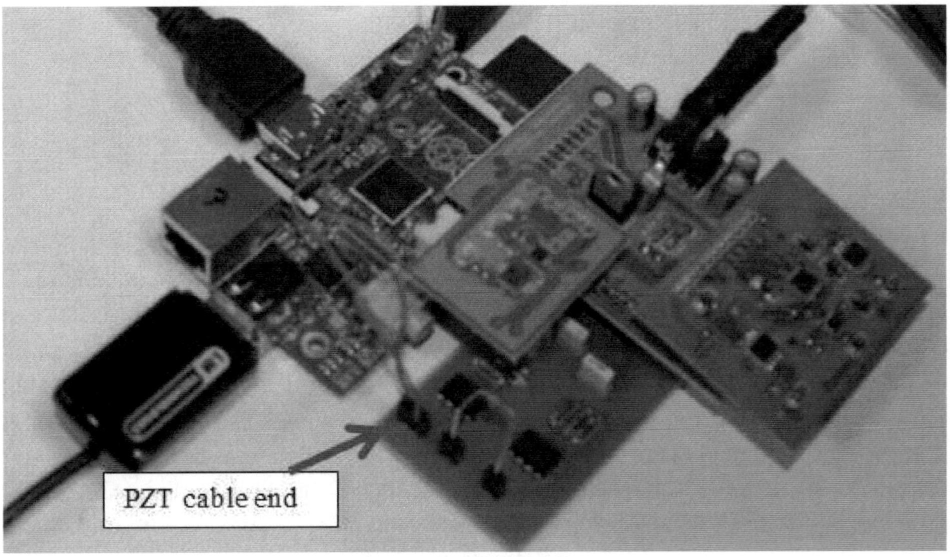

Figure 1. WiAMS

According to the EMI approach a PZT transducer is employed to excite the structure with a high frequency excitation and monitor the changes in the electrical impedance of the transducer. If a PZT transducer placed on a structure is excited with a sinusoidal voltage the area near the PZT (influence radius) will start to vibrate (converse piezoelectric effect). The structure will respond to the vibration by producing electric current on the PZT due to the inverse piezoelectric effect.

The output data of this procedure are the spectra of the voltage versus frequency for a range of frequencies from 160 – 250 kHz. The specific range of frequencies is the appropriate range for reinforced concrete structures as the Voltage – Frequency spectra exhibit peak values of the Voltage at frequency values of 185-195 KHz (1st resonant frequency) and 235-245 KHz (2nd resonant frequency). The resonant frequency is the frequency at which the ceramic element vibrates most readily, and most efficiently converts the electrical energy input into mechanical energy. The WiAMS has a control unit based on a Raspberry pi single board computer which can transfer data without a base station, perform processing-hungry operations and most important to connect to the internet via WiFi. The last feature is very important as it provides the ability to obtain measurements simply by adding the device to the home network and perform structural health monitoring.

In the experimental procedure the output data were uploaded on a database and then downloaded through MySQL (an add-in for Excel). The ability to control the system remotely was achieved by using the free open source application PUTTY. A more detailed description of the system and the EMI method applied in this experimental research is provided as mentioned above by [1].

3. Experimental Procedure

All six reinforced concrete specimens (four beams and two columns) retrofitted with CFRP were subjected to monotonic loading. The beam specimens were subjected to four point bending while the column specimens to uniaxial compression. In this paper the experimental procedure and results of only one specimen (Beam 4) will be presented. More information can be found in [2] concerning the rest of the specimens. Beam 4 was strengthened in flexure with CFRP laminates in the tension zone, tested with the WiAMS. The geometry of the specimen is presented in figure 2 (dimensions in mm).

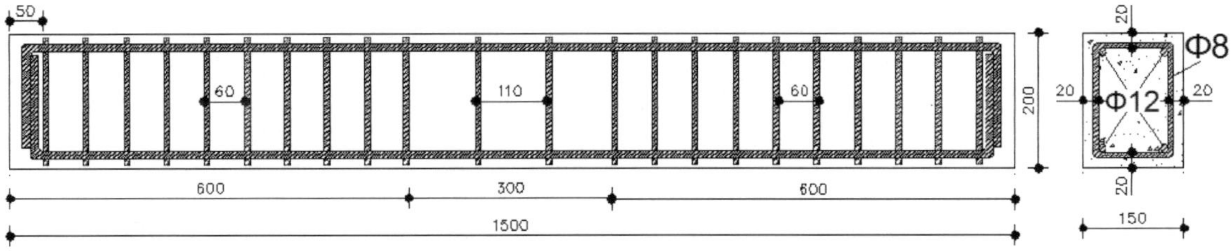

Figure 2. Geometry of the specimen

The average compressive strength of the concrete was 26.76 MPa. The longitudinal and transverse reinforcement had a yield stress of 545 MPa and tensile strength of 600 MPa. The CFRP laminates applied on the beam had a mean tensile strength and modulus of elasticity equal to 3100 MPa and 165 GPa, respectively. The width of each laminate was 50 mm and the thickness was 1.4 mm. Two laminates were applied the one next to the other. The length of application was 900 mm. The epoxy resin used was a two-component epoxy matrix with tensile strength equal to 72.4 MPa and tensile modulus 3.18 GPa respectively. Figure 3 shows the arrangement of the transducers on the beam and the failure of the beam. Debonding occurred at the right edge (as seen it can be seen in figure 3) of the laminates as it was expected. The vertical FRP strip at the left edge of the laminates was placed in order to make sure that debonding will occur at the other edge.

(a) (b)

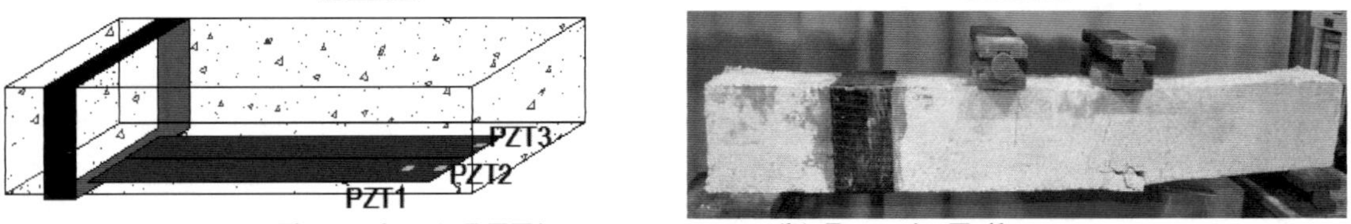

Figure 3. (a) PZT's arrangement, (b) Beam's Failure

The PZTs were placed at the edge of the laminates where the debonding was more intense. The main properties of the PZTs are presented in Table 1.

Table 1

Property	Remarks
Density	7.80 (g/cm^3)
Electric Permittivity $\varepsilon_{33}^{T}/\varepsilon_0$	1750
Piezoelectric Strain Coefficient d_{31}	-180 (10^{-12} C/N)
Elastic Compliance Coefficients S_{11}^{E} S_{33}^{E}	16.1 (10^{-12} m^2/N) 20.7 (10^{-12} m^2/N)
Dielectric loss factor $\tan\delta$ (10^{-3})	20

4. Experimental Results and Discussion

As already mentioned, the output data of the measurements obtained with the WiAMS were the spectra of the voltage versus frequency for various displacement or strain values. A representative graph of voltage – frequency is illustrated in figure 4(a). In figure 4(b) the Load-displacement curve of the beam is presented.

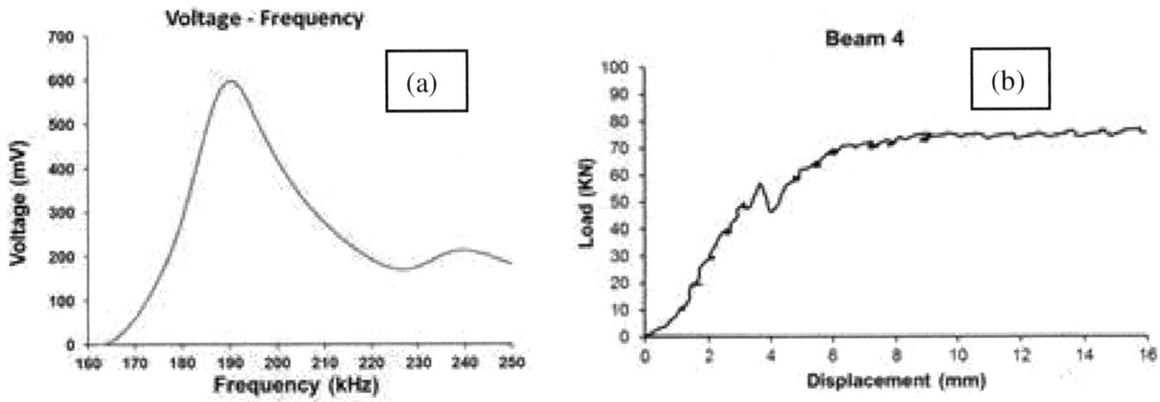

Figure 4. (a) Indicative Voltage – Frequency graph, (b) Load – Displacement graph

The quantification of the damage is achieved by employing the RMSD index, defined by the following equation:

$$RMSD = \left(\sqrt{\frac{\sum_{k=0}^{n} [V_0(\omega_k) - V_i(\omega_k)]^2}{\sum_{k=0}^{n} [V_0(\omega_k)]^2}} \right) \times 100 \quad (\%) \qquad (1)$$

where n is the number of frequency points, $V_0(\omega k)$ is the initial measurement of the voltage of the *kth* frequency point (undamaged condition) and $V_i(\omega k)$ is the measurement of the voltage at the *ith* displacement value of the *kth* frequency point . Figure 5 provides the graphs of the RMSD index for Beam 4 and the final image of the area close to the PZTs after the experiment.

The variation of this index in relation to the displacement or strain value during the test is an easy and understandable way for quantifying the damage. The increase of the RMSD index indicates that the damage also increases with the increase of the applied displacement on the specimens. All six experiments were displacement – controlled. As the displacement increased the experiment was paused at various displacement values in order to obtain the voltage measurements.

As it can be seen in Figure 5 the transducers PZT2 and PZT3 were placed at the edge of each laminate. PZT1 was placed on the same laminate with PZT2 at a distance of 100 mm. The RMSD index for PZT2 has an increased value since the beginning of the experiment. This fact suggests that the specific area where PZT2 was placed was debonded since the beginning. PZT1 placed at a distance of 100 mm from PZT2 exhibits a different behavior with a smooth variation of its values until it reaches a maximum value at the end of the experimental procedure. The variation of the RMSD index for PZT3 is similar to the one of PZT1 but with one order of magnitude higher almost since the beginning (for low displacement values). This is possibly due to the fact that PZT1 was placed in an area where debonding occurred later for the other transducers. Hence the debonding was more gradual, leading to a higher variation of the RMSD index.

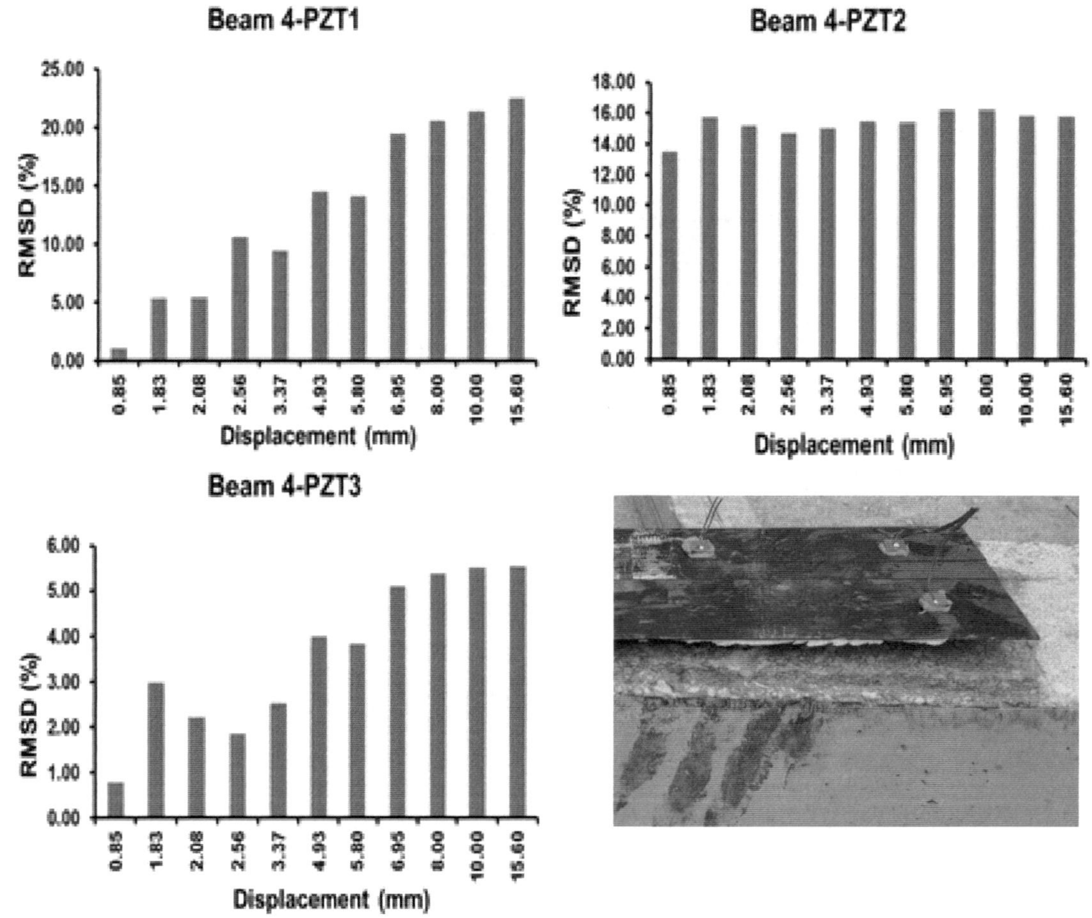

Figure 5. RMSD index and final image of the area close to the PZTs after the experiment

5. Conclusions

According to the experimental results the following conclusions can be drawn.

• When the PZT transducers are placed close to the area where the failure occurs the damage can be detected at an early stage.

• The variation of RMSD increases with the load. The specific measurement system is quite sensitive to damage detection at an early stage and much more sensitive than the one used in the previous study by [3].

Although the results of the research presented are highly promising, more experimental work should be carried out in the specific research field.

Acknowledgments

This research has been co-financed by the European Union (European Social Fund-ESF) and Greek National Funds through the Operational Program "Education and Lifelong Learning" of the National Strategic Reference Framework (NSRF) – Research Funding Program: THALES – Investing in knowledge society through the European Social Fund. The authors wish to thank the students S. Koukiainis and N. Trochoutsou, for their assistance in the experimental program.

References

[1] C.P. Providakis, S. Tsistrakis, M. Voutetaki, J. Tsompanakis, M. Stavroulaki, J. Agadakos, E. Kampianakis, and G. Pentes, 'An innovative active sensing platform for wireless damage monitoring of concrete structures', Structural Control and Health Monitoring (under review).

[2] E.A. Tzoura, E. G. Choutopoulou, T.C. Triantafillou, C.P. Providakis, A.V. Tsantilis, C.G. Papanicolaou and D.L. Karabalis, 'Damage Detection of Reinforced Concrete Elements Retrofitted with FRP by Using a Wireless Measurement System', Structural Monitoring and Maintenance, (under review).

[3] E.A. Tzoura, T.C. Triantafillou, C.P. Providakis, A.V. Tsantilis, C.G. Papanicolaou and D.L. Karabalis 'Damage detection of reinforced concrete columns retrofitted with FRP jackets by using PZT sensors', Structural Monitoring and Maintenance, Vol 2, No 2, pp 165-180, June 2015.

Some Aspects of Signal Processing for Nonlinear Pulse Ultrasonic Mixing Spectroscopy

Karel HAJEK

Department of Electrical Engineering, University of Defence, Brno, Czech Republic
Phone: +420 776298417; e-mail: karel.hajek@unob.cz

Abstract
Principles of nonlinear ultrasonic mixing impulse spectroscopy (NUMIS) show great potential advantages in comparison with other nonlinear ultrasonic methods. It is first of all from the point of view of sensitivity and possibility of quick, simple and precise localization of a defect.
This paper is focused on problems for improve a signal processing, because the sensitivity of nonlinear ultrasonic methods are limited on principle by various noise signals. The problem of sensitivity limits and connected problem of usable dynamic range is discussed in general. One of the important problems consists in optimisation of envelope shape of the exciting tone-bursts. The optimization criterion is maximum suppression of the spectrum of the excitation signal for the frequency band detection of difference frequency.
Next important task was directed to realisation of optimum digital filtration of asked signal with differential frequency and maximum rejection of excitation signals and other disturbing signals inclusive thermal noise. Finally, time dependences of measured signals for detection of a defect place is shown.

Keywords: Nonlinear ultrasonic spectroscopy, mixing principle, sensitivity, burst, dynamic range

1. Introduction

The non-linear ultrasonic spectroscopy (NUS) is developed in order to obtain more appropriate methods in comparison with conventional ultrasonic methods. Specifically, the goal is to achieve a particularly high sensitivity to small defects, less sensitivity to parasitic reflections of the ultrasonic signal and capabilities for complex shapes of test objects [1]. These nonlinear methods are considered in different directions; mainly this is for case with one [3], two or more excitation signals, excitation by continuous signal or by impulse signal. The most published method uses the modulation principle [4-7]. The Time-Reversal method is considered as a most progressive in this time [2].

Published results show that these methods did not achieve expected results, mainly in terms of higher sensitivity and accuracy of localization. These problems were analyzed in [8] and different principles were compared. This article is aimed at perfecting new methods NUMIS, based on the mixing principle, because this principle allows the use of efficient analog prefiltration and thus increase the equivalent dynamic range and sensitivity. The basic block diagram of this method is shown in Fig. 1.

The main idea consists in sending two harmonic bursts with different frequencies f_1 and f_2, see Fig. 1 and 3. These bursts are mutually time-shifted in successive steps. When two waves come across in the place of defect with nonlinear properties, the new frequency component with different frequency f_d will be created. The example with exciting frequencies f_1 = 1.5 MHz and f_2 = 1 MHz and the difference frequency f_d = 500 kHz was shown in [8]. This choice of the frequency values fulfils two aims. It enables the use of sufficient analog pre-filtration, and it also allows the sufficiently precise localization because the high of difference frequency corresponds to a short wavelength. A solving of this problem was published in [9].

Further analysis discusses a problem of limits of the sensitivity as the one of main property of this new NUS method. Following there are analyzed some aspects of the signal processing.

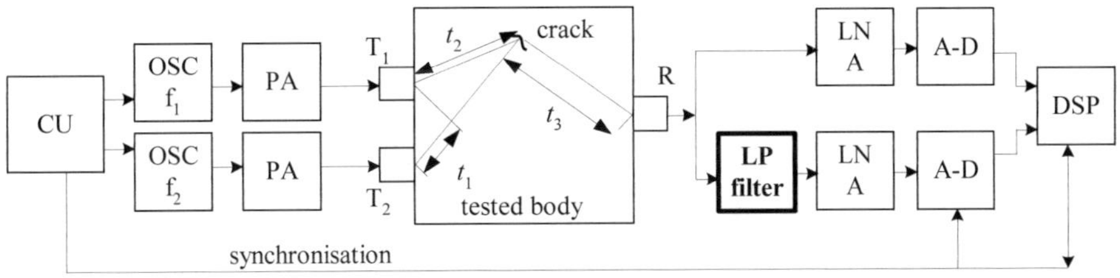

Figure 1. Principal block diagram of nonlinear ultrasonic mixing impulse spectroscopic method (CU – control unit, OSC- oscillator, PA - power amplifier, T_1,T_2 – transmitters, R – receiver, LNA – low noise amplifier, DSP – digital signal processing) [8].

2. Analysis of sensitivity limits for NUS

Various methods of the NUS have different sensitivities. The sensitivity of a NUS method is limited by ratio of exciting ultrasonic signals level and noise level in passband for evaluated signal. It is necessary to discuss all important sources of this noise. These noise sources have various origins and they depend on principle of NUS method. They can operate in both in the part of generation of excitation signals, and in part measured signal processing. The noise sources can be divided to these main groups:

- Higher harmonic components of the ultrasonic exciting signal (generator, exciter, contact between exciter and DUT).
- Additive and differential harmonic components of the ultrasonic exciting signals (mixing and modulating methods with two or more exciting frequencies for one common signal way).
- Equivalent thermal noise of signal processing string (preamplifier, A-D converter).
- Equivalent higher harmonic components and additive and differential harmonic components created by nonlinearity of all signal processing string (e.g. analog preprocessing, A-D converter).
- Other special noise sources (e.g. spectrum spreading of exciting pulses and bursts).

It is evident that the sensitivity is limited by both borders, thermal noise for minimum of processed signal and creating of new parasitic harmonic components for maximum of processed signal. This property is generally expressed by size of possible dynamic range for processed signal. Theoretically, the dynamic range of the signal chain is not directly dependent on the absolute signal level. But on the other hand, it should be noted that the sensitivity depends on the absolute level of the excitation signal and the level of thermal noise.

The basic comparison of various NUS methods from this point of view shows that group of methods with one exciting signal has greatest dynamic range and sensitivity limitation. In this case we evaluate level of created higher harmonic components (e.g. 3th) for some level of exciting first harmonic component, see (2) and Fig. 2.

$$f_v = n f_1 \Big|_{n = 0, 1, 2, \ldots} \tag{1}$$

On the other hand the excitation circuit (generator, exciter) product the same parasitic components without possibility of substantial rejection of this effect. By this way we can obtain dynamic range c. 30-40 dB.

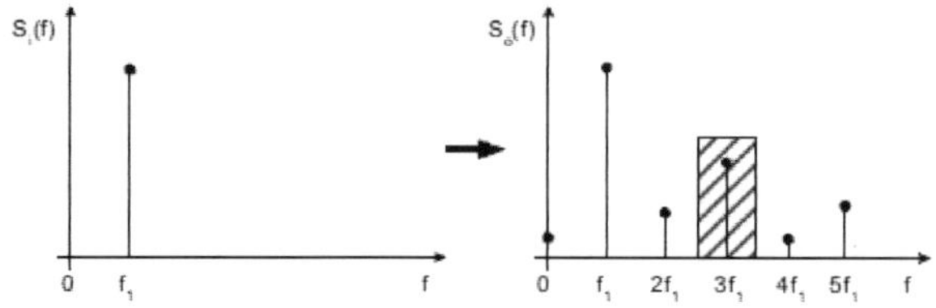

Figure 2. Principle of NUS with one exciting harmonic component.

Therefore the NUS methods with 2 or more exciting signals with different frequencies have basic advantage that frequency of evaluated signal is different that frequencies of excitation signals and their higher harmonics (2), see Fig. 3. A necessary condition, there are separate paths of excitation for each excitation signal.

$$f_r = \left| \pm m f_1 \pm n f_2 \right|_{\text{m, n} = 0, 1, 2..\infty} \quad \rightarrow f_r = \left| \pm f_1 \pm f_2 \right| \tag{2}$$

It is evident that sensitivity of NUS method is based on frequency filtration of evaluated frequency components. This frequency filtration enables suppression of spurious frequency components in the signal evaluation, especially for systems with two or more excitation signals as it was mentioned above. On the other hand, the frequency filtration has various real limits. Therefore various NUS methods with 2 or more exciting frequency components have different abilities to use the real frequency filtering.

The frequency filtering in the signal chain can be divided to three possibilities with different properties:

- Passive ultralinear analog signal prefiltering (RLC filters)
- Analog signal filtering (RLC and ARC filters)
- Digital signal filtering

The passive ultralinear analog signal prefiltering has maximum dynamic range with minimal danger of rise of the parasitic harmonic components. The condition is that it is in front of analog electronics chain, especially in front of the amplifier. On the other hand it is not simple to obtain high filtration selectivity. The analog signal filter is used in the analog electronics chain with amplifiers. Therefore the dynamic range of this part is limited by dynamic range of used amplifiers but his filtration selectivity is higher than previous block. It also provides the anti-aliasing filtration for the AD converter. The third type, the digital filtering has the greatest possibility for filtration selectivity, but it has the lowest dynamic range, due to the limited dynamic range of used AD converter.

Various methods NUS with more excitation signals can be compared with regard to the above frequency filtering options. We can divide these methods to two main groups, see fig. 3. In case of use exciting with one signal with low frequency f_1 and second signal with high frequency f_2 the methods are based on so-called amplitude modulation (AM), Fig. 3 a. The second case uses frequencies of exciting signals with little difference, see Fig. 3 b. These methods are usually called as mixing.

As we can see, the mixing principle enables higher attenuation of excitation signals by a simpler using of the analog linear prefiltration than AM methods. Therefore the mixing methods can obtain higher sensitivity, because they apply evaluated signals with lower dynamic range to chain of analog and digital signal processing. In other words, under the condition of the same dynamic range of signal processing of both systems, the mixing principle allows to process of input sensed signals with higher dynamic range. Therefore the equivalent dynamic range of all system is higher than dynamic range of the analog and digital processing chain.

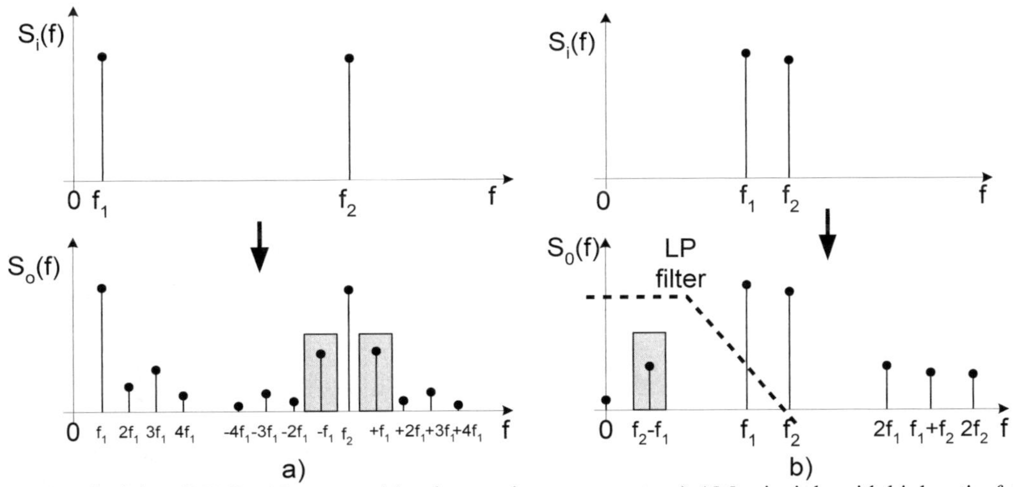

Figure 3. Principle of NUS with two exciting harmonic components: a) AM principle with high ratio f_2/f_1, b) mixing principle with low ratio f_2/f_1 and indicated filtration for rejection of signals with exciting frequencies.

The time-reverse nonlinear methods have a little different principle. Nevertheless, the problem limitation of dynamic range is the same, because the dynamic range of all system is lower than dynamic range of the analog and digital processing chain.

The use of pulse (burst) NUS methods brings special problems for sensitivity limitation. These burst exciting signals enables to solve problem of defect localization. On the other hand the use of the exciting burst brings spreading of monochromatic spectrum to wide band. By this way, the spectrum of exciting signals can take noise effect for detected band in case of NUS methods with two or more exciting signals. The aim is a solution to minimize this effect, equivalent to reducing dynamic range of the system and thus the achievable sensitivity. It should take advantage of the two possibilities, optimizing the shape of the excitation signal and the subsequent optimization of digital filtering of the measured signal for obtaining the most accurate form of envelope detected pulse.

To solve the model digital frequency filtering and subsequent envelope detection of the burst was chosen modeling in Matlab. Because the goal is to model the process nonlinearity, the input signal was modeled as a sum of the detected pulse of 800 kHz with a unit height and both excitation pulses of 1.5 MHz and 2.3 MHz with a correspondingly greater height (up to 80 dB). But confirmed the assumption that the signal at 2.3 MHz is detected pulse due to the increased frequency distance less influence than the 1.5 MHz signal. Therefore, it was subsequently used a simpler model with no input pulse 2.3 MHz.

3. Optimisation of the exciting burst shape

As it was discussed above, the use of burst brings spreading of monochromatic spectrum to wide band. Both signals were modeled with the desired sampling rate of 50 MHz. In the first step, they were considered approximately estimated pulses with a rise time of about 15 μs. However, the first modeling showed that spreading the spectrum of the excitation pulse with a finite length and slope of the leading and trailing edges extending to the detection band (800 kHz). Therefore, spectral analysis was made of the excitation pulse of 1.5 MHz and finding suitable envelope function minimizes this effect. In doing so, it was used well-known theory of window functions used for FFT or IIR filtering.

The spectrum of pulse shape without limitation (rectangular window) was compared with the originally contemplated shape $\sin^2$ and available alternatives windows in Matlab. This comparison showed a window having the smallest parasitic influence of the detected band 800 kHz window called Nuttall (Fig. 4), which is one of the variants Blackman-Harris window. The result of comparison of the rectangular window, and the window pulse $\sin^2$ Nuttall is shown in Fig. 4 b. Logically greater spread spectrum near the fundamental frequency us in this case does not matter, because the band does not interfere evaluated frequency band.

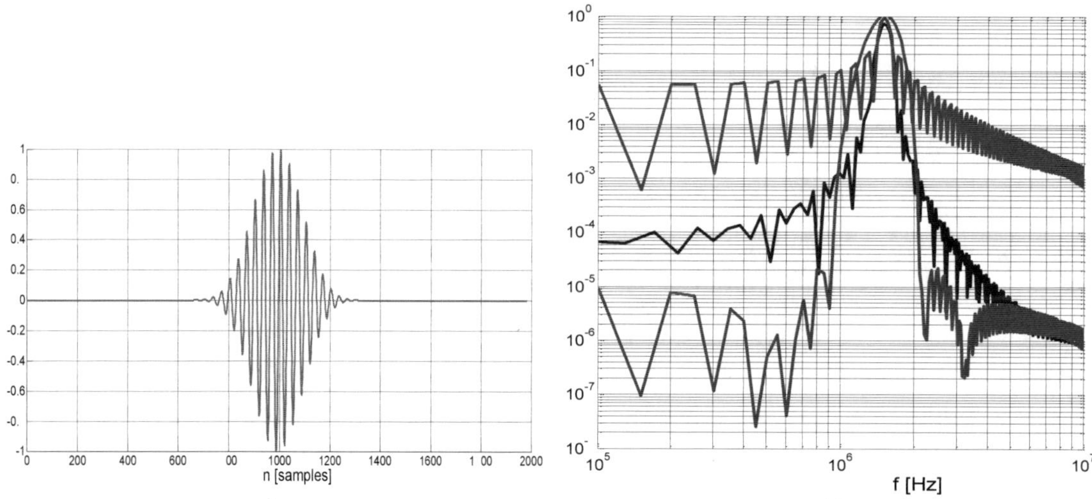

a) b)

Figure 4. a) Example of burst with Nuttal shape. b) Comparison of spectrum of bursts with various shapes: rectangular shape (red), $\sin^2$ shape (blue), Nuttall shape (green).

4. The optimum digital signal processing for frequency filtering

We can consider different variants of digital frequency filters for filtration. It was chosen IIR filter due to the simple configurability and modifiability of the parameters (order, quality factor Q, resonant frequency, filter type). For ease of experimental modeling technology was selected cascading band-pass 2nd order blocks with a choice of quality factor Q and the same resonant frequency f_0. This frequency was set to $f_2 - f_1$, therefore 800 kHz. The approximation resulting from a cascade connection of 2nd order block with the same parameters for the specified purpose quite comfortable with the group delay is reasonable dependence and leads to minimal change in the shape of the detected pulse. More complex variants of approximation (eg. Bessel) would not bring a significant improvement and optimization process would be considerably more complicated.

During this optimization were considered excitation pulses with a final length and shape of a given window function Nuttall (Fig. 4a), as is evident from the previous chapter. The influence of the excitation signal has been tested only for less distant frequency of 1.5 MHz, because the influence of the pulse is below 2.3 MHz and nonlinear effects are modeled. For consistent optimization is needed to monitor the effect of multiple variables. An important parameter is the level of the excitation signal. In all these cases, the maximum level is applied excitation signal U_2= 10 000×U_1 so U_2/U_1=80 dB.

For the pulse was chosen 20 signal periods, which at a frequency of 1.5 MHz, corresponding to 13.3 μs (667 samples), so the rise time is about 6.6 μs (334 samples). The individual options figure shows the envelope of the detected pulse of 800 kHz (blue), the total signal as sum of excitation and detected signal (1.5 MHz and 800 kHz) after filtration IIR (green) and the detected envelope of the filtered signal (red).

In terms of the properties of IIR filtering and the 800 kHz signal detection is necessary to evaluate influence of the order of the filter, the filter type and quality factor. To detect the envelope of the pulse and the subsequent influence of filtering it is necessary to select other parameters. As the most important and most sensitive proved choice of values order and filter quality factor, so the parameters that determine the selectivity. Due to the mutual relation of both parameters in influencing the selectivity there were considered variations of values of quality factor and different variations of the order of the filter.

The example in Fig. 5 a) - c) shows the influence of chosen values of the quality factor Q. The effect of filtration of IIR filter is insufficient and the influence of the excitation pulse disproportionately increases for low Q. Excessive increase in the value of Q leads to an excessive decline in the level of the detected pulse.

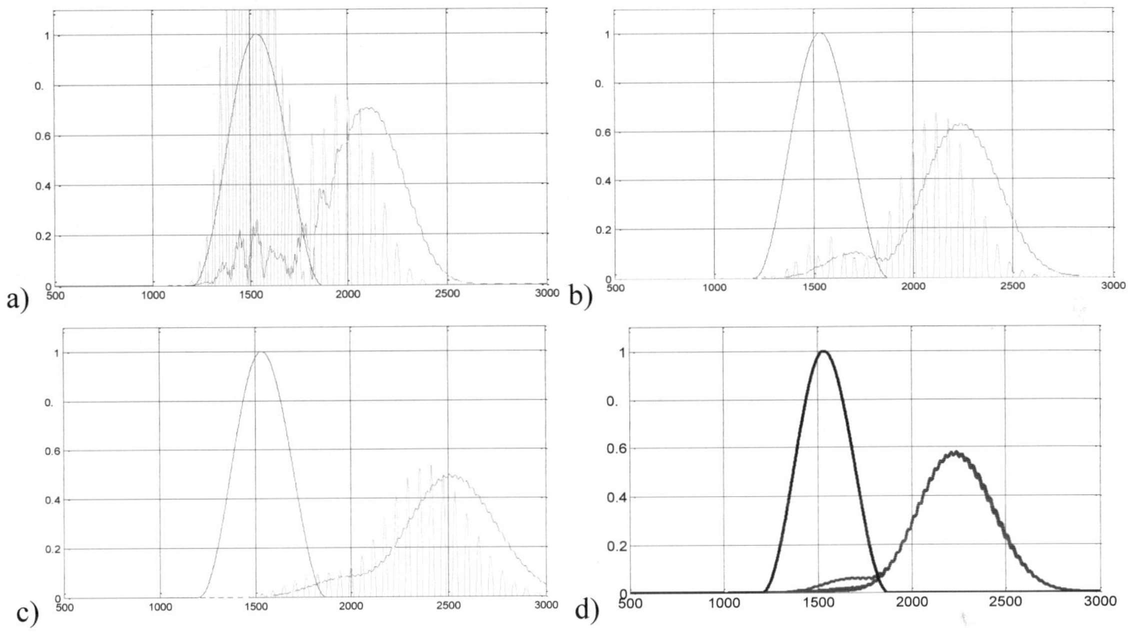

Fig. 5. Variation of Q value: a) Q = 0.75, b) Q =1, c) Q = 1.5, d) Comparison of filtration efficiency for U_2/U_1=60 dB, 70 dB a 80 dB.

Filtration efficiency can be checked for various levels of the interfering excitation signal, using the optimal values of the selected parameters. Fig. 5 d) shows a comparison for the ratios U_2/U_1 = 60 dB, 70 dB and 80 dB. As shown, the influence of the excitation signal is negligible up to 70 dB, only when it increases to 80 dB it is starting to show.

5. Practical verification

The results of shown optimization were used in the practical measurement on a real sample with crack. We scanned the series of time responses for gradually shifted burst with frequency 1.5 MHz. Example of one measured signal (both excitation and one scan) for time delay 16 μs is shown in Fig. 6. It is apparent that the sensor signal has value almost 1 V after amplification 60 dB. Excitation bursts were scanned through a voltage divider 1:100.

After digital processing according to the above-mentioned results we obtain waveforms of the envelope signal with the difference frequency of 800 kHz as shown in Fig. 7. To test the purity of excitation signals were also measured waveforms for generating excitation signals themselves. This was tested parasite transmission of these signals to 800 kHz band by any of the above principles. They are shown under the designation 29u15a (1.5 MHz), and 98u20a (2.3 MHz) in Fig. 7 and are drawn dashed. It is evident that the transfer of parasitic excitation signals to the measuring band of 800 kHz is minimal and practically does not affect the actual measured waveforms of differential signals. Thus it determines the basic equivalent sensitivity to the value of about 0.5 μV at the amplifier input.

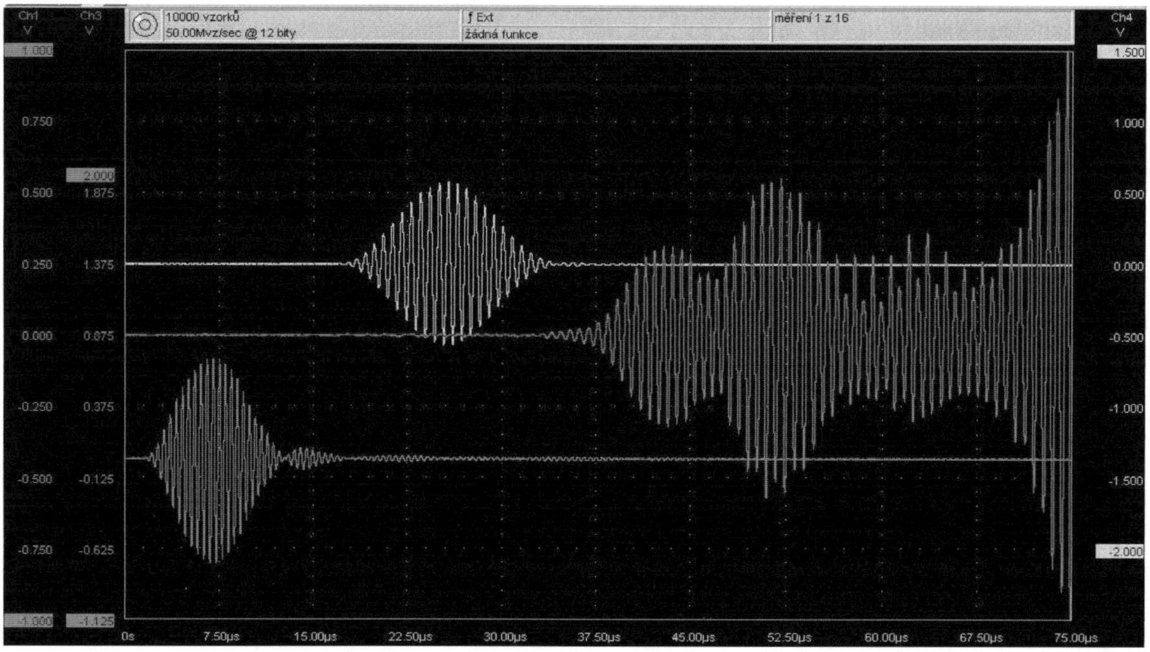

Figure 6. Example of one measured signal (both excitation and one scan) for time delay 16 μs

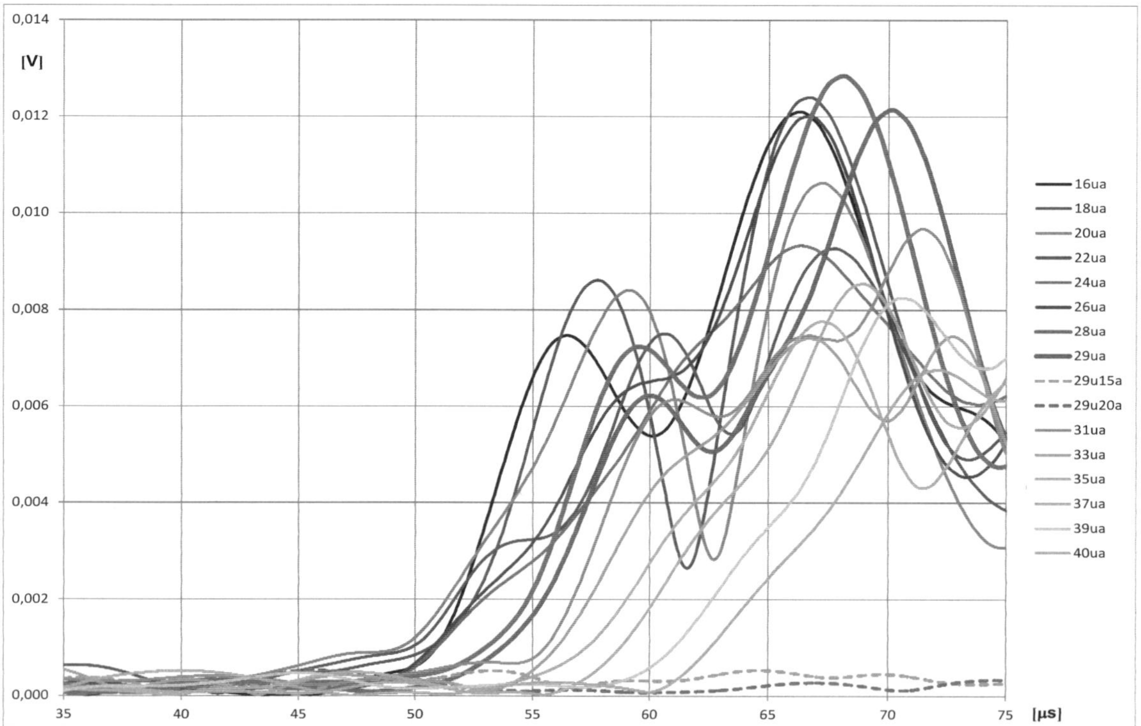

Figure 7. Series of time responses for gradually shifted burst with frequency 1.5 MHz

It is possible to determine the time offset between both bursts for the defect site as about 27μs from the set of curves in Fig. 7. It corresponds to the shift 13.5 μs from middle in direction to exciter 1.5 MHz. This is a starting point for locating of the defect [9].

6. Conclusion

This article discusses the problems of sensitivity for methods NUS and specifically deals with some aspects influencing the sensitivity of the new method NUMIS. It shows that this method should be among the most sensitive. Specifically, it is focusing to some of the parameters and factors limiting the sensitivity of this method. It deals with parasitic transmission of excitation signals to the evaluated bandwidth for the signal with the difference frequency. It shows that the excitation pulse bursts spread spectrum in principle and therefore it is looked for a shape of the burst excitation so that the effect was minimal.

It also deals with the optimization of the subsequent digital processing so as to make the most suppression of excitation spectra of bursts on the shape of the evaluated differential signal with frequency. This is realized in an optimum manner as frequency filtering, as well as means of obtaining the envelope function of the detected pulse. Verification of these processes to practical measurements shows that the proposed procedures are successful and allow us to achieve high sensitivity of this new method.

Acknowledgement

The work presented in this paper has been supported by the grant TA CR No. TA04010165.

References

1. VAN DEN ABEELE, K. E.-A, JOHNSON, P. A., SUTIN, A. Nonlinear Elastic Wave Spectroscopy (NEWS) Techniques to Discern Material Damage, Part I: Nonlinear Wave Modulation Spectroscopy (NWMS). Res Nondestr Eval (2000) © 2000 Springer-Verlag New York Inc. DOI: 10.1007=s001640000002.
2. DOS SANTOS, S. VEJVODOVA S., PREVOROVSKY, Z. Nonlinear Elastic Wave Spectroscopy in Symbiosis with Time Reversal for Localization of Defects: Tr-News. In 5th International Workshop NDT in Progres, Prague 2009. http://www.ndt.net/article/Prague2009/ndtip/proceedings/DosSantos_10.pdf
3. HAJEK, K., KORENSKA, M., SIKULA J., Nonlinear Ultrasonic Spectroscopy of Fired Roof Tiles, In 16th World Conference on Nondestructive Testing, Montreal, Canada, August 30-September 3, 2004, Book of Abstracts, pp. 42, CD
4. ZHAO, X., QI, X., ALEXANDER S, ABERNETHY, E. Damage Detection in Composite Cylinders Using Modulated Guided Wave Vibration, NDCM 2011, Blacksburg (USA), Available from: http://www.ndt.net/article/ndcm2011/papers/A-03-1.pdf
5. PIECZONKA, L., STASZEWSKI, W. J., AYMERICH, F., UHL, T. Analysis of Nonlinear Vibro-Acoustic Wave Modulations Used for Impact Damage Detection in Composite Structures, In 6th European Workshop on Structural Health Monitoring EWSHM 2012, Materials and Qualification. Available from: http://www.ndt.net/article/ewshm2012/papers/th2a2.pdf
6. ZAITSEV, V., MATVEEV, L., MATVEEV, A. On the ultimate sensitivity of nonlinear modulation method of crack detection, NDT&E International, 2009, Vol. 42. Available from: http://vladimir.zaytsev.net/articles/2006_nmod_NDT_zaits.pdf
7. KAZAKOV, V., JOHNSON P. A. Modulation Acoustic Method of Crack Location. 16th International Symposium on Nonlinear Acoustics Moscow, Russia, August 2002.
8. HÁJEK, K., ŠIKULA, J. New possibilities to increase sensitivity of the ultrasound non-linear modulation methods. Int. J. Microstructure and Materials Properties, Vol. 6, Nos. 3/4, 201, p. 283-292. DOI: 10.1504/IJMMP.2011.043223
9. HÁJEK, K., NENAKHOVA, V. Quick Simple and Sensitive Tomography by Nonlinear Ultrasonic Mixing Spectroscopy. 11th European Conference on Non-Destructive Testing (ECNDT 2014), October 6-10, 2014, Prague. ISBN 978-80-214-5018-9

Nonlinear Time Reversal for Non Destructive Testing of complex medium : a review based on multi-physics experiments and signal processing strategies

Serge Dos Santos[1], Martin Lints[1,2], Nathalie Poirot[3], Zuzana Farova[1,4], Andrus Salupere[2], Michael Caliez[5] and Zdenek Prevorovsky[4]

[1] INSA Centre Val de Loire, Unité Mixte de Recherche "Imagerie et Cerveau", Inserm U930, 3 Rue de la Chocolaterie, CS 23410, 41034 Blois cedex, France, serge.dossantos@insa-cvl.fr

[2] CENS, Institute of Cybernetics at Tallinn University of Technology, Akadeemia tee 21, 12618 Tallinn, Estonia

[3] GREMAN CNRS, IUT de Blois, 15 rue de la chocolaterie 41000 Blois, France

[4] Institute of Thermomechanics AS CR, v.v.i., Dolejskova 5, CZ-18200, Prague 8, Czech Republic

[5] INSA Centre Val de Loire, Blois Campus; Laboratoire de Mécanique et Rhéologie, 3 rue de la Chocolaterie, CS23410, 41034 Blois, France

Abstract The interaction between an acoustic wave and a complex media has an increase interest for nondestructive testing (NDT) applications. For specific applications using nonlinear imaging, inversion procedure needs specific signal processing techniques and new coding schemes. This paper tends to show how symmetry analysis can help us to define new methodologies and new experimental set-up involving modern nonlinear signal processing tools. Generalized TR (Time Reversal) based NEWS (Nonlinear Elastic Wave Spectroscopy) methods and their associate symmetry skeleton will be taken as an example with some new spatio-temporal signal processing tools. The purpose of this paper is to present the extension of TR-NEWS for skin ageing characterization. Hysteresis behavior coming from the complex loading of the skin has been identified with PM-space statistical approach, usually associate to aging process in NDT. Phenomenological hysteretic parameters extracted will be presented and associated to standard parameters used for skin characterization. As another application of mixing properties in a wide frequency range, new broadband techniques are needed in the domain of the preservation of cultural heritage. The use of TR-NEWS based analysis (10 MHz) combined with a FTIR-based system (wavenumber range from 3650-650 cm-1) has shown a specific property of the tuffeau limestone which is a mix of calcite, SiO2 and CaSO4,2H2O whereas other damaged sample is calcite.

Key words: Ultrasonic Testing (UT), Other Methods, signal processing, TR-NEWS, nonlinear ultrasonic imaging, nonlinear acoustic testing, nonlinear time reversal

1 Introduction

Nondestructive Testing (NDT) of complex media is of increasing importance, not only in aeronautic industry, but also in various areas of industry also including modern biomedical media. A general definition of NDT is a examination, test, or evaluation performed on any type of test object without changing or altering that object in any way, in order to determine the absence or presence of conditions or defects that may have an effect on the usefulness or serviceability of that object[1].

Nondestructive tests are usually conducted in order to measure various characteristics such as size, dimension, configuration or structure, including alloy content, hardness and grain size. NDT, as a technology, has seen significant growth over the past 30 years and is considered today to be one of the fastest growing technologies from the standpoint of uniqueness and innovation.

With the increase of modern signal processing tools observed during the last 20 years, the area of interest of NDT based innovations can be explored in such domain like old or new materials for biomedical, or within the increasing demand of inspection of ancient materials in the domain of cultural heritage. Among these modern signal processing tools, nonlinear signal processing is a new field of research with the aim to optimize the excitation of information coming from nonlinear effects[2, 3]. Within the wide range of possibilities of signal processing methods, those using symmetry properties seems to be more efficient for extraction of a specific information. Various nonlinear methods have been investigated in recent years for the detection of faults and fatigue in carbon-fibre reinforced composite materials and structures. However, there is no universally-agreed rationale for which technique is best suited to the detection of which type of defect. Furthermore, even if such a rationale were to exist, real-world samples potentially contain a variety of defects (e.g. micro-cracking, delamination and disbonding) induced by various damage mechanisms (stress, impact, heat) such that no single nonlinear testing technique can offer the optimum inspection choice in all circumstances[4, 5].

Various nonlinear methods have been developed in recent years for defect detection in complex materials[6, 7]. The most common include harmonic and overtone generation, inter-modulation product generation and resonant frequency shift, also known as Nonlinear Elastic Wave Spectroscopy (NEWS) methods. Consequently, new optimized excitations are needed and, thanks to the analysis of symmetry properties of the system such as reciprocity, nonlinear time reversal[8] and other pulse-inverted (PI) techniques, a wide range of innovation can be proposed[9]. This is the case of TR-NEWS methods that use pulse-inversion and chirp-coded coding schemes[10, 11]. The interaction between an acoustic wave and a complex media has an increase interest for NDT applications, but also for biomedical ultrasound. For specific applications using nonlinear imaging, inversion procedure needs specific signal processing techniques and new coding schemes[12]. Nonlinear techniques use the fact that microcracks and delaminations generate harmonic and/or subharmonic tones of the frequency at which they are excited. As an example, intermodulation product generation which is based on the monitoring of nonlinear wave mixing in the material[7] is an alternative and a potentially more sensitive nonlinear method for defect detection in composites. In order to confirm these results, this paper tends to show how symmetry analysis can help us to define new methodologies and new experimental set-up involving modern nonlinear signal processing tools[13].

There is, however, still a need for applying NEWS in a complex medium where an adequate knowledge of the initial excitation and the geometry of the medium should be known precisely in order to predict the ultrasonic propagation with high efficiency. Consequently, using symmetry invariance, TR-NEWS methods are supplemented and improved by new excitations having the intrinsic property of enlarging frequency analysis bandwidth and time domain scales, with now both medical acoustics and electromagnetic applications[12]. Among the family of pulse coded excitation, solitonic coding constitutes a new scheme in the sense that solitary waves are the best candidates for pulse propagation in nonlinear and dispersive media[14]. Such spatio-temporal focusing may enable efficient delivery of solitonic optimized excitations to a specific target location within a complex enclosure. This method has the potential to be applied in many real-word contexts where the wish to focus acoustic energy at a location is impractical or impossible. Their robustness during propagation could inform aeronautic end-users during monitoring process of layered; granular, lightweight or functionally graded materials.

The purpose of this paper is also to present the extension of TR-NEWS for skin aging characterization using the nonlinear time reversal signal processing tool known to localize, in a complex medium, sources of nonlinearity potentially responsible of complex material aging. One other aspect related to the problem of complex medium aging is the multiscale aspect of the nonlinear signature which can be studied using PM-space. Since the validation of nonlinear time reversal methods within the NDT community[15], another fields of applications have been investigated recently. One can cite landmine detection, composite and echodentography[16]. The NEWS methods have been shown to improve cracks detection and might, therefore, also be advantageous in medical diagnostic ultrasound applied to echodentography. All of theses applications concern aging characterization or degradation in a complex medium. Linear and nonlinear behavior of skin elasticity is measured locally thanks to an acousto-mechanical loading of the skin conducted with INSTRON loading machines specifically optimized for biomaterials. Hysteretic behavior coming from the complex loading of the skin is identified with PM-space statistical approach, usually associate to ageing process in NDT. Phenomenological hysteretic parameters extracted is presented and associated to standard parameters used for skin characterization.

Finally as another application of mixing properties in a wide frequency range, new broadband techniques are needed in the domain of the preservation of cultural heritage. This is an interdisciplinary challenge which needs the expertise of many scientific disciplines as well as chemistry, physic, optics, etc. Advanced signal and image processing for art and cultural heritage investigation is a topic of major interest in order to provide a quantitative source of information to art historians. The Loire Valley in France has a diversified cultural heritage, including monuments where the mainly construction units is stone. Most castles were built with two French highly porous limestone called Richemont and tuffeau limestones. The objective of this paper is to investigate TR-NEWS for skin aging[17] and for tuffeau limestone characterization (Fig.3) which are both subjected to a growing area of research.

2 The Strategy of Symmetry Analysis

Invariance with respect to time is one of the properties of a more general algebraic approach that is applied in physics which uses intrinsic symmetries for the simplification and the analysis of complex systems. Symmetry analysis is a framework of the bases of a systemic approach aimed at using absolute symmetries like Time Reversal (TR), reciprocity between emitters and receivers, and others. These methods are highly strategic in the sense that they conduct improvement of the measurements so as to optimize the determination of nonlinear properties extracted with coded excitations (e.g. pulse-inversion PI or chirp-coded processes, etc.) with the virtual increase of the effective bandwidth of the excitation. Coded excitations and signal processing is now considered as an efficient way for imaging the complexity in bio-materials with hierarchical properties[18]. Accurate analysis of nonlinear time reversal systems needs the use of new methods of signal processing[19]. Most of systems used in engineering presents a level of nonlinearity that was considered negligible and included in the small stochastic part of the noise. Modern engineering is developed now by considering this stochastic part of the nonlinear signature as a new vector of information coming from the complex system under study[9]. The huge variety of information extracted from this small stochastic part of the response coming from a complex system induces an increase of uncertainty associate to the linear part. This linear part, with its underlying hypothesis of stationarity and determinism, should be consequently associated to a grater uncertainty if the system under study presents intrinsically a complex structure with mesoscopic properties, memory effects, preconditioning, and aging. Of course, these properties are breaking now the stationarity hypothesis implicitly assumed in any linear signal processing, since linear systems theory dominates the field of engineering. This consequence completely justifies this strategy to use nonlinear signal processing tools.

3 Generalized TR based NEWS multimodal instrumentation

Recent years there has been a considerable development of TR based NEWS methods using invariance with respect to TR and reciprocity, both in numerical and experimental aspects. These methods were practically elaborated as the well-known TR-NEWS methods[20]. Chirp-coded pulse excitation[10, 11] is proposed for improving the SNR of ultrasound imaging. Nevertheless, the chirp-coded pulse elongated that the axial resolution of an ultrasound image. Thus, pulse compression technique is utilized to improve the axial resolution of the images. However, it is still lacking the information about the effects of chirp-coded pulse, which was modulated with different window functions, in conjunction with different strain analysis algorithms upon the qualities of imaging. The aim of this study is development of advanced TR-NEWS methods with chirp-coded excitation for assessing biomechanical properties of porcine skin and for NDT of limestones studied within cultural heritage applications. Improvement of TR-NEWS has been conducted with coded excitation using chirp frequency excitation and the concept was presented and validated in the context of NDE imaging[9]. The chirp-coded TR-NEWS method uses TR for the focusing of the broadband acoustic chirp-coded excitation. The method consist in the successive steps :

- emission of a linear frequency sweep signal (the chirp-coded excitation)
- recording of the response to the emitted signal (the chirp-coded coda)
- computation of the pseudo-impulse response which is the correlation between the chirp-coded excitation and its response

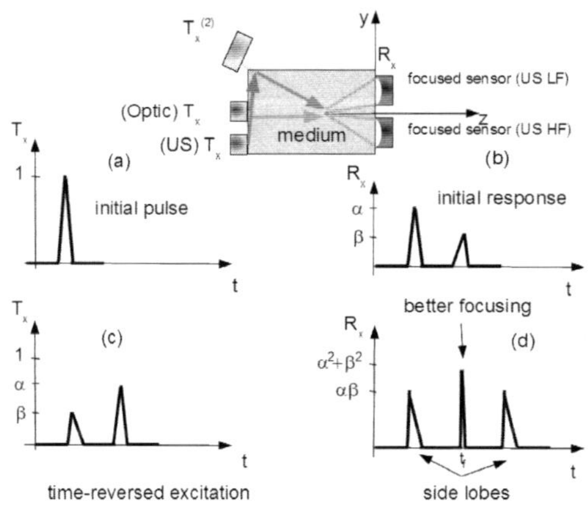

Figure 1: *The four elementary steps of the multi-modal TR-NEWS method : (a) after the excitation of a broadband initial excitation (a short pulse here), the recorded initial response (b) is averaged, filtered, time reversed (c) and rebroadcasted as a time-reversed excitation with the same experimental conditions. The optimized excitation based on this TR-NEWS approach is used for extraction of the local nonlinear signature located in the time domain with respect to the maximun and the side lobes of the crosscorrelation function given by the focused signal (d). The multimodal property come from the fact that the focusing is increase thanks to the use of a broadband multi-physics modality (US LF and HF, optics, etc.).*

- recording of the response to the time-revered emitted pseudo-impulse excitation (chirp-coded TR-NEWS coda)

In order to describe chirp-coded TR-NEWS process in terms of a time-invariant system, we consider a chirp excitation $c(t)$ send in a medium with impulse response $h_{21}(t)$. As described schematically for a simple medium (Fig.1), side lobes could be interpreted with cross-correlations functions and present a symmetry property with respect to the focusing time t_f. The chirp-coded TR-NEWS method uses TR for the focusing of the broadband acoustic chirp-coded excitation. In order to describe chirp-coded TR-NEWS process in terms of a time-invariant system, we consider a chirp excitation $c(t) = A\cos(2\pi f(t)t)$ where $f(t) = At + f_0$ send in a medium with impulse response $h_{21}(t)$. The chirp-coded coda response $y(t,T)$ recorded for a finite time duration T is given by $y(t,T) = h_{21}(t) * c(t) = \int_{\mathbb{R}} h(t-t',T)c(t')dt'$, where $h(t-t',T)$ is an approximation of the (linear) Green's function that satisfy the linear wave equation. The correlation $\Gamma(t)$, computed during Δt, is given by

$$\Gamma(t) = \int_{\Delta t} y(t-t',T)c(t')dt' \simeq h(t) * c(t) * c(T-t,T), \tag{1}$$

and is called the pseudo-impulse response. $\Gamma(t)$ is also proportional to the impulse response $h(t)$ if $\Gamma_c(t) = c(t) * c(-t) = \delta(t)$. Under these hypothesis, $\Gamma(t)$ can be considered proportional to the impulse response (referred as the coda) of the medium and used for enhancing the TR-NEWS focusing. If $\Gamma(t)$ is time reversed and used as a new excitation, the response $y_{TR}(t)$ of the medium (called chirp-coded TR-NEWS coda) is then given by

$$y_{TR}(t,T) = \Gamma(T-t) * h_{21}(t) = \Gamma_h(T-t,T), \tag{2}$$

and provides the linear autocorrelation of the system which peaks at $t = T$ and induces a spatial focusing at the receiver. All this theory is valid under linear behavior of the medium represented by its impulse response $h_{21}(t)$. Any source of nonlinearity in the system will result to a perturbation of this method, and will induce additional terms in Eqs.(1-2). When focusing is performed inside a reverberating medium, like the skin or the Tuffeau limestone, the effective aperture of the TR-NEWS process becomes virtually infinite. Indeed, the impulse response computed using previous equations become sharp. This corresponds to an infinite aperture (sources surrounding the focus) producing a focal area used for analyzing the medium.

4 TR-NEWS for skin imaging monitoring

The measurement of skin's mechanical properties (nonlinearity, anisotropy, and viscoelasticity) is important in several fields, including medicine and cosmetics, and present a huge dispersion depending on age, gender, physical size of individual and location on the body (forearms, face, etc). Small changes in the mechanical properties are very sensitively reflecting many diseases. Their influence on the appearance of aging, and their role indicating disease and pathologies is also a societal goal of biomechanical research[17, and ref. therein]. Hysteresis is a phenomenon that occurs in ferromagnetic and ferroelectric materials, as well as in the deformation of some mesoscopic materials, which are flexible or compressible. For example, sand rock[21], which is one of the example of nonclassical nonlinear materials, for which hysteresis behavior is one of the key properties. In electronics, hysteresis is produced by positive feedback to avoid an oscillation. Today hysteresis and memory based modeling is one of the most interesting and challenging fields of innovation in many engineering applications such as actuators. It seems to be a promising way to understand mesoscopic properties of biomechanical materials. Like porous materials characterized by anisotropy, nonlinearity, hysteresis and susceptibility to fluid sorption, we propose in this paper to study human skin with a multiscale model where each state is defined by a set of internal parameters acting on the element. The other consequence is a pragmatic analysis swarming by phenomenological approaches in the family of PM models. The accurate extracted information coming from such systems needs to be associated to the symmetry hypothesis of the underlying mesoscopic structure with respect to scaling effects responses.

The behavior of PM space responses with respect to the optimized various types of signals with the PM space theory was studied. The PM space model has a prominent future in the modeling of viscoelastic behavior of complex materials. It has been recently proposed for understanding physical mechanism including porosity, amorphous or crystalline bonding material and pre-existing cracks, magnetics, magnetostrictive and shape memory, and smart structure components. With this representation and its identification and classification, the physical response (electrical for electromagnetic relays, or acousto-mechanical biomaterial like the skin) can

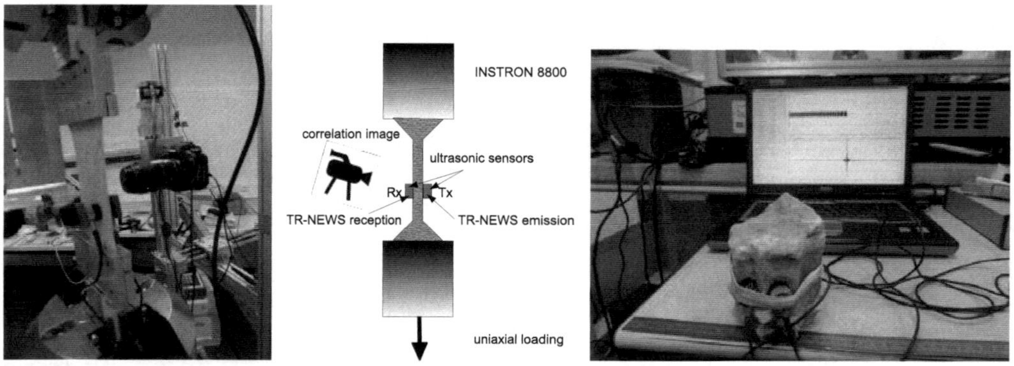

Figure 2: *TR-NEWS experimental set-up for acousto-mechanical imaging of the porcine skin aging (a,b) and for the aging characterization of Tuffeau limestones*

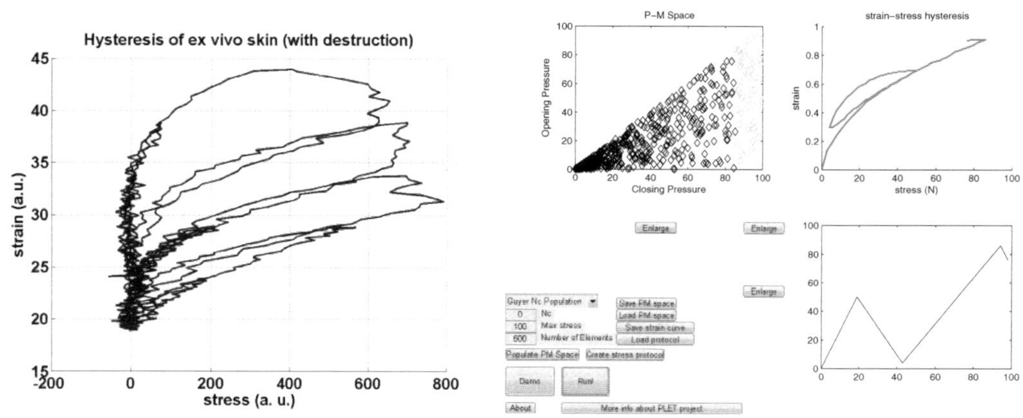

Figure 3: *Skin PM space for memory analysis extracted from hysteretic strain-stress measurements conducted on ex vivo porcine skin. Optimization of the strain-stress curve induces a PM space with 600 hysterons distributed along a Guyer distribution[17]. This original result constitutes the first PM-space identification of viscoelastic and/or hysteretic properties of the skin under uniaxial loading.*

be predicted. Since the strain-stress characteristic of porcine skin under increasing loading is characterized by hysteretic behavior, the preconditonning signature will be investigated with the objective of extracting the memory of the skin[18].

5 Cultural heritage and new NDT multimodal imaging

Electromagnetic waves with Terahertz (THz) frequencies or corresponding sub-millimeter wavelenght are associated to the range between infrared and microwaves. These properties give to the THz frequency range an interesting future for new applications in NDT. Besides security applications, THz technologies for NDT are about to enter markets as pharmaceutics and the composite industries[22], but also for studying NDT on historical heritage. Structural maintenance of historical buildings requires knowledge based strategies to keep under control all parameters governing the structural integrity of these historical buildings, and particularly a robust mon-

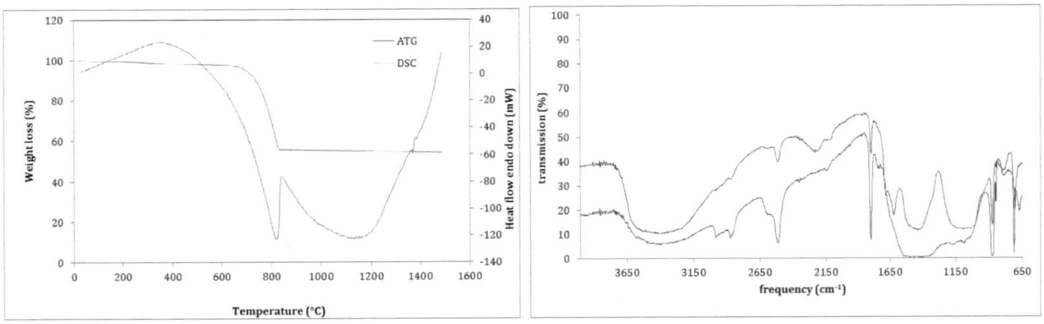

(a) (b)

Figure 4: *(a): ATD/DSC analysis of the Tuffeau limestone. The slope temperature was chosen to 10°C/mn. The transition observed at 800°C corresponds to the CaCO3 decomposition. (b): FTIR analysis performed in the frequency range [3650:650]*cm^{-1} *where CaCO3 vibration modes are identified.*

itoring. In order to test TR-NEWS approach for NDT of the fine art objects and the cultural and historical heritage, some preliminary tests have been conducted on Tuffeau limestones. The objective is to confirm the interest of using multi-physics laser based imaging methods for low-cost ultrasound sensor for the detection of defects[23].

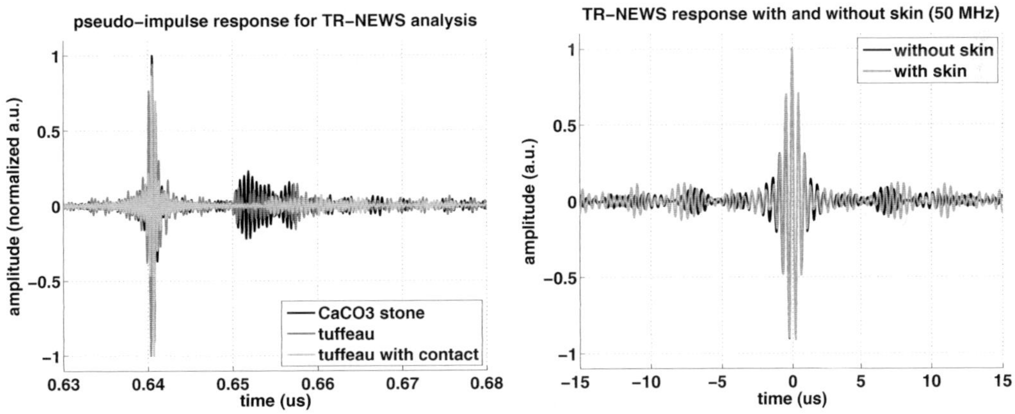

(a) Tuffeau limestone (b) Porcine skin

Figure 5: *TR-NEWS pseudo impulse response (a) and cross correlation function (b). The skin signature is located near the maximun of the crosscorrelation function.*

As seen in Fig.5, the chirp-coded excitation $c(t)$ with $\Delta f \in [0.1 : 1.5]$ MHz is plotted in the time domain with $\Delta t = \frac{N}{f_e} = \frac{64000}{50e6} = 1.28$ ms. The pseudo-impulse response given by Eq.1 is plotted in the time domain after filtering with a 20.75 MHz low-pass filter and $N_a = 148$ averages. The pseudo-impulse properties has been compared in the time and the frequency domains. The reverberant properties (in the time domain) of the skin and the Tuffeau limestone have been reached as observed in (a).

6 Conclusions and perspectives

The use of TR-NEWS based analysis (50 MHz) combined with a FTIR-based system (wavenumber range from 2000-800 cm-1) has shown a specific property of the tuffeau limestone which is a mix of calcite, SiO2 and CaSO4,2H2O whereas other damaged sample is calcite. In the future, the ultrasound nonlinear imaging system should be used on human subjects to demonstrate the feasibility of collecting *in vivo* data. The objective to extend this modern approach to skin and human brain, from whose memory effects are currently admitted, gives to this approach a promising future for modern engineering, and medical biomechanical imaging. Concerning cultural heritage applications, we believe that TR-NEWS and its multi-modal aspect including US and FTIR techniques will benefit art collection community as a whole, and further emphasize scientist's contribution to the cultural heritage as well.

7 Acknowledgments

This work is supported by the Région Centre-Val de Loire under the PLET project.

References

[1] C. Hellier, *Handbook of Nondestructive Evaluation, Second Edition.* McGraw-Hill Professional, 2012.

[2] E. N. Janssen *et al* , "Dual energy time reversed elastic wave propagation and nonlinear signal processing for localisation and depth-profiling of near-surface defects: A simulation study," *Ultrasonics*, vol. 51, no. 8, pp. 1036 – 1043, 2011.

[3] S. Park, *et al* , "Evaluation of residual tensile strength of fire-damaged concrete using a nonlinear resonance vibration method," *Magazine of Concrete Research*, vol. 67, pp. 235–246, 2014.

[4] J. Jingpin, *et al* , "Micro-crack detection using a collinear wave mixing technique," *NDT & E International*, vol. 62, p. 122, 2014.

[5] J. Jiao, *et al* , "Evaluation of the intergranular corrosion in austenitic stainless steel using collinear wave mixing method," *NDT & E International*, vol. 69, pp. 1–8, 2015.

[6] K. Van Den Abeele, *et al* , "Quantification of material nonlinearity in relation to microdamage density using nonlinear reverberation spectroscopy: Experimental and theoretical study," *The Journal of the Acoustical Society of America*, vol. 126, no. 3, pp. 963–972, 2009.

[7] P. Armitage and C. D. Wright, "Design, development and testing of multi-functional non-linear ultrasonic instrumentation for the detection of defects and damage in CFRP materials and structures," *Composites Science and Technology*, vol. 87, pp. 149 – 156, 2013.

[8] M. Frazier, *et al* , "Nonlinear time reversal in a wave chaotic system," *Phys. Rev. Letters*, vol. 110, p. 063902, 2013.

[9] S. Dos Santos, *et al* , "Nonlinear signal processing for ultrasonic imaging of material complexity," *Proc. of the Estonian Academy of Sciences*, vol. 59, pp. 301–311, 2010.

[10] J. Liu and M. F. Insana, "Coded pulse excitation for ultrasonic strain imaging," *Ultrasonics, Ferroelectrics, and Frequency Control, IEEE Transactions on*, vol. 52, pp. 231–240, 2005.

[11] H. Peng and D. C. Liu, "Enhanced ultrasound strain imaging using chirp-coded pulse excitation," *Biomedical Signal Processing and Control*, vol. 8, pp. 130–141, 2013.

[12] S. Dos Santos, *et al* , "Time reversal based news methods for localization and classification of nonlinear vibrations in bio-mechanical media," in *20th International Congress on Sound & Vibration : ICSV20, Bangkok : International Institute of Acoustics and Vibration*, M. Crocker and M. Pawelczyk, Eds., 2013.

[13] V. Sanchez-Morcillo, *et al* , "Spatio-temporal dynamics in a ring of coupled pendula: Analogy with bubbles," in *Localized Excitations in Nonlinear Complex Systems*, Springer International Publishing, 2014, vol. 7, pp. 251–262.

[14] M. Lints, *et al* , "Simulation of solitary wave propagation in carbon fibre reinforced polymer," *Proc. of the Estonian Academy of Sciences*, vol. 64, pp. 297–303, 2015.

[15] X. Guo, *et al* , "Detection of fatigue-induced micro-cracks in a pipe by using time-reversed nonlinear guided waves: A three-dimensional model study," *Ultrasonics*, vol. 52, p. 912, 2012.

[16] S. Dos Santos and Z. Prevorovsky, "Imaging of human tooth using ultrasound based chirp-coded nonlinear time reversal acoustics," *Ultrasonics*, vol. 51, no. 6, pp. 667 – 674, 2011.

[17] S. Dos Santos, *et al* , "Skin hysteretic behavior using acousto mechanical imaging and nonlinear time reversal signal processing," in *Proceedings of the 22nd International Congress on Sound and Vibration, Malcolm J. Crocker Marek Pawelczyk, Francesca Pedrielli, Eleonora Carletti, Sergio Luzzi*, Eds., 2015.

[18] S. Dos Santos and J. Chaline, "Symmetry analysis for nonlinear time reversal methods applied to nonlinear acoustic imaging," in *to be published in proc. of the 20th International Symposium on Nonlinear Acoustics*, A. C. Proceedings, Ed., 2015.

[19] E. N. Janssen and K. Van den Abeele, "Dual energy time reversed elastic wave propagation and nonlinear signal processing for localisation and depth-profiling of near-surface defects: A simulation study," *Ultrasonics*, vol. 51, pp. 1036 – 1043, 2011.

[20] T. Ulrich, *et al* , "Time reversal and non-linear elastic wave spectroscopy (TR-NEWS) techniques," *International Journal of Non-Linear Mechanics*, vol. 43, no. 3, pp. 209 – 216, 2008.

[21] R. A. Guyer, *et al* , "Hysteresis, discrete memory, and nonlinear wave propagation in rock: A new paradigm," *Phys. Rev. Lett.*, vol. 74, pp. 3491–3494, 1995.

[22] P. U. Jepsen, *et al* , "Thz spectroscopy and imaging – modern techniques and applications," *Laser Photonics Rev.*, vol. 5, pp. 124–166, 2011.

[23] F.-J. García-Diego, *et al* , "Development of a low-cost airborne ultrasound sensor for the detection of brick joints behind a wall painting," *Sensors*, vol. 12, pp. 1299–1311, 2012.

INSTITUTE OF THERMOMECHANICS AS CR, v. v. i.

VIIIth NDT in Progress Prague 2015

VIIIth International Workshop NDT In Progress

October 12 – 14, 2015, Prague, Czech Republic

Characterization of the Anisotropy and Directional-Dependent Nonlinearity of the Human Skin In-vivo Using Ultrasound

Daniel TOKAR Zdenek PREVOROVSKY

Institute of Thermomechanics
Academy of Sciences of the Czech Republic
Dolejskova 1402/5
Prague, Czech Republic

daniel.tokar@it.cas.cz

Outline

What is this presentation about…

VIIIth NDT in Progress Prague 2015

1. Introduction to mechanical behaviour of the human skin
2. Motivation for the research and background
3. Measuring method description
   - multi-directional flexible circular probe
   - measuring system: composition, calibration
   - anisotropy, NEWS, Time-Reversal
4. Results
   - human skin *in-vivo*
   - carbon-fabric fiber composite (CFRP)
5. Conclusion
6. Perspective and future work

2/16

Introduction

Human skin's mechanical behaviour

- **Human skin anatomical structure:**
  - epidermis
  - dermis – elastin and collagen fibres
  - hypodermis – fatty tissue
- **Human skin mechanical properties:**
  - nonlinear stress-strain diagram
  - viscoelasticity, mechanical hysteresis
  - anisotropy

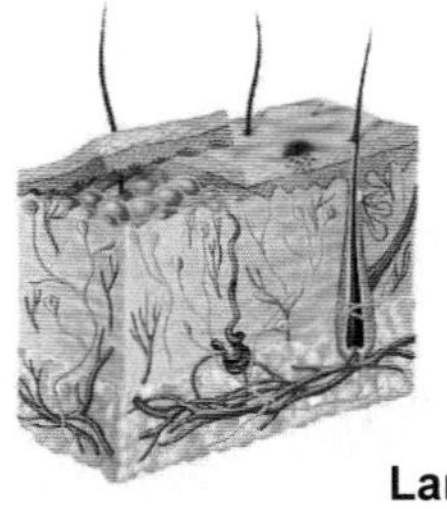

Langer's lines

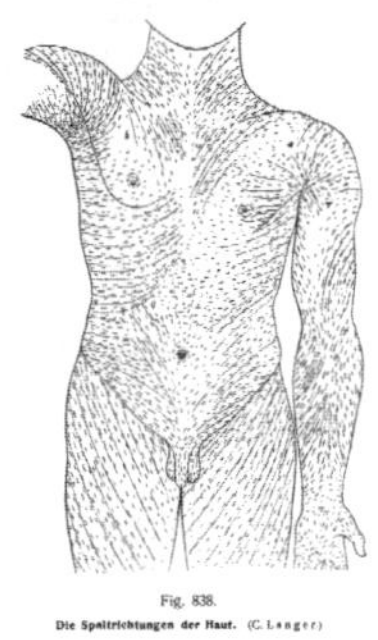

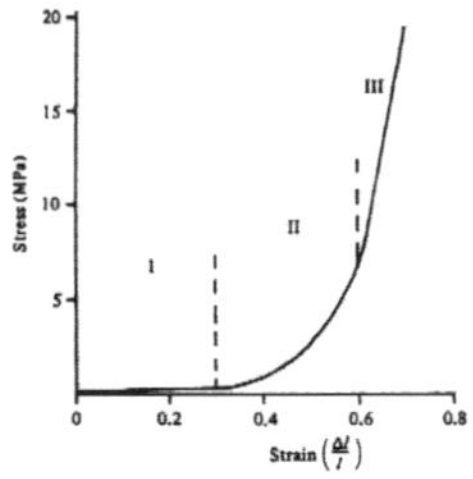

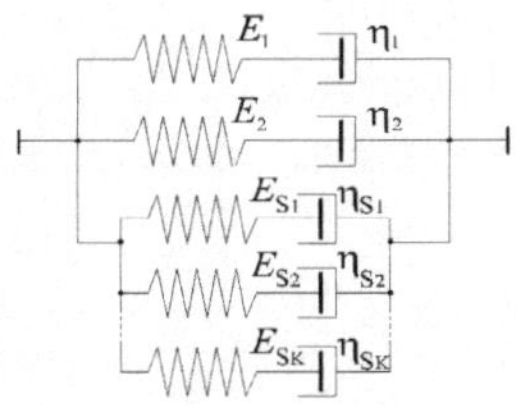

HENDRIKS, F. M.: *Mechanical behaviour of human epidermal and dermal layers in vivo.* Proefschrift. Eindhoven: Technische Universiteit Eindhoven, 2005. ISBN 90-386-2896-X.

Introduction

Motivation for the research and background

- Biological aspects of the skin are well known, mechanical behavior is not yet fully investigated
- Skin mechanical properties are affected by environmental and other conditions: age, temperature, humidity, UV radiation, etc.
- **Measurement of human skin anisotropy**
  - development of fast and easy to use method for local anisotropy investigation *in-vivo*
  - contribution to knowledge of global skin properties
  - plastic surgery – implants localization
- **Nonlinear methods**
  - investigation of changes in mechanical properties of the skin related to environmental conditions, skin diseases, ageing, etc.

Methods

System for local anisotropy evaluation

- Multi-directional flexible circular ultrasonic probe

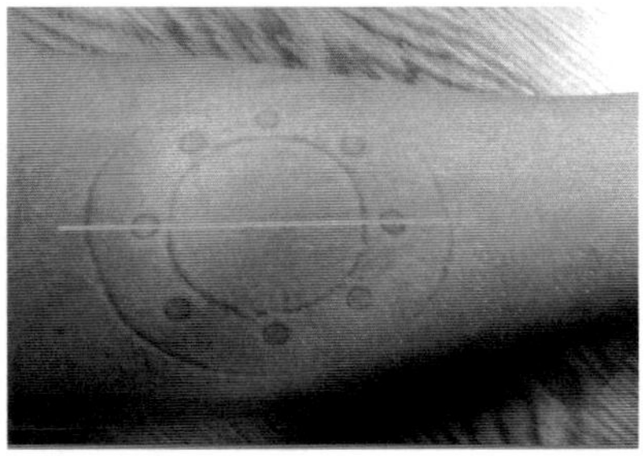

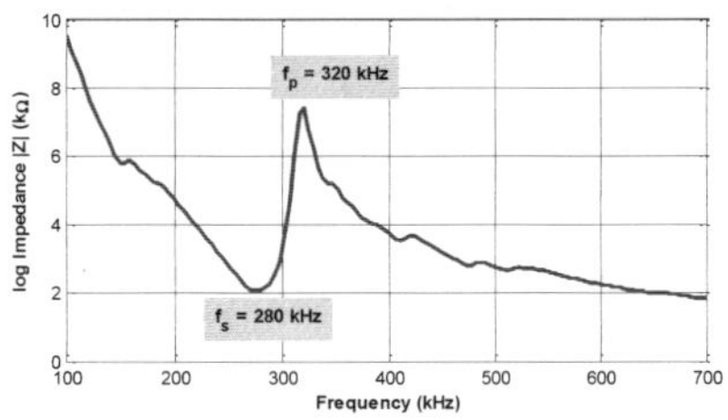

VIII th NDT in Progress Prague 2015

Methods

System for local anisotropy evaluation

- **Measuring system:** AWG + power-amplifier, high-speed digitizer + preamplifiers (20 dB gain, HP 20 kHz), PXIe, 8-ch multiplexer

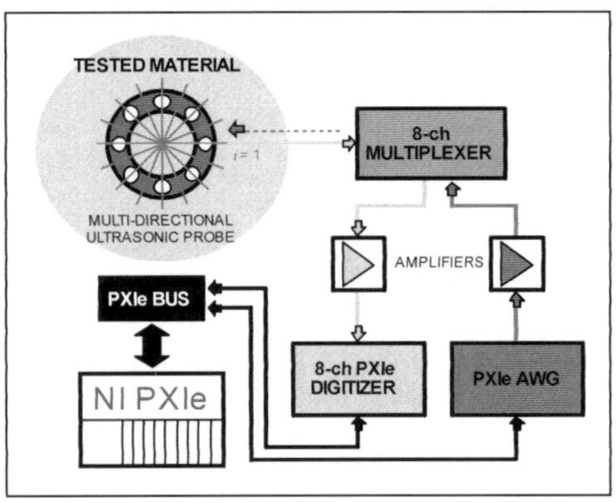

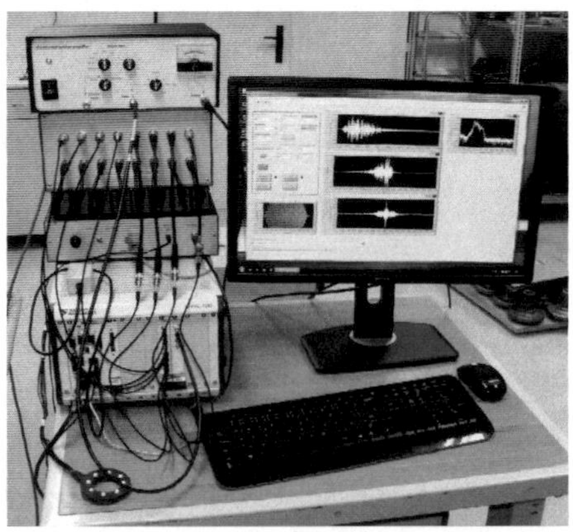

Methods

System for local anisotropy evaluation

- System's calibration on isotropic polymethylmethacrylate sample

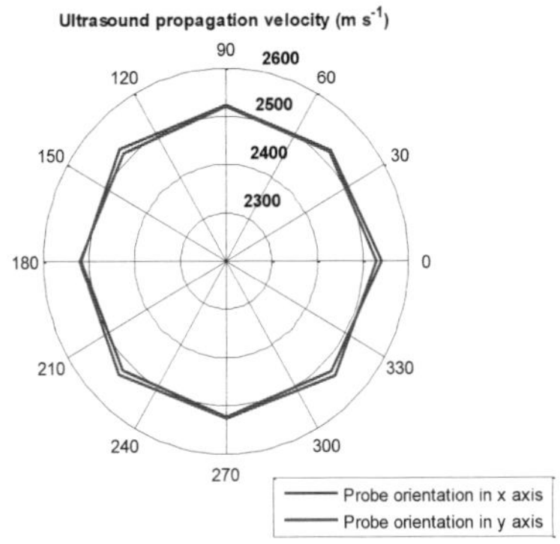

- signal averaging
- threshold-value detection

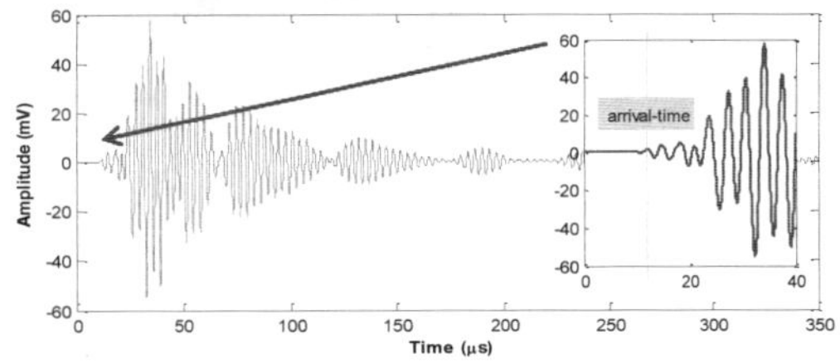

Results

Forearm skin anisotropy measurement

VIIIth NDT in Progress Prague 2015

- Volunteer: female, 25 years, skin thickness 1.3 mm

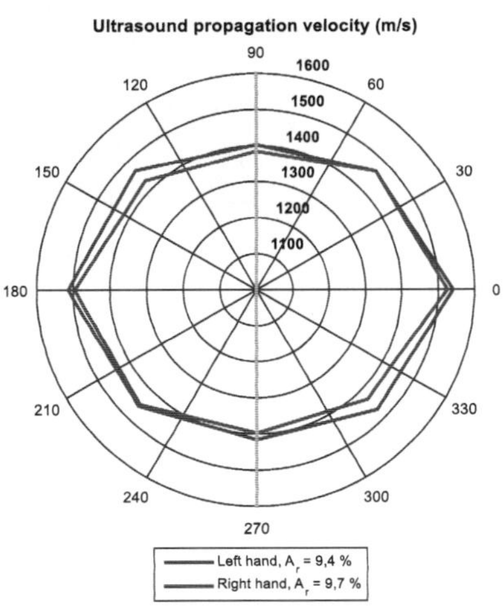

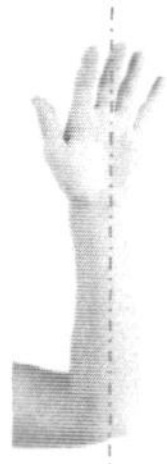

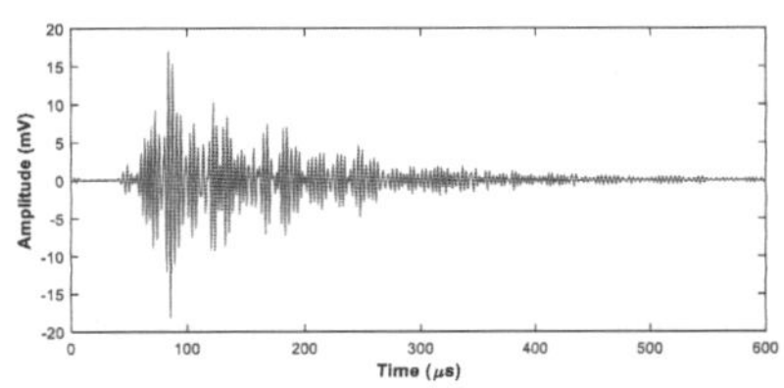

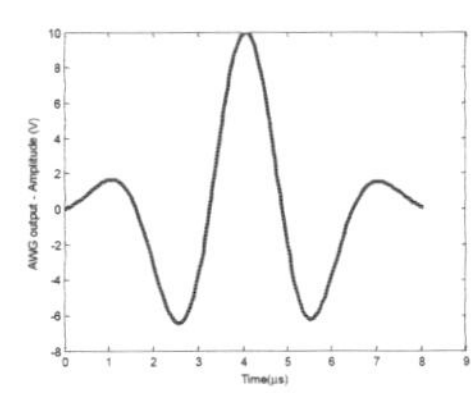

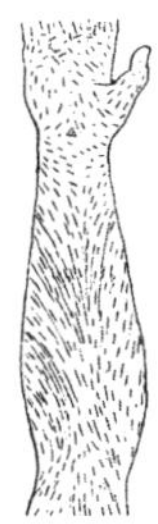

$$A_r = \frac{c_{max} - c_{min}}{\bar{c}} \cdot 100 \quad (\%)$$

Results

Forearm skin anisotropy measurement

VIIIth NDT in Progress

Prague 2015

- Volunteer: male, 28 years, skin thickness 1.9 mm

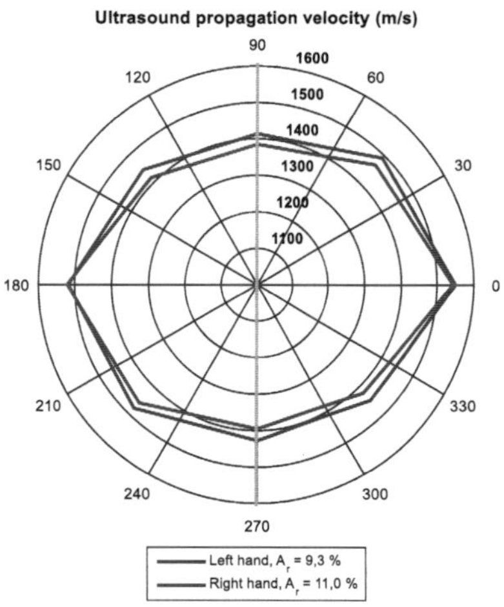

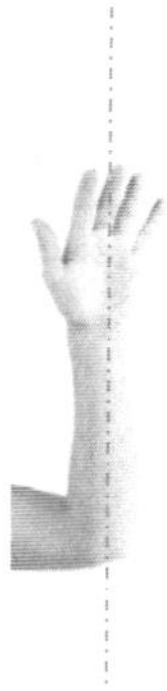

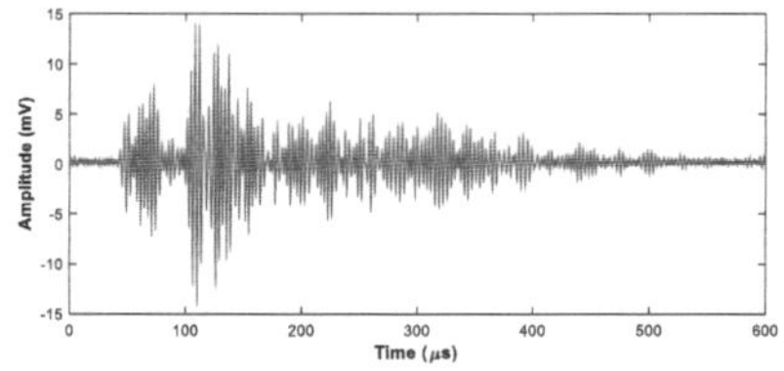

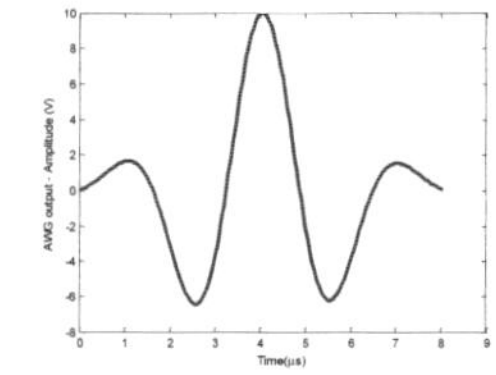

$$A_\mathrm{r} = \frac{c_\mathrm{max} - c_\mathrm{min}}{\bar{c}} \cdot 100 \ \ (\%)$$

Results

Carbon fabric-reinforced plastic - composite

VIIIth NDT in Progress

Prague 2015

Intact specimen

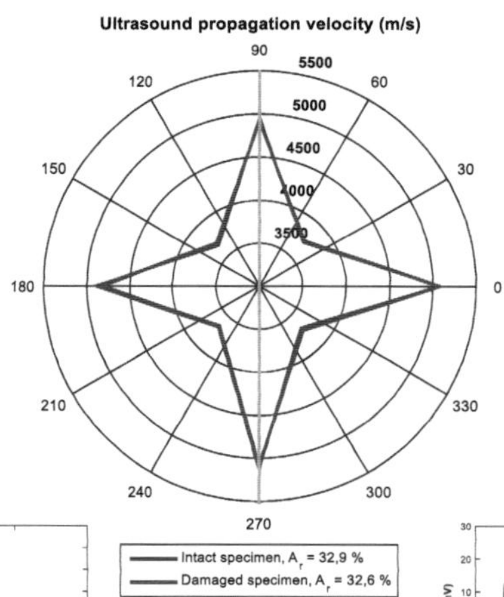

Damaged specimen

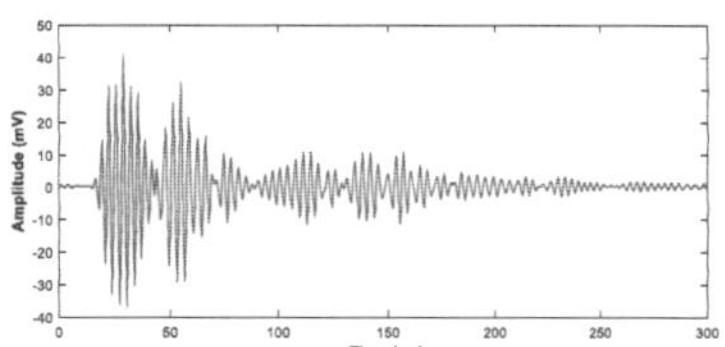

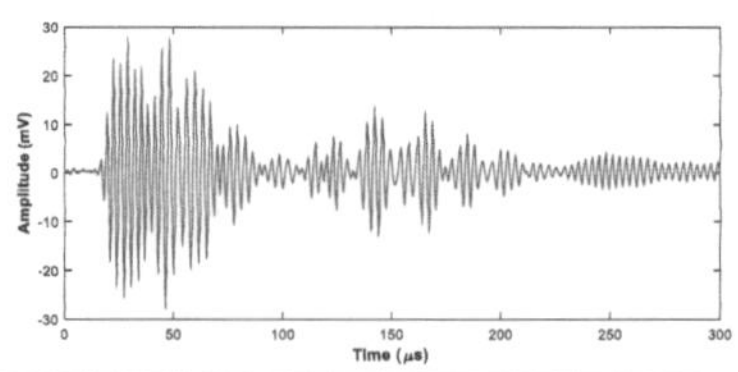

Results

NEWS methods applied on the forearm skin

VIIIth NDT in Progress Prague 2015

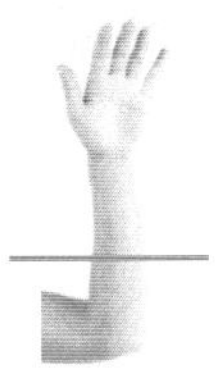

Sine-train, f = 308 kHz

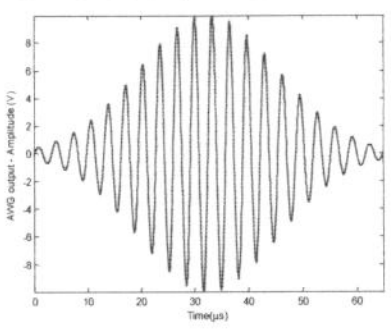

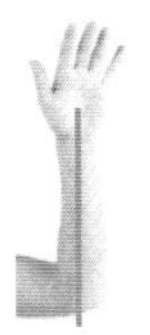

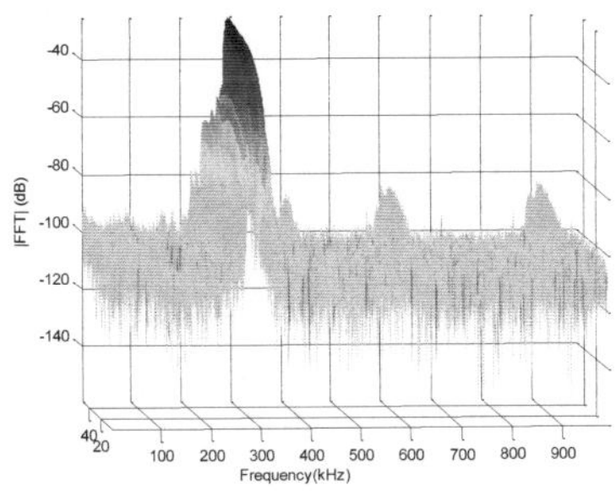

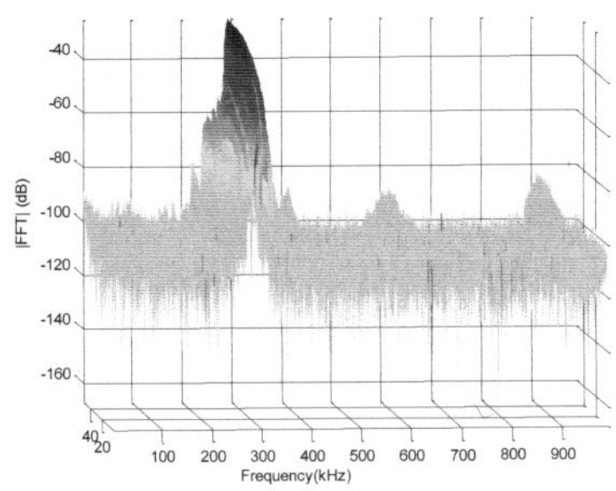

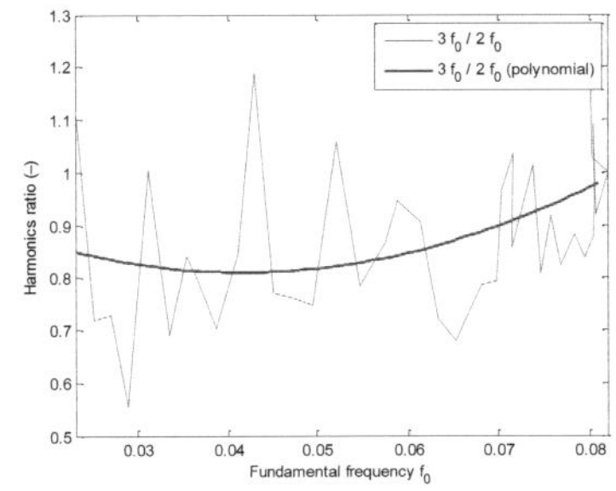

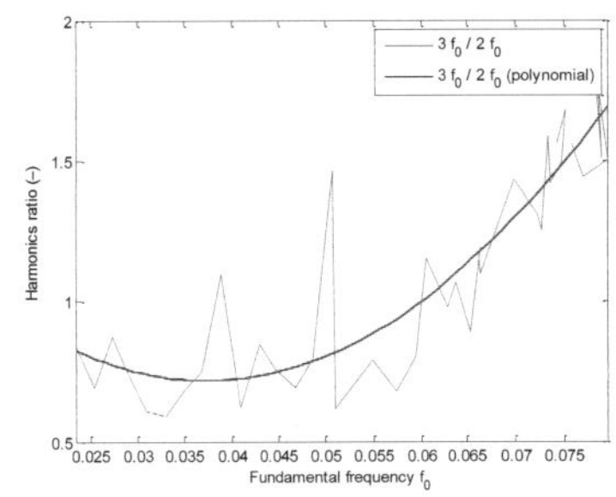

11/16

Results

NEWS methods applied on CFRP composite

VIIIth NDT in Progress Prague 2015

Sine-train, f = 308 kHz

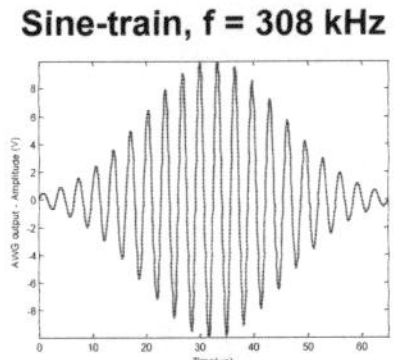

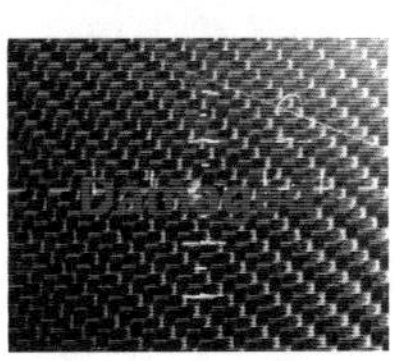

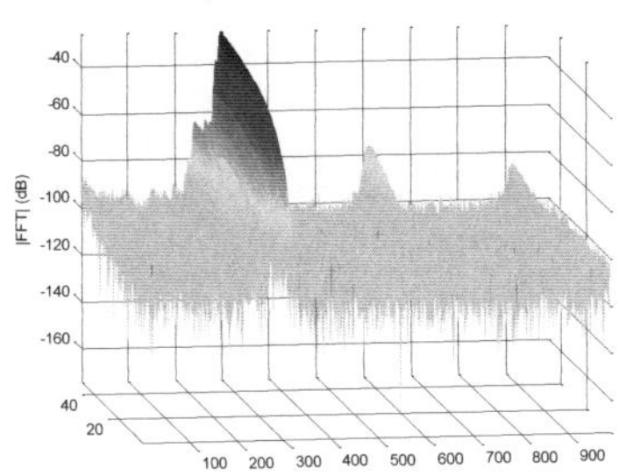

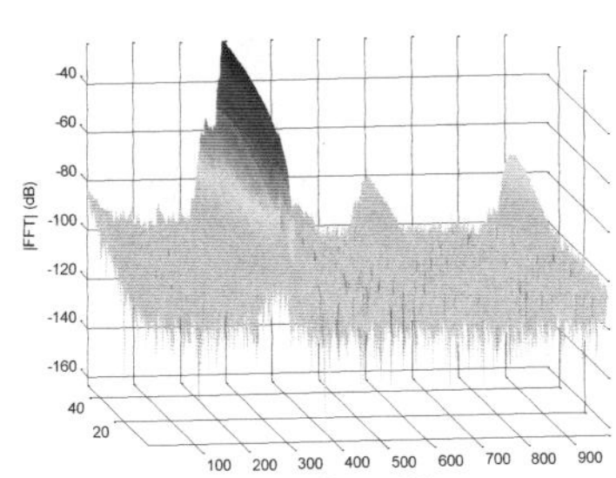

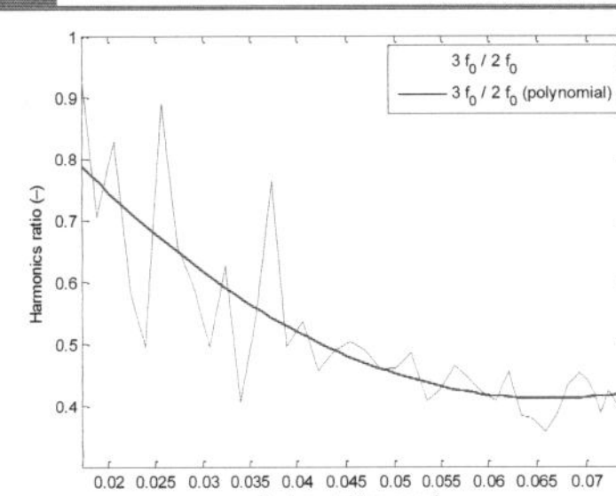

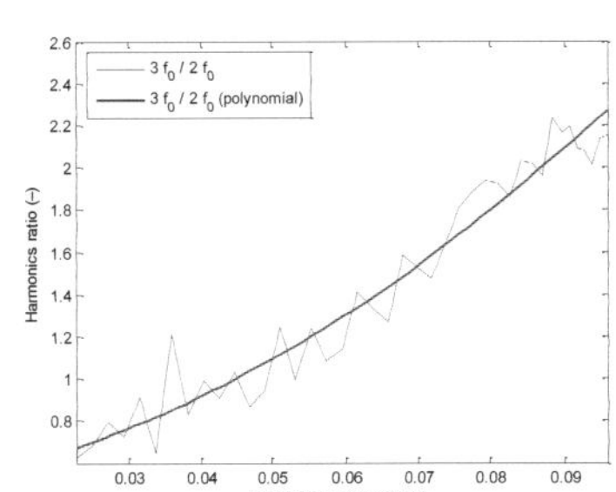

12/16

Results

Chirp-coded TRA applied on the forearm skin

VIIIth NDT in Progress

Prague 2015

- Volunteer: female, 25 years, skin thickness 1.3 mm

Linearly swept sine, (200–400) kHz

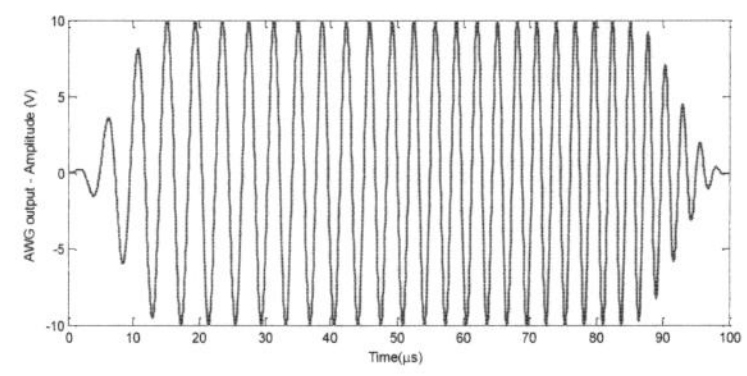

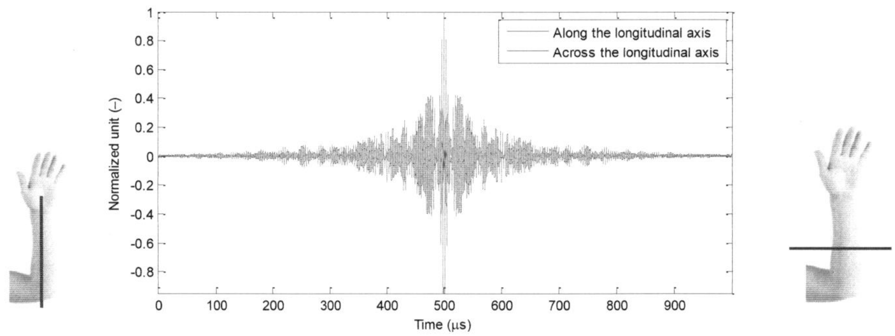

13/16

Results

Chirp-coded TRA applied on the forearm skin

VIIIth NDT in Progress

Prague 2015

- Volunteer: male, 28 years, skin thickness 1.9 mm

Linearly swept sine, (200–400) kHz

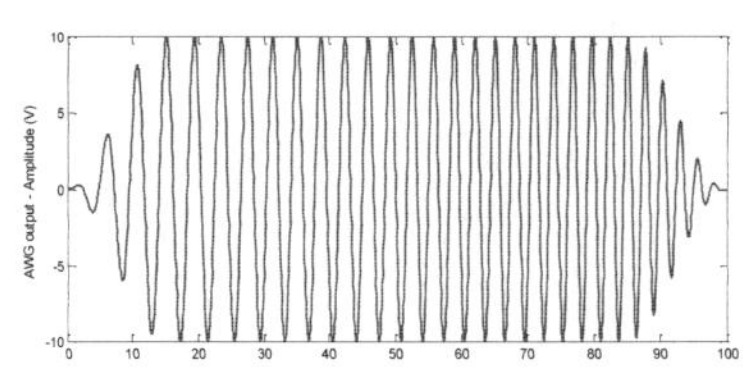

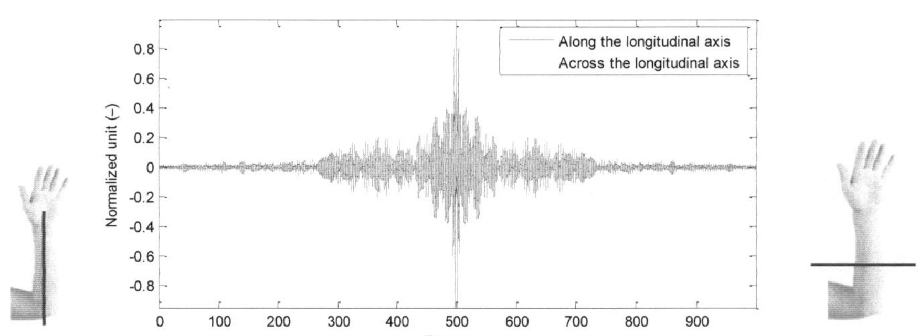

14/16

Results

TRA reconstruction during mechanical loading

VIIIth NDT in Progress Prague 2015

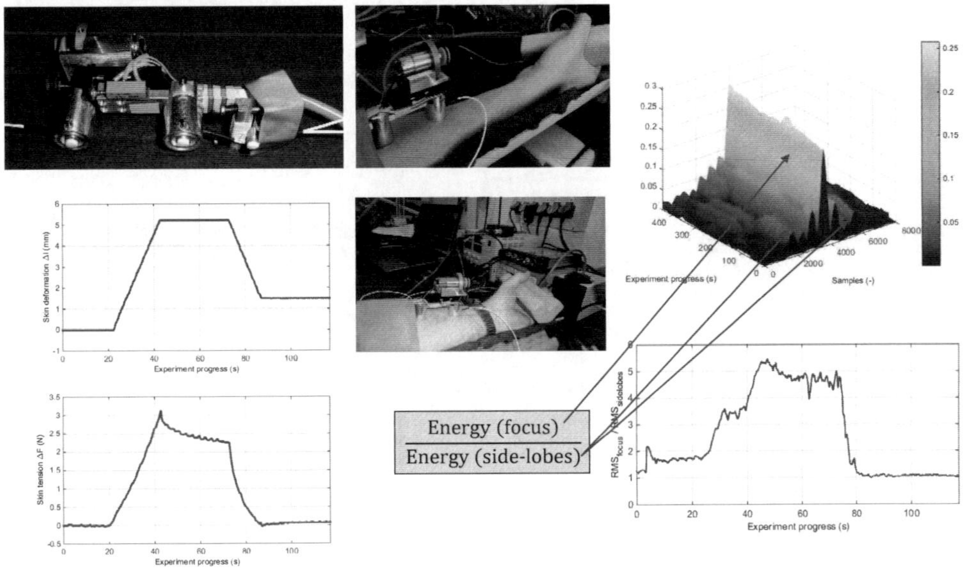

15/16

Summary

Conclusion and future work

VIIIth NDT in Progress Prague 2015

- New measurement method for local anisotropy evaluation of various biological and industrial materials; tested on human skin in-vivo, carbon fabric-fiber composite, and polymethylmethacrylate
- Anisotropy diagrams for the forearm skin corresponds with Langer's line orientation but embody individual differences – useful in plastic surgery
- Sufficiently accurate results: calculated variation coefficients < 1 %
- Results of NEWS and TRA methods applied on the skin show differences in the nonlinear signatures related to the direction of propagation through the skin (orientation of collagen fibers)
- Sensitive TRA method is promising for detection of changes in the mechanical behavior caused by tensile loading and external environmental influences
- Presented results and approaches should help to comprehend complex mechanical behavior of the human skin

16/16

INSTITUTE OF THERMOMECHANICS AS CR, v. v. i.

VIIIth NDT in Progress Prague 2015

VIIIth International Workshop NDT In Progress

October 12 – 14, 2015, Prague, Czech Republic

THANK YOU FOR YOUR ATTENTION

Daniel TOKAR Zdenek PREVOROVSKY

daniel.tokar@it.cas.cz

A Reference Free Method for Damage Detection and Location Based on Geometric Similarity

M.S. Salmanpour*[1,a], Z. Sharif Khodaei[2,a], M.H. Aliabadi[3,a]

[a]Department of Aeronautics, Imperial College London, South Kensington Campus, Roderic Hill Building, Exhibition Road, SW72AZ, London, UK. [1]msv09@imperial.ac.uk, [2]z.sharif-khodaei@imperial.ac.uk, [3]m.h.aliabadi@imperial.ac.uk

Abstract: This paper outlines a method exploiting geometrically similar areas as mutual baselines. The proposed method eliminates the need for previously recorded baseline sets and the pitfalls of changing environmental conditions between recording of signals. Through the use of instantaneously recorded baselines at the same time as the current interrogation signal and transducer calibration, it is possible to accurately predict the location of damage. The proposed transducer calibrated instantaneous baseline (TCIB) method was first applied to numerically simulated results. It was then validated experimentally for two different locations of artificial damage.

Keywords: Structural Health Monitoring, Lamb wave, Damage detection, Baseline free, Non-destructive testing, delay and sum algorithm

1. Introduction

Structural health monitoring (SHM) is an emerging branch of non-destructive testing promising real time, in-situ monitoring of structural integrity with minimal intrusion to the normal operation schedule. A network of attached ultrasonic transducers, typically piezoelectric Lead Zirconate Titanate (PZT), generate and sense diagnostics ultrasonic signals which are influenced by the material and geometric properties of the host structure.

Presence of damages can affect the ultrasonic wave propagation compared to that of pristine cases, however the signal is highly convoluted and features are difficult to interpret. Baseline comparison algorithms can isolate the damage scattered signal from other signal features, and in doing so detect and locate damage [1, 2]. The elegance of the baseline comparison methods lies in the fact that they can deal with the highly convoluted signals from complex geometries. Baseline signals are recorded in the pristine state and current signal at a later time. The current and baseline signals are subtracted, the Hilbert envelope of this is then referred to as the residual signal. Ideally the residual should only contain damage scattered signal. If the baseline is recorded at a different condition to that of the current interrogation response, the accuracy of damage detection may be significantly reduced. Non-damage related environmental factors can impact the structures' response, for example a temperature rise reduces propagation velocity [3], resulting in non-damage related residual. A library of baselines recorded at various temperatures increments could be used but the library of baselines may become less effective if the operation profile changes or modifications are made to the geometry [4].

Not relying on previously recorded baselines is a possible solution to changing environmental conditions. Strictly speaking for damage detection there often needs to be a comparison between response at two states of the structure or at least of key parameters [5]. "Baseline free" is a subjective term: what exactly constitutes a baseline? For instance Yeum et al [6] and Sohn [7] proposed reference or baseline free methods, exploiting dual transducers arrangements, in which modes after the first arrival where associated to damage. Wave propagation and arrival times (which will also change under varying environment) need to have been characterised previously which could arguably constitute as a required reference input.

Another promising class of reference free methods are non-linear detection approaches. It is argued that unlike normal wave propagation, damage scattered signals are nonlinear, hence not time reversible [8] or scalable [9]. In the case of time reversal this often results in the loss of temporal information and so damage cannot be accurately located [10], although complex hybrid methods have been proposed [11]. Scaled subtraction was successfully applied to composite panels to detect nonlinear delaminations [12]. But it is not clear what propagations are linear and non-linear, for example to prevent false alarm features due to step changes in thickness (i.e. stiffener foot) would have to be scalable but those due to damage not. Anton et al [13] proposed a baseline free method in which transducer paths would act as instantaneous baselines for each other and highlight damage affected paths. This does not directly locate damage from the wave propagation features but uses the affected paths to find a coarse probable area.

Taking works by Anton et al [13] as inspiration, a new method is proposed were instantaneous baselines are used for damage detection and location. The novelty is in using an entire area as baseline to locating damage with wave propagation features for good spatial resolution, not just using superposition of affected paths. This may result in applicability to complex stiffened panels and allow accurate damage location predictions. The TCIB method aims to provide a calibrated residual signal only containing damage scatter to a delay-and-sum algorithm. Geometrically similar areas are used as mutual baselines allowing for accurate prediction of damage location; this was demonstrated with numerical results. However, for practical applications; even under laboratory conditions, the Lamb wave signals differ between seemingly identical transducers and geometries, due to slight inconsistencies in the geometry, transducers attachment and positioning. These discrepancies, similar to environmental factors, may drown out the effects of damage. Accounting for these phenomena in the TCIB method, the transducer responses are calibrated allowing the instantaneously recorded signals to predict damage location with good spatial accuracy. The method was validated experimentally by locating an artificial damage at different locations.

2. Proposed Method Outline

The developed transducer calibrated instantaneous baseline (TCIB) method allows for the use of baselines recorded at the same time as the current/damaged signal. That is, the sparse transducer array could be arranged to form repeating cells, and each cell can act as the baseline for other cells The transducers' should be arranged in such a way so as to not break any existing symmetry in the structure. This would allow the sensors to be mirrored and reflected in space, to be used as baselines for other sections, see Figure 1. To select which cell is the interrogation area and which are baseline cells, there has to be many repeating cells to allow adequate comparison. For a structure with many repeating cells it is possible to identify a cell with large levels of discrepancy with respect to the other cells acting as a baseline in a round robin fashion, assuming damage is confined to one cell. As with most baseline free approaches, the developed TCIB method loses the integral link with the geometric uncertainties, naturally built into baseline comparison strategies. For example it may be possible that the bonding layer is inconsistent between the transducers, edges are not perfectly straight or the transducer position is slightly off. As the same inconsistencies are present in both the baseline and the current signal they would cancel out in classical baseline comparison methods. This is not the case for baseline free approaches.

It can be anticipated that the high level of detection accuracy obtained from FEA results, in the following section, can be hard to attain experimentally with data from an actual damaged panel. This is because the 'sterile' environment in which residuals are only caused by damage (or even temperature for that matter) is contaminated by noise and manufacturing tolerances.

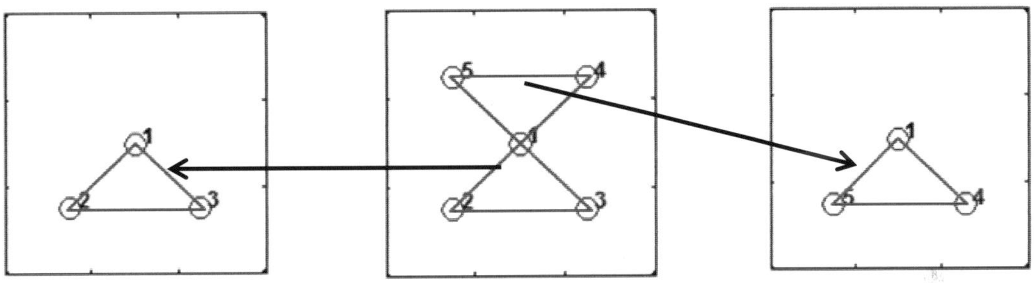

a b c

Figure 1. Transducer calibrated instantaneous baseline sensor mapping to generate baseline and current signal: inspection area (a) and baseline mapped to inspection area (c).

The solution proposed here is to use a calibration signal to account for the geometric discrepancies not related to damage, which are present even in the undamaged plate. For example signal for sensor path 1-5 is not identical to that for sensor path 1-3 and subtracting the two signals leaves a sizable residual, see Figure 1 and 2. The calibration is performed as follows: (1) in the initial state the signal would be recorded, envelope of the residual between each sensor pair retained as the calibration envelope. (2) Data is then recorded for the current damaged case. (3) Transducers are mapped with suitable instantaneous baselines used, and the residual envelope calculated. (4) The calibration envelopes are subtracted from the corresponding residual envelope depending on the mapping used. For further details the reader is referred to [14]. It should be stressed that the calibration envelope will not be affected significantly by environmental effects, as all phase information is discarded by taking the envelope. This will be demonstrated by using calibration signal recorded at significantly different temperature. To minimise the impact of the boundary, the inspection area will be limited to the area where round path between the inspection point and sensors pair is smaller than that to the closest boundary. The calibrated residual signals are then used as inputs to a delay-and sum algorithm, probability based damage index method (PBDIM) [2].

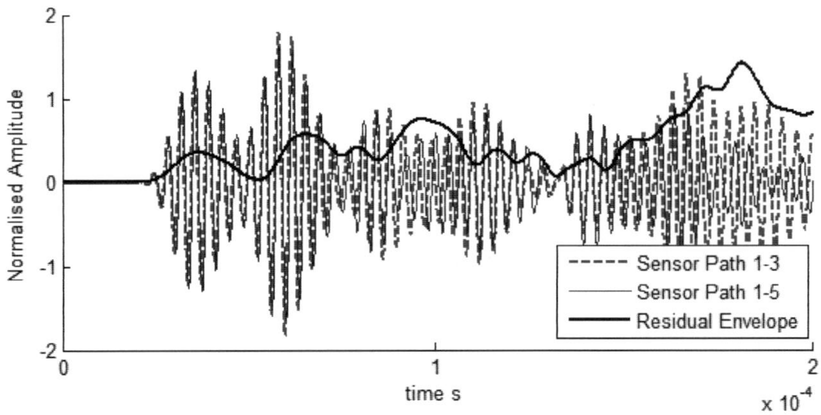

Figure 2: Experimental signal for two similar sensor paths and the residual envelope between them.

In the current form the method can only be used for isotropic plates where Lamb wave velocity is uniform in all direction. However, it may be possible to map sensors in such a way to also match the directional velocity profiles by taking into account the composite layup directions.

3. Experimental and Numerical Procedure

Experiments were conducted on two aluminium 6082 T6 plate of thickness 1.2 mm, with 5 disc PIC255 piezoelectric transducers bonded with cyanoacrylate adhesive, see Figure 3. Transducers were excited with a five cycled tone burst signal with central frequency of at 300 kHz. This was to ensure the actuation energy was concentrated around the central frequency, without using a continuous sine wave so a as to allow wave packets due to damage scattering to be distinguished. Actuation voltage V applied to the transducers was of the form:

$$V = \left(1 - cos\left(2\frac{\pi f t}{n}\right)\right)\sin(2\pi f t)\, H(\tau) \qquad (3.1)$$

Where f is the central frequency, n is the number of cycles, t is the time and $H(\tau)$ is the Heaviside function.

The acquisition hardware included a National Instrument (NI) arbitrary signal generator (PXIe5412), a MAA high frequency amplifier and the response was captured with a PXIe5105 high speed oscilloscope. The plate was heated in an Instron heating chamber. The plates were allowed to reach an isothermal temperature profile over 30 minutes and temperate was measured with K-type thermocouple. The plate was heated from the ambient lab temperature of 18 ^{o}C in steps of 5 degrees upto 55 ^{o}C. Damage was introduced in the form of glued on mass (1.3 kg 4.3 cm diameter) to cause scattering.

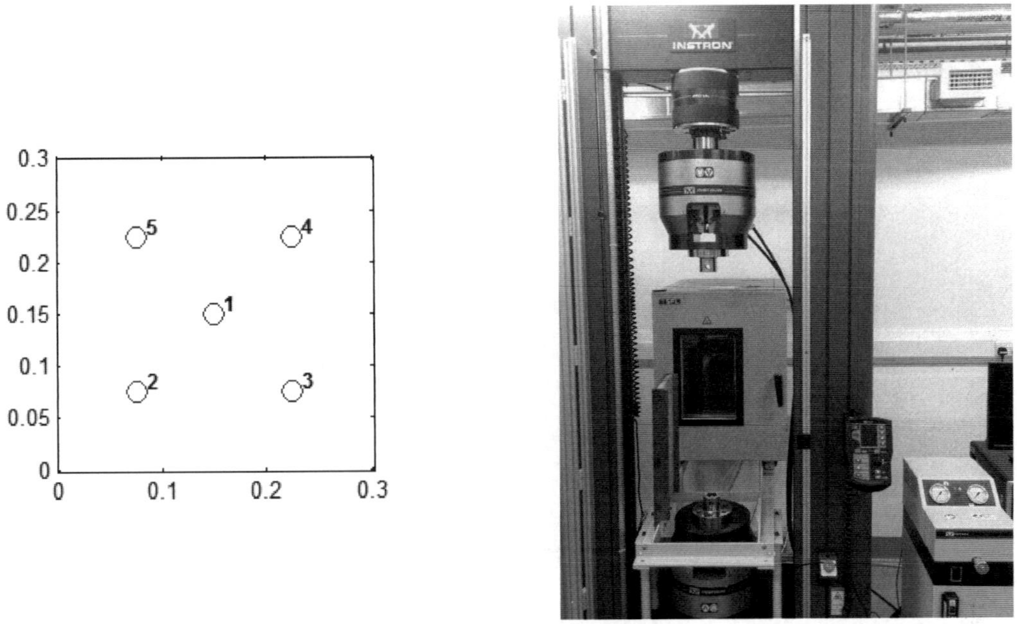

Figure 3. Schematic of aluminium plate (left), Dimension in meters. Instron Environmental chamber (right)

As mentioned before the damage scatter and its associated parameters including amplitude and time of arrival (TOA) are integral to damage detection. Numerous researchers have used means induce wave scatter to represent damage scatter, including holes [15], clamps [16], weights and putty [13]. A preliminary study was conducted to assess scattering features (time of flight and amplitude) of difference artificial damage types in the direct transducer path. Scattering from a 10 mm hole, 1 x10 mm crack (perpendicular to direct sensor path), clamp, 1.3 kg mass with diameter 4.5 cm glued to the plate and a G-clamp were collected. It was found that in the amplitude of scatter was largest for the glued on weight/mass, Figure 4. The artificial damages were located at the same position and in all cases the time of flight was the same. This suggests that the artificial damages are a suitable development tool for damage detection algorithms.

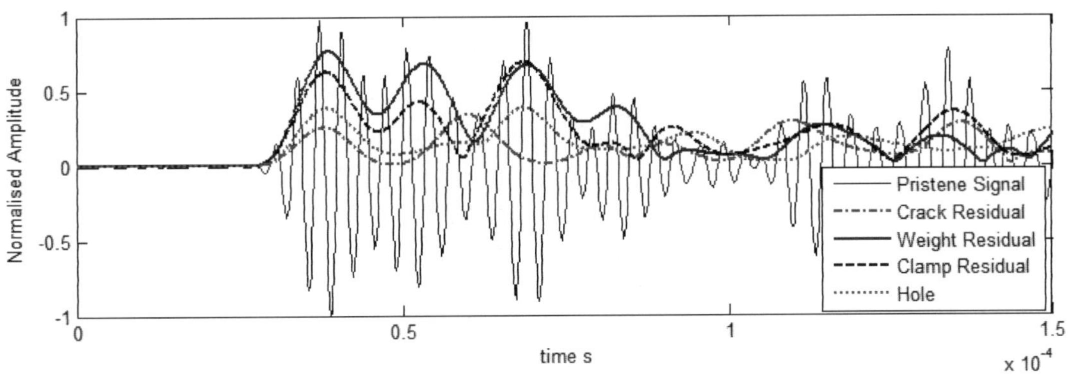

Figure 4: Scattering caused by different actual and artificial damage types on an aluminium plate.

Finite element analysis (FEA) is a cost effective method in terms of computation and modelling time, for simulating Lamb wave propagation. An aluminium plate in this paper was modelled with the ABAQUS explicit solver with shell elements, using the same strategy as Sharif Khodei and Aliabadi [2]. The transducers were modelled with brick elements with the actuation signal applied at the upper circumference, see Figure 5, a full overview of the modelling strategy used can be found in works by Su and Ye [17].

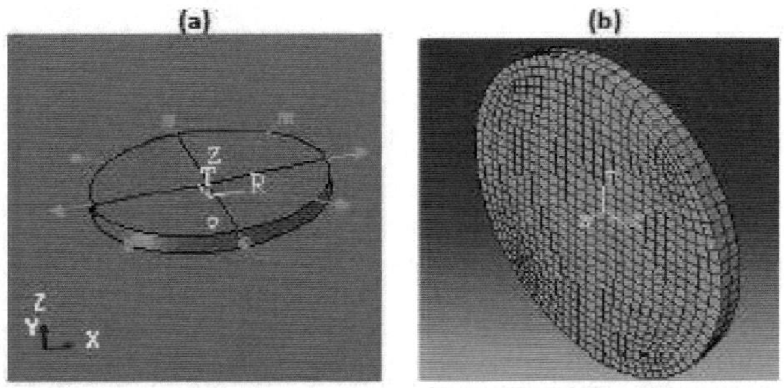

Figure 5: Actuation of ultrasonic signals on edge top circumference via radial displacement (a) meshed transducer (b).

4. Numerical Results

The TCIB method was first tested with data obtained with FEA results. This was done by simulating a damaged aluminium plate with an array of five transducers, see Figure 1. The damage was modelled as a through thickness square hole of size 1 cm^2.

It was found that the damage location was predicted accurately with the peak damage index (DI) confined to the area near the actual damage, see Figure 6.

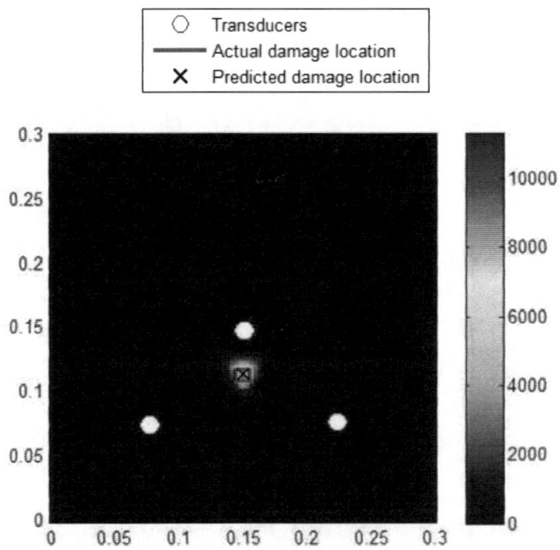

Figure 6. PBDIM for FEA results with instantaneous baseline.

The accurate detection achieved is numerically simulated signals maybe difficult to achieve experimental, due to transducer position error bonding and edge discrepancies. To arbitrary these, the signal can be time gated in the data fusion step with the time of flight to the nearest boundary as the limit, hence eliminating boundary reflections. This would create areas with different levels of coverage, and only areas with highest level of coverage were retained as shown in Figure 7.

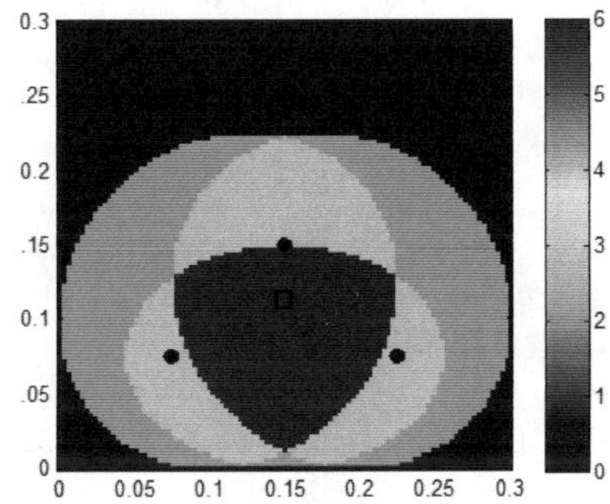

Figure 7. Coverage limited by boundary, contours indicated number of sensor covering a point.

To validate the calibrated instantaneous baseline method the same obtained FEA results were used in the PBDIM method, but this time the position of the middle transducer was passed onto the PBDIM algorithm with error. The position was shifted by 10 mm in the y and x direction. It was found that the detection prediction was still accurate, offset only by a small margin from the actual location.

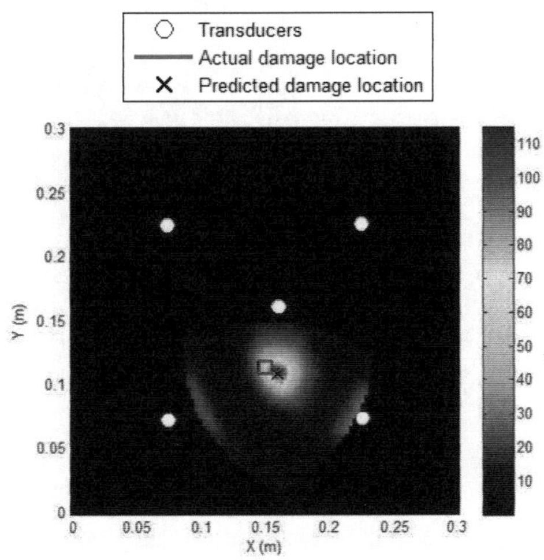

Figure 8. PBDIM for FEA results with instantaneous baseline, middle sensor has 10 mm error in position in detection algorithm compared to model case.

5. Experimental Results

The TCIB method was experimentally demonstrated by gluing a mass to the plate at various simulated damage location. The current signal was recorded at 18 ^{o}C. Highlighting the fact that the calibration in no way acts as a "baseline" like classical baseline comparison methods, the calibration data used was recorded at a different physical location (different lab) and most importantly recorded in the temperature oven at a temperature of 55 °C. it can be observed that at this temperature the non-damage related residual is significant and at times nearly twice the magnitude of the original signal, see Figure 9. For this level temperature difference normal baseline comparison methods simply cannot predict damage location accurately as shown in Figure 10.

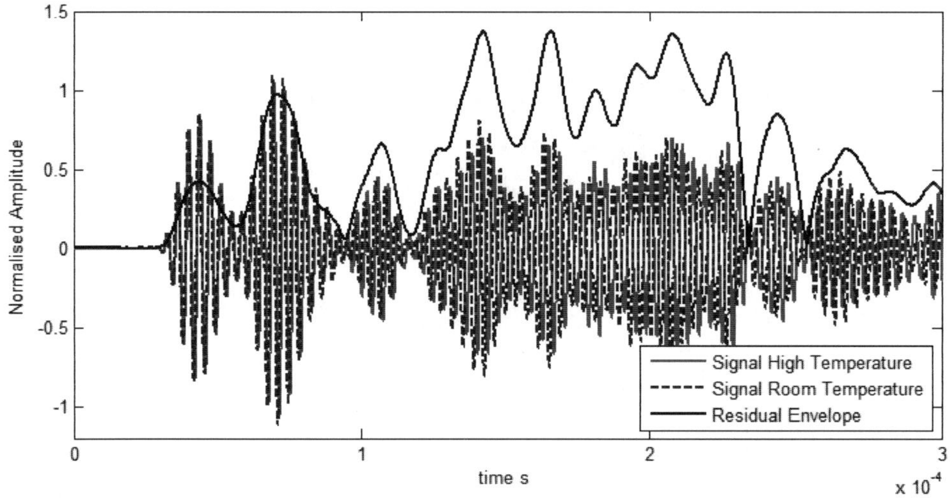

Figure 9: Non-damage related residual envelope for aluminium plate at 18 and 55 °C.

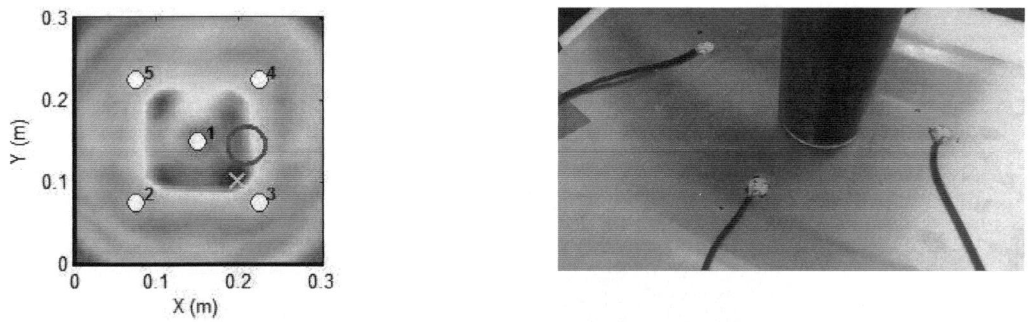

Figure 10. Simulated damage (right) damage detection algorithm applied to experimental signals recorded at 18 °C and at 55 °C.

By applying the TCIB method, the peak damage index was located near the centre of the first damage position (Figure 11a), while for the second case the peak damage index area was located near the edge (Figure 11b). Both positions of damage were predicted accurately within the radius of actual damage, see Figure 11.

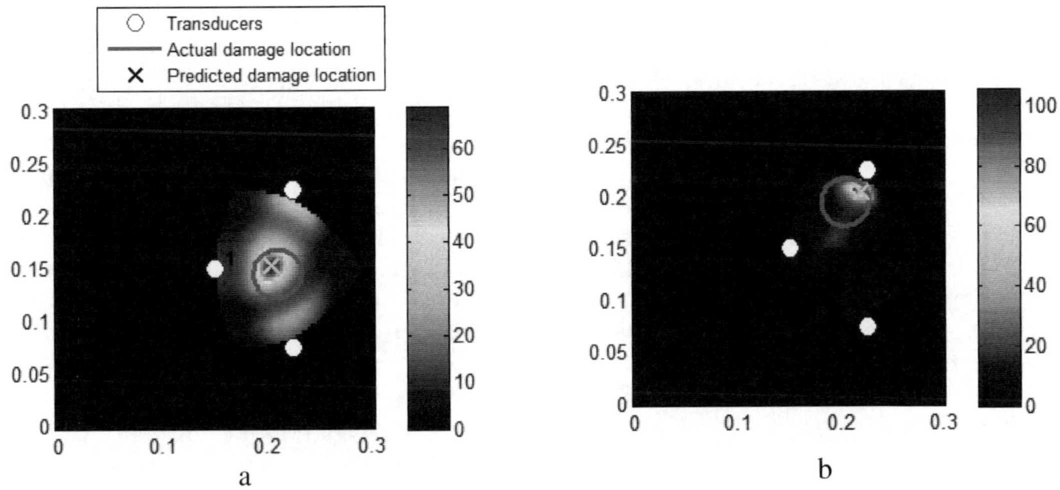

Figure 11. Instantaneous baseline damage detection for experimental aluminum plate, with damage located centrally (a) and near a sensor (b).

6. Conclusions

The developed instantaneous baseline method was demonstrated first with FEA simulation. The method performs well even when a transducer location was passed inputted incorrectly into detection algorithm. The method was then validated for two different locations with artificial damage. This ensured good detection accuracy was not linked to the position of damage chosen.

In future works the method should be applied to actual damage cases. The developed scheme will be applied to more complex structures including composite stiffened panels for detecting impact induced delamination.

Acknowledgment

This research was funded by the UK Engineering and Physical Sciences Research Council (EPSRC), Doctoral Training Account.

References

[1] Michaels, J.E., Detection, localization and characterization of damage in plates with an in situ array of spatially distributed ultrasonic sensors. Smart Materials and Structures, 2008. **17**(3): p. 035035.

[2] Sharif-Khodaei, Z. and M.H. Aliabadi, Assessment of delay-and-sum algorithms for damage detection in aluminium and composite plates. Smart Materials and Structures, 2014. **23**(7): p. 075007.

[3] Marzani, A. and S. Salamone, Numerical prediction and experimental verification of temperature effect on plate waves generated and received by piezoceramic sensors. Mechanical Systems and Signal Processing, 2012. **30**: p. 204-217.

[4] Putkis, O. and A.J. Croxford, Continuous baseline growth and monitoring for guided wave SHM. Smart Materials and Structures, 2013. **22**(5): p. 055029.

[5] Worden, K., et al., The fundamental axioms of structural health monitoring. Proceedings of the Royal Society A: Mathematical, Physical and Engineering Science, 2007. **463**(2082): p. 1639-1664.

[6] Yeum, C.M., et al., Reference-free delamination detection using Lamb waves. Structural Control and Health Monitoring, 2013: p. n/a-n/a.

[7] Sohn, H., Reference-free crack detection under varying temperature. KSCE Journal of Civil Engineering, 2011. **15**(8): p. 1395-1404.

[8] Park, H.W., et al., Time reversal active sensing for health monitoring of a composite plate. Journal of Sound and Vibration, 2007. **302**(1-2): p. 50-66.

[9] Scalerandi, M., et al., Nonlinear acoustic time reversal imaging using the scaling subtraction method. Journal of Physics D: Applied Physics, 2008. **41**(21): p. 215404.

[10] Cai, J., et al., High spatial resolution imaging for structural health monitoring based on virtual time reversal. Smart Materials and Structures, 2011. **20**(5): p. 055018.

[11] Jun, Y. and U. Lee, Computer-aided hybrid time reversal process for structural health monitoring. Journal of Mechanical Science and Technology, 2012. **26**(1): p. 53-61.

[12] TABATABAEIPOUR, M., et al. Reconstruction Algorithm for Probabilistic Inspection of Damage (RAPID) in Composites. in 11th European Conference on Non-Destructive Testing (ECNDT 2014). 2014. Prague, Czech Republic.

[13] Anton, S.R., D.J. Inman, and G. Park, Reference-Free Damage Detection Using Instantaneous Baseline Measurements. AIAA Journal, 2009. **47**(8): p. 1952-1964.

[14] Salmanpour, M.S., Z.S. Khodaei, and M.H. Aliabadi, Baseline Free Impact Damage Localisation Based on Geometric Similarity. In Progress.

[15] Giurgiutiu, V., Piezoelectric Wafer Active Sensors for Structural Health Monitoring of Composite Structures Using Tuned Guided Waves. Journal of Engineering Materials and Technology, 2011. **133**(4): p. 041012.

[16] Raghavan, A. and C.E.S. Cesnik, Guided-wave signal processing using chirplet matching pursuits and mode correlation for structural health monitoring. Smart Materials and Structures, 2007. **16**(2): p. 355-366.

[17] Su, Z. and L. Ye, Identification of damage using Lamb waves : from fundamentals to applications. Vol. 48. 2009: Springer Science & Business Media.

Interpolation Analysis of Ultrasonic Guided Wave Signals in the Temperature Domain

Ren-Jean LIOU [1] and Jennifer E. MICHAELS [2]

[1] Department of Computer and Communications, National Pingtung University, Taiwan, ROC;
Phone: +886 8 7663800; e-mail: renjean@mail.nptu.edu.tw

[2] School of Electrical and Computer Engineering, Georgia Institute of Technology, Atlanta, GA, USA;
e-mail: jemichaels@gatech.edu

Abstract

To determine the state of damage in structural health monitoring using guided ultrasonic waves under various operational conditions, such as temperature, moisture, stress, etc., the measured signal is often evaluated by subtracting a damage-free baseline signal. However, it is not possible to obtain baselines for all possible conditions. Applying interpolation to the collected baselines provides an effective way to approximate the real baseline. This paper studies techniques of interpolation in the temperature domain. We first analyze the sampling criteria for the data used. Conventional interpolation methods that are described in most numerical analysis books are applied. The interpolation study proceeds into the frequency domain where approaches in the *f-x* and *f-k* domains are implemented. The performance of different interpolation methods is compared and their efficacy is discussed.

Keywords: Structural health monitoring, interpolation, *f-x*, *f-k*, temperature domain, ultrasonic guided wave

1. Introduction

A structural health monitoring (SHM) system is inevitably subjected to environmental and operational variations because of its *in situ* and in-service nature of operation [1][2]. To determine the structural state under current conditions, such as temperature, moisture, stress, etc., the measured signal is often compared to a baseline signal recorded in the damage-free state. Normally structural problems can be detected by subtracting the baseline signal from the measured signal. However, it is not possible to obtain baseline signals for all possible operational conditions. If the current condition is within the range of conditions of a set of baseline signals, applying interpolation to these baseline signals can approximate the baseline for the current condition and thus eliminate the need for acquiring all possible baselines.

Another motivation for investigating interpolation is that it is desirable to sample data no more finely than is actually needed for a specific application. Since digitizers with high sampling rates are readily available and inexpensive, the minimum allowable sampling frequency is typically exceeded to provide improved resolution in the time domain. Oversampling in other domains, however, may not be so readily accomplished. For example, sensor size can limit receiver spacing in the spatial domain. In addition, acquisition time requirements may dictate spatial increments for scanned systems. Interpolation for the purpose of spatial up-sampling has been addressed for both geophysics [3] and biomedical ultrasonic applications [4], and has also been considered to address compensation for nonfunctioning receivers [5].

It is also useful to consider interpolation in other domains, such as temperature or applied stresses. Interpolations with respect to these conditions are applicable to SHM where it is desired to match current signals to baseline data that might have been acquired under different environmental and operational conditions. Existing methods to address temperature variations require stretching signals to compensate for temperature mismatch and thus achieve better baseline matching [6][7]. However, the accuracy is still subject to the presence of noise and a more general method is desired for other factors such as applied stresses.

It is common to use conventional interpolation techniques [8] for all kinds of ultrasonic signals. An in-depth study was presented in [9]. In this paper we comprehensively investigate ultrasonic signal interpolation in the temperature domain using conventional sinc, linear, spline and FFT methods. Two other advanced methods, *f-x* and *f-k* domain approaches, are also applied and discussed. These two methods are frequently used in seismic data processing. Simulated data are utilized to evaluate interpolation performance.

2. Sampling theorem and conventional interpolation methods

The traditional sampling theorem is based on the Nyquist criterion. A bandlimited temporal signal $x(t)$ can be perfectly recovered from an infinite sequence of samples if the sampling rate exceeds $2f_{max}$, where f_{max} is the highest frequency of $x(t)$.

Reconstruction of a signal can be readily performed by sinc function multiplication, which is equivalent to applying a perfect lowpass filter. If $x(nT)$ represents the samples of $x(t)$, where T is the sampling period, the reconstructed signal $\tilde{x}(t)$ is obtained as

$$\tilde{x}(t) = \sum_{n=-\infty}^{\infty} x(nT)\frac{\sin[\pi(t-nT)/T]}{\pi(t-nT)/T}. \tag{1}$$

However, it can be problematic in practice to use the sinc function. The first difficulty is the computational requirements because summing a large number of samples is required. The primary difficulty, however, is related to edge effects because signals are windowed in all domains. The multiplicative windowing operation generates high frequency content, which means that the Nyquist criterion is not satisfied and interpolation performance may be poor. There are other interpolation methods available that have different characteristics. Two of the more commonly used schemes are considered here.

For spatial sampling, consider a simple 1-D waveform in space that represents a traveling plane wave captured at a specific time. The well-known relationship between wavenumber k, wavelength λ, frequency f, and wave speed c is

$$k = \frac{2\pi}{\lambda} = \frac{2\pi f}{c} \quad \text{and} \quad \lambda = \frac{c}{f}. \tag{2}$$

If Δx is the spatial sampling interval, then the Nyquist criterion for spatial sampling is

$$\Delta x < \frac{\lambda}{2}. \tag{3}$$

This criterion only applies to signals that are sampled for an infinite spatial extent; a space-limited signal cannot be band-limited and perfect reconstruction is therefore not achievable. Only an approximation can be obtained, which may not be sufficiently accurate.

By examining the characteristics of ultrasonic signals recorded at different temperatures, it can be seen that multiple echoes are analogous to multiple waves propagating at different wave speeds. The so-called "temperature domain," where signals are expressed as a function of temperature, is analogous to the spatial domain, where signals are a function of position. Since the speed of propagation changes linearly with temperature change [6][7], individual echoes shift linearly as a function of temperature. We define a "wave speed" $\tilde{c} = \Delta T/\Delta t$ in the temperature domain, where ΔT is the temperature change (i.e., temperature sampling interval), and Δt is the time shift of an echo associated with the temperature change ΔT. The corresponding wavelength is $\tilde{\lambda} = \tilde{c}/f$, where f is frequency. The Nyquist criterion for temperature sampling thus becomes

$$\Delta T < \frac{\tilde{c}_{min}}{2f_{max}}, \qquad (4)$$

where $\tilde{c}_{min}$ occurs at the maximum time t_{max}, and f_{max} is the maximum frequency. A simple illustration is depicted in Figure 1.

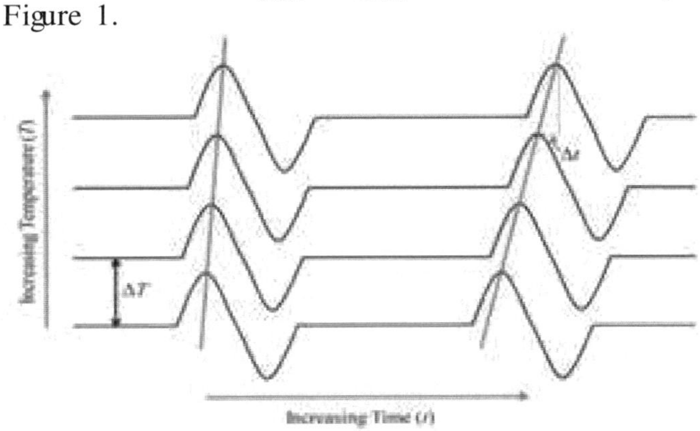

Figure 1. Ultrasonic signals in the temperature domain.

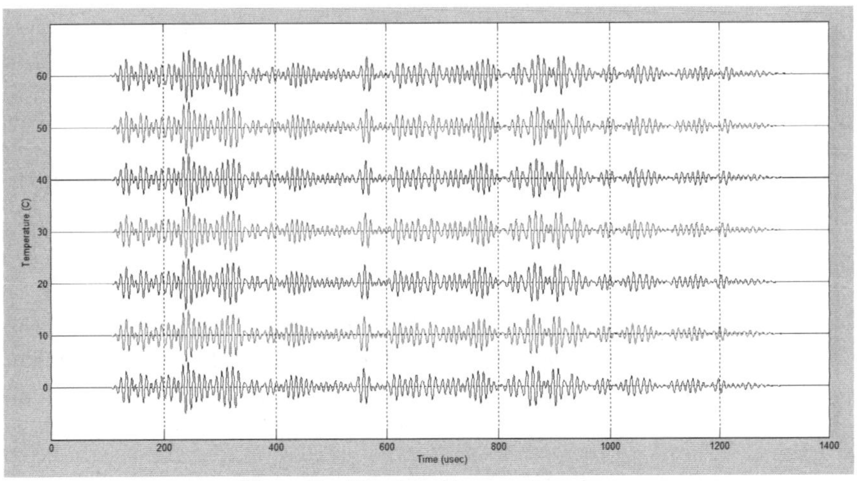

Figure 2. Simulated temperature data.

Figure 2 shows simulated temperature signals from 0°C to 60°C with a 10°C increment based upon the A_0 guided wave mode and a 100 KHz tone burst excitation. The time shifts on the right side of the plot can be clearly seen.

Suppose there are M signals available. We can define the equally sampled signal $x_k(n)$ at fixed temperature increment for $k = 0,\ldots,M-1$. For each signal with a total of N samples, the discrete Fourier transform (DFT) is

$$X_k(f) = \sum_{n=0}^{N-1} x_k(n) e^{-j2\pi fn/N}, \tag{5}$$

where $f = 0, \ldots, N-1$.

Let us assume that $x_0(n)$ is recorded at the lowest temperature. From [9], the next temperature increment waveform $x_1(n)$ is approximately stretched (or compressed) with a time shift ξn, i.e., $x_1(n) = x_0(n + \xi n)$, where ξ is a function of the temperature change. The signal at temperature k has the recursive form

$$x_k(n) = x_{k-1}[(1+\xi)n] = x_0[(1+k\xi)n], \tag{6}$$

where $k\xi n$ is the shift in time domain. The shift yields frequency scaling in the frequency domain, i.e.,

$$X_k(f) = X_0(f/(1+k\xi)). \tag{7}$$

When $1 + k\xi$ is greater than one, the corresponding frequency response is stretched.

3. Interpolation in the frequency domain

Interpolation can also be performed in the frequency domain. Two of the commonly seen techniques, *f-x* and *f-k* domain processing in geophysics research, are applied.

Interpolation is a type of prediction problem. Prediction filters are a very useful way to obtain data interpolation. The technique of prediction filter in adaptive filtering [9] can be utilized. Prediction in the time domain has been discussed for many decades. However, signal features are not well represented in time domain. Hence many approaches use feature extraction first to concentrate the signal energy, reduce the computational requirements, and increase the accuracy of results. Frequency domain processing is among the most popular techniques.

Spitz [10] presented an interpolation operator in the *f-x* domain for seismic data by solving a set of linear equations whose coefficients depend only on the spectrum of the spatial prediction filter defined by the recorded signal.

Let us assume that the length of the prediction is L, which can be referred to as the order of an AR process. The coefficients in the frequency domain of the forward and backward one-step prediction filter are $P_i(f)$. For M signals, each signal $X_k(f)$ at frequency f may be modeled as

$$X_k(f) = \sum_{i=1}^{L} P_i(f) X_{k-i}(f), \quad k = L+1, \ldots, M \tag{8}$$

$$X_k^*(f) = \sum_{i=1}^{L} P_i(f) X_{k+j}^*(f), \quad k = 1, \ldots, M-L \tag{9}$$

where "*" denotes the complex conjugate. The prediction filter $P_i(f)$ can be quickly obtained by matrix inversion and multiplication; sometimes the pseudoinverse operation is required. If interpolation is to be performed between every adjacent increment, then the length of the DFT can be doubled to fill up the empty data positions. The output of the interpolated waveform in the frequency domain is denoted as $X_k'(f)$. Since it is reasonable that the same prediction filter can be applied to the new interpolated data, equations (8) and (9) are accordingly modified with double length as

$$X_k'(f) = \sum_{i=1}^{L} P_i'(f) X_{k-i}'(f), \quad k = L+1, \ldots, 2M-1 \tag{10}$$

$$X_k'(f) = \sum_{i=1}^{L} P_i'^{*}(f) X_{k+i}'(f), \quad k = 1, \ldots, 2M-L-1 \tag{11}$$

where $X_k(f) = X_{2k-1}'(f)$; i.e., the odd-indexed X' is equivalent to the original data and the even-indexed X' is the interpolated data. The relations between unknown and known signals are hence provided. By joining the above two equations and rearranging terms, we derive the following expression,

$$\mathbf{A}(\boldsymbol{P}'(f)) \begin{bmatrix} X_2'(f) \\ X_4'(f) \\ \cdot \\ \cdot \\ X_{2M-2}'(f) \end{bmatrix} = \mathbf{B}(\boldsymbol{P}'(f)) \begin{bmatrix} X_1'(f) \\ X_3'(f) \\ \cdot \\ \cdot \\ X_{2M-1}'(f) \end{bmatrix} \tag{12}$$

where **A** and **B** are matrices composed of the prediction filter. The interpolated data in the frequency domain can be calculated by multiplying both sides of Eq. (12) by the inverse matrix $\mathbf{A}^{-1}(\boldsymbol{P}'(f))$. Again pseudoinverse operation may be required. The prediction vector can be determined using the properties of phase shift so that $\boldsymbol{P}_j'(f) = \boldsymbol{P}_j(f)$. The interpolated data can be readily obtained by computing the inverse FT.

Another adaptive data interpolation method is designed in the DFT domain (*f-k*) for spatially aliased data [11]. The method utilizes the cyclic properties of the DFT. Although it is best to illustrate the algorithm graphically, the main principles are briefly described. Four signals in the original data domain are needed.

1. The original data to be interpolated with length *N*.
2. Data with zeros filled at every other location of the original signal. The data length is now become 2*N*-1.
3. Data with zeros appended to the original data till the total length equals 2*N*-1.
4. Data by periodically zeroing every other data of the third signal. The total length is still 2*N*-1.

After preparation of the above four signals, their DFTs are computed. The DFT of the second signal is actually the spectrum of the expected interpolated signal with aliasing and about half of the peak amplitude. We need to find the operator to boost the amplitude. The operator can be obtained by calculating the ratio between the DFTs of the third and the fourth signals. By multiplying this operator with the DFT of the second signal, the results will be very close to the desired spectrum. The interpolated data can be obtained by computing the inverse DFT.

4. Simulation results

The temperature signals were simulated using an aluminum plate with thickness 6.8 mm at 25°C and dimensions of 500 mm × 500 mm. The exciting signal is a tone burst at 100 kHz. The sampling frequency is 20 MHz. The locations of the two transducers are (451.8, 194.2) and (175.4, 375.4). The plate is homogeneous and isotropic, and it is assumed that the edges are perfect reflectors. All temperature signals from 0°C to 60°C at a 1° increment were simulated. The full time duration of each signal is 1.4 ms. We applied six different interpolation methods: spline, linear, *f-x*, *f-k*, sinc, and FFT via zero-padding. For each temperature increment, the waveform is approximately stretched (or compressed) with a time-dependent time shift; i.e., all adjacent signals are directly related. Hence, a first order predictor is sufficient for prediction in *f-x* interpolation.

Figure 3 shows, for a fixed time duration of 1 ms, the minimum signal-to-noise-ratio (SNR) versus temperature interpolation increment using the six different approaches. We can see that the spline method has the best performance followed by *f-x*. In general, the performance decreases as the temperature increment increases.

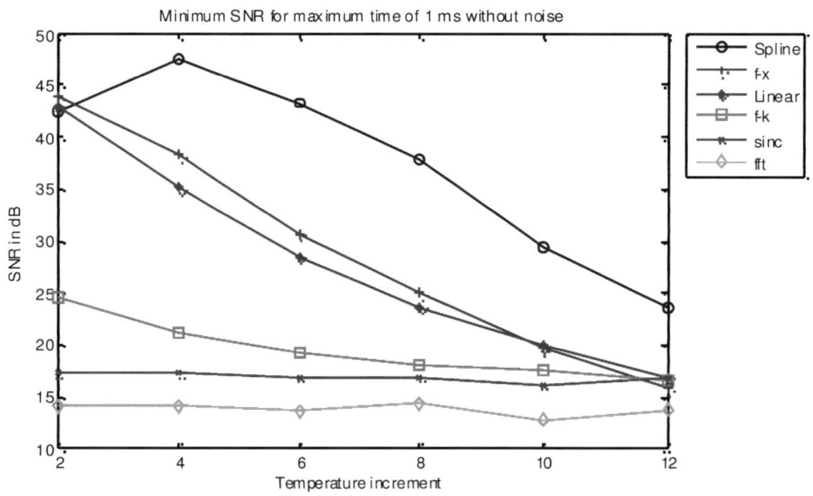

Figure 3. SNR for a fixed time duration of 1 ms and different temperature increments.

Figure 4 shows, for a fixed temperature increment of 4°C, minimum SNR versus signal duration. As before, the spline and *f-x* methods have better performance. The SNR generally decreases as the time duration increases. Due to the characteristics of the temperature dependence, the time shift increases as time evolves, which yields bigger interpolation errors.

In order to observe the interpolation performance at each temperature, we set the total time duration to 1.4 ms and the temperature increment to 4°C. The original signals are at 0°C, 4°C, ..., 60°C, and the interpolated signals are hence at 2°C, 6°C, ..., 58°C. The results for each temperature are shown in Figure 5. We can see that the performance using linear interpolation and the *f-x* method with order of 1 are quite stable. Spline interpolation still has the best performance for most of the time with fluctuating SNR and without apparent edge effects. All other methods demonstrate lower SNR on the two edges of the temperature range. In particular, the FFT method has excellent performance in the center but very poor performance on the edges.

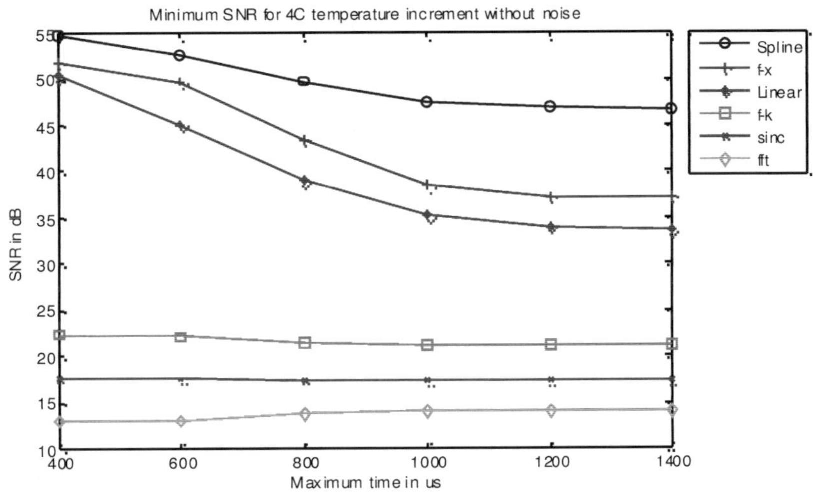

Figure 4. SNR at a fixed temperature increment of 4°C and varying time durations.

For further analysis, we eliminated the methods of sinc, FFT and *f-k*, which had poor performance in previous simulations. Gaussian noise is added to the signal with the SNR maintained at 30 dB. Similar evaluations as for Figures 3 and 4 are performed and the results are shown in Figures 6 and 7, respectively. The spline method no longer has the best performance and is even worse than the linear method at some times. The *f-x* method is more effective in tolerating noise.

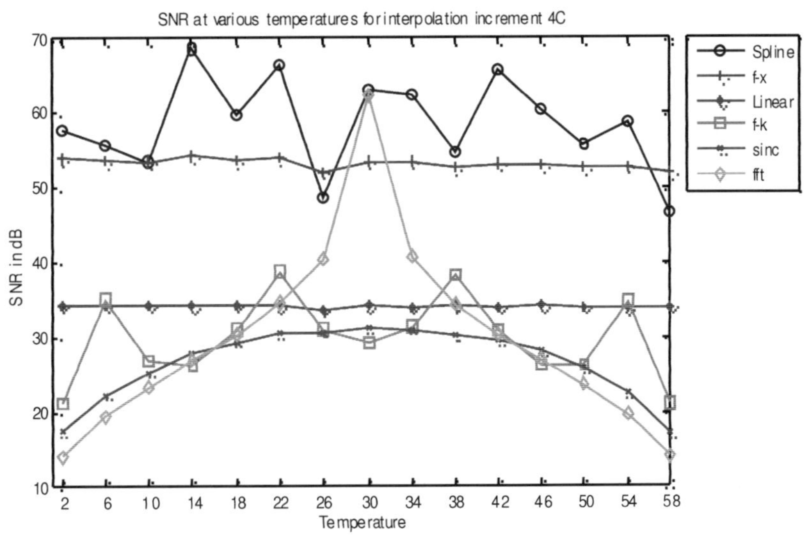

Figure 5. Interpolation performance at different temperatures.

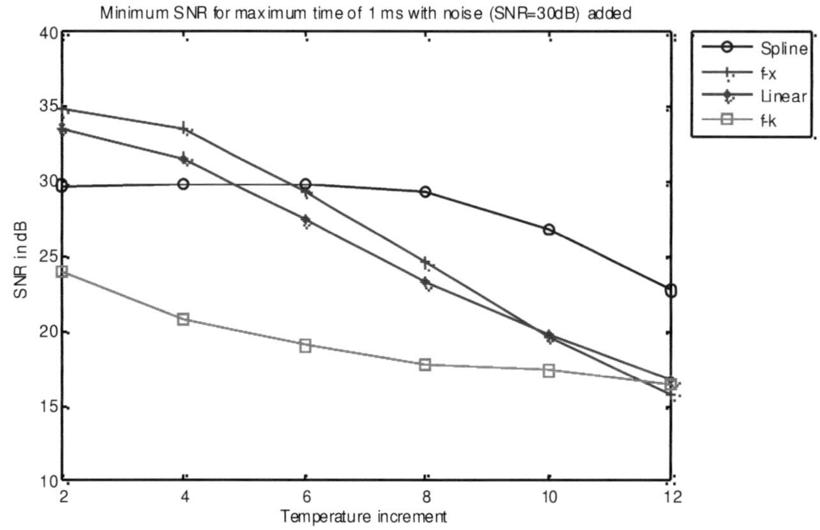

Figure 6. Same as Figure 3 but with noise added.

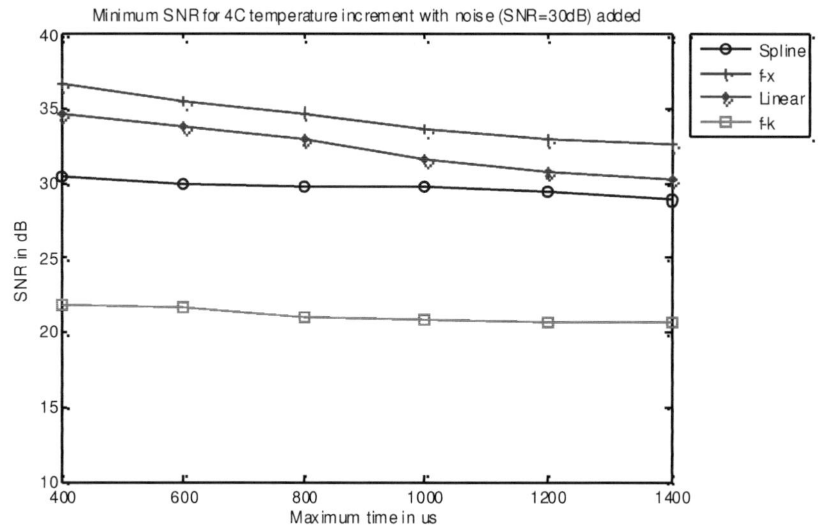

Figure 7. Same as Figure 4 but with noise added.

5. Conclusions

This paper presents results from a numerical study on the interpolation of ultrasonic guided wave signals in the temperature domain. We utilized the conventional methods of linear, cubic spline, sinc and FFT based interpolation. We also adopted interpolation in the *f-x* and *f-k* domains, which are popular techniques used in seismic data processing. We applied these methods on simulated data in the temperature domain. The data were interpolated for different temperature increments and different time durations. In general, the interpolation performance decreased as the temperature increment and time duration increased. Linear, spline and *f-x* domain methods demonstrated better performance in most cases. We also showed that by evaluating interpolation performance at every temperature within the range considered, spline, linear and *f-x* domain methods do not have edge effects. In the presence of noise, *f-x* domain interpolation has more tolerance and provides better results, which can be more clearly seen in simulated data since no measurement variations are introduced. The results of this paper provide guidance for obtaining ultrasonic measurements in a variable temperature environment in terms of tradeoffs between sampling rates and interpolation performance.

Acknowledgements

The first author would like to thank Ministry of Science and Technology of Taiwan, ROC, for partial support of this work under Grant Number NSC102-2918-I-251-001 and MOST103-2221-E-251-003.

References

1. J. E. Michaels and T. E. Michaels, "Detection of structural damage from the local temporal coherence of diffuse ultrasonic signals," *IEEE Transactions on Ultrasonics, Ferroelectrics, and Frequency Control*, vol. 52, no. 10, pp. 1769-1782, 2005.

2. Y. Lu and J. E. Michaels, "Feature extraction and sensor fusion for ultrasonic structural health monitoring under changing environmental conditions," *IEEE Sensors Journal*, vol. 9, no. 11, pp. 1462-1471, 2009.

3. B. Liu and M. D. Sacchi, "Minimum weighted norm interpolation of seismic records," *Geophysics*, vol. 69, pp. 1560-1568, 2004.

4. A. Trucco, S. Curletto, and M. Palmese, "Interpolation of medical ultrasound images from coherent and incoherent signals," *IEEE Transactions on Instrumentation and Measurement*, vol. 58, no. 7, pp. 2048-2060, 2009.

5. A. J. W. Duijndam, M. A. Schonewille, and C. O. H. Hindriks, "Reconstruction of band-limited signals, irregularly sampled along one dimension," *Geophysics*, vol. 64, no. 2, pp. 524-538, 1999.

6. Y. Lu and J. E. Michaels, "A methodology for structural health monitoring with diffuse ultrasonic waves in the presence of temperature variations," *Ultrasonics*, vol. 43, pp. 717-731, 2005.

7. A. J. Croxford, J. Moll, P. D. Wilcox and J. E. Michaels, "Efficient temperature compensation strategies for guided wave structural health monitoring," *Ultrasonics*, vol. 50, pp. 517-528, 2010.

8. C. Gerald and P. Wheatley, *Applied Numerical Analysis*, Addison-Wesley, 1994.

9. J. E. Michaels, R.-J. Liou, J. P. Zutty, and T. E. Michaels, "On the interpolation of ultrasonic guided wave signals," *Review of Progress in Quantitative Nondestructive Evaluation*, 31A, D. O. Thompson and D. E. Chimenti (Eds.), American Institute of Physics, pp. 679-686, 2012.

10. S. Haykin, *Adaptive Filter Theory*, 4th ed., Prentice Hall, 2002.

11. S. Spitz, "Seismic trace interpolation in the F-X domain," *Geophysics*, vol. 56, no. 6, 785-796, 1991.

12. N. Gulunay, "Seismic trace interpolation in the Fourier transform domain," *Geophysics*, vol. 68, no. 1, 355-369, 2003.

Laser-ultrasonic measurement of local elastic moduli

Alexandr A.KARABUTOV [1,2],Elena B. CHEREPETSKAYA [1], Natalia B. PODYMOVA [1,3]

[1]Laboratory of Laser Ultrasonic Nondestructive Testing, National University of Science and Technology MISiS , Moscow, Russia Phone: +7 499 2309593; [2] International Laser Center of M.V.Lomonosov Moscow State University; [3]Faculty of Physics of M.V.Lomonosov Moscow State University; e-mail: aak@edu.ilc.ru, npodymova@mail.ru, echerepetskaya@mail.ru

Abstract Laser excitation of ultrasonic wave enables one to measure the velocity of longitudinal, shear and surface waves with high precision (~ 0.5%) locally (travel distance - several mm). Knowledge of material density makes it possible to determine the local value of total set of elastic modulus. The Poisson's ratiois determined from the ratio of shear and longitudinal acoustic wave velocity, as is no need no to know exact thickness and the density of the sample under testing. Usage of this technology as a new tool of material evaluation is discussed.

Keywords Laser ultrasonic, local modulus, Poisson's ratio, Young's modulus

1. Introduction

The continuous decrease of the characteristic dimensions of electronic and micromechanical systems, the thickness of the film coatings and other products requires actual development of adequate methods of investigating the characteristics of their physical and mechanical properties at nano-, micro- and mesoscale [1-3]. The most important characteristics of the elastic properties of such materials are the Young's modulus and Poisson's ratio for isotropic materials or components of the stiffness matrix for anisotropic media.

There have been several attempts to determine Young's moduli of films and nanomaterials by micro- or nanoindentation techniques [4-7].However, the complex physical and chemical processes involving elastic and plastic deformation of the microinjections, the dissipation of elastic energy through various processes, local heating, changes in the crystal structure, mechano-chemical reactions and chemical modification of the surface, electrical and emission phenomena greatly complicate the interpretation of the results of microindentation. The lateral displacement of a contact body relative to another, passing an electric current through the contact, the effect of the environment, electric and magnetic fields can dramatically change the course of these processes and lead to significant errors in the measurement of elastic moduli [7].

An alternative to the indentation measurements for thin films and other nanostructures is the ultrasonic techniques. It is known that the elastic properties of materials affect the phase velocities of acoustic waves propagating therein. Therefore, at present, wide use is made of ultrasonic methods for determining the elastic moduli of structural materials from measured phase velocities of longitudinal and transverse (shear) acoustic waves [8]. These methods are rapid and can be used for samples of arbitrary shape and size and in regions whose size does not exceed a few millimeters.

The aim of this work was to develop a pulse acoustic method of quantifying the influence of porosity on the local Young's modulus of isotropic composite materials and measuring Poisson's ratio using a laser source of ultrasound. The method is based on a laser thermooptical method of ultrasound generation [9] combined with measurement of the phase velocity of longitudinal and shear acoustic waves in samples. The main advantage of this method compared with traditional ultrasonic research methods is the efficient generation of short and powerful probing acoustic pulses required for reliable testing of ultrasound absorbing and scattering heterogeneousmaterials.Using as an example samples of a metal matrix composite based on silumin with reinforcing microparticlesof silicon carbide, SiC (CM composite), in various concentrations, it is shown that to provide an effective increase inYoung's modulus with increasing concentration of SiC, the porosity of the final sample should notexceed 2%.

2. Method of calculating of elastic modulus of isotropic composite materials

The porosity P (the volume concentration of gas pores averaged over its volume) of a CM sample is defined by the formula

$$P = (1 - \rho / \rho_0) \cdot 100\% \quad (1)$$

where ρ is the density of the sample calculated from the known volume of the sample and the results of its hydrostatic weighing or weighing in air, ρ_0 is the calculated density of the solid phase of the sample determined from the known densities of the matrix ρ_m and the filler ρ_f and their volume concentrations n_m and n_f in this sample

$$\rho_0 = n_m \rho_m + n_m \rho_m \quad (2)$$

According to the theory of elasticity, the Young's modulus E_0 of an isotropic composite without pores (P is zero) can found the formula

$$E_0 = \rho_0 C_S^2 \frac{3C_{L0}^2 - 4C_S^2}{C_{L0}^2 - C_S^2} \quad (3)$$

where ρ_0 is calculated by formula (1,2); C_{L0} is the theoretically calculated phase velocity of longitudinal acousticwaves in the CM without pores, and C_S is the measured phase velocity of shear acoustic waves in the sample. Thepossibility of using the experimentally obtained value of C_S to calculate Young's modulus is due to the fact thatporosity does not affect the shear stiffness of the sample, and a reduction of C_S due to the scattering of shear wavesby pores for sufficiently small porosity (less than 5%) can be neglected. To calculate C_{L0}, it is proposed to usethe following model of propagation of longitudinal acoustic waves in a two-phase (two-component) medium

$$C_{L0}^2 = \frac{1}{\rho_0} \left(\frac{n_m}{\rho_m C_{Lm}^2} + \frac{n_f}{\rho_f C_{Lf}^2} \right)^{-1} \quad (4)$$

Here the phase velocities of longitudinal acoustic waves in the matrix C_{Lm} and filler C_{Lf} of the composite are assumedto be known.Young's modulus of the samples are calculated by formula (3) in which ρ_0 and C_{L0} are replaced by themeasured densities ρ and the phase velocity of longitudinal acoustic waves C_L for each CM sample.

The Poisson's ratio ν_0 is defined by the formula

$$\nu_0 = \frac{C_{L0}^2 - 2C_S^2}{2(C_{L0}^2 - C_S^2)} \quad (5)$$

3. Method for measuring phase velocities of longitudinal and shear ultrasonic waves in an isotropic composite

Thermoopticalsources of ultrasound produce acoustic signals of short duration about 100 ns, which provides high accuracy in measuring the travel time of an acoustic wave even in samples of small thickness of 100 μmand more. An ultrasonic beam with a typical diameter of a few millimeters can be used to produce and measure the phase velocity of transverse (shear) acoustic waves.

The scheme of the laser optoacoustic system for measuring elastic modulus of isotropiccomposite materials shown in Fig.1.

The ultrasonic pulse generated in an OA source (the reference signal) propagates in the sample and is detected using a broadband piezoelectric transducer having acoustic contact with the sample. To ensure this contact, the OA source, sample, and piezotransducer are separated by layers of an immersion fluid (in this case, distilled water, not shown in Fig. 1) and are mounted into a cell, which is an optoacoustic measuring cell. The transverse size of the test region is determined by the transverse size of the generated ultrasonic beam equal to the characteristic diameter of the laser beam. Electrical signals from the piezotransducer are sent to a Tektronix type dual-channel digital storage oscilloscope; the time of start of the oscilloscope is synchronized with the moment of laser pulse generation. The signal-to-noise ratio of the recorded electrical signals is 50-60 dB. The laser OA system has the following characteristics: operating frequency range 0.2-50.0 MHz, pressure amplitude of ultrasonic pulses 0.01-10.00 MPa, sample thickness 0.1-70.0 mm, and transverse size of the testing region 1-2 mm.
The samples of composites based on a matrix of AK12M2MgN alloy (silumin) reinforced with particles of silicon carbide, SiC, with an average diameter of 14 μm and various volume concentrations were studied. The density of the solid phase of each sample ρ_0 was calculated by expression (2) using the known density of the filler ρ_{SiC}= 3200 kg/m3 [10], the density of the silumin matrix ρ_A= 2740 kg/m3, and the specified volume concentrations of the matrix n_A and the filler n_{SiC}. The density of silumin was calculated from the known densities and mass concentrations of the alloy components. The density of each sample $\rho$was calculated from the mass measured by weighing in air and the calculated volume. Accordingly, the porosity of each sample P averaged over its volume was determined by formula (1.1). Parameters of the investigated samples of the CM are given Table 1.

Table 1. Parameters of CM samples

Sample number	H, mm	n		ρ_0,kg/m^3	ρ,kg/m^3	P, %
		AK12M2MgN	SiC			
1	11,20	1,00	0	2,74	2,72	0,7
2	10,20	0,97	0,03	2,75	2,71	1,5
3	7,85	0,93	0,07	2,72	2,67	3,6
4	4,75	0,86	0,14	2,80	2,67	4,6

From Table 1, it follows that the averaged porosity of the sample increases with increasing concentration of the SiC filler. Presumably, this is due to features of the manufacturing technology of the material (nonuniformreinforcement). For the same reason, the local porosity in different parts of the same sample may be different,which affects the local Young modulus. Since the porosity of the samples does not exceed 5%, in calculating the Young's modulus of the composite without pores by formula (3), we can use the measured phase velocity of shear acoustic waves in the sample.All tested samples were plane-parallel disks with a diameter of 40 mm, whose surfaces were polished by acorundum abrasive powder having an average particle diameter of 20 μm. It should be noted that in samples withthis ratio of the thickness and diameter, it is impossible to create strictly uniaxial tension required for measuringYoung's modulus using the mechanical method of tension or compression.The mechanical and acoustic properties of the obtained CM samples allowed them to be considered isotropicsince the shape of the filler particles is quasi-spherical and the particles are almost uniformly distributed over the volume of each sample.

3. Results and discussion

The absolutevalue of phase velocity of longitudinal acoustic waves is given by the formula

$C_L = 2H / \Delta T_L$ (6)

where H is the thickness of the sample, ΔT_L is the difference in time of arrival at the piezotransducer between theultrasonic pulse passed one times through the sample and the pulse passed thrice through the sample and reflectedfrom the interfaces between the sample and the immersion fluid layers. This method of measuring the phase velocityis called the time-of-flight method. Figure 2 shows the typical temporal shape of the longitudinal acoustic wave pulse passed through sample No. 4 of a silumin based composite with silicon carbide particles (Table 1) and thepulse passed three times through the sample. Figure 3 shows the temporal shape of the ultrasound signalgenerated upon absorption of the laser pulse in sample No. 3 of a composite based on silumin with silicon carbide.The recorded shape of the wave *S* (see Fig. 3) is determined by the acoustic fieldof the shear wave transformed into a longitudinal wave in passing from the sample to the immersion fluid. In this case, the shape of the wave *S* is influenced by the direction of the shear wave, determined by the transverse sizeof the longitudinal wave beam *L*, the dependence of the transformation coefficient on the angle of incidence of theshear wave on the interface between the sample and the immersion fluid, and the finiteness of the piezotransduceraperture. Therefore, the recorded pulse *S*, determined by the shear wave, is highly extended compared with thelongitudinal wave pulse *L*, and the arrival time of the negative peak of the signal *S* is determined by the timeof shear wave propagation in the sample. The pulse following this signal (see Fig. 3) is the pulse of longitudinalwaves *L* reflected in the immersion liquid layer between the sample and the piezotransducer.

The measured difference ΔT_{LS} in time of arrival at the piezotransducer between the maximum of the pulse *L*and the minimum of the pulse *S* and the measured phase velocity of longitudinal acoustic waves C_L are used todetermine the phase velocity of shear waves C_S in the sample

$$C_S = H(\Delta T_{LS} + H / C_L)^{-1} \quad (7)$$

The error of time-of-flight measurements of phase velocities is fairly small: about 0.5% for longitudinal waves andabout 2% for shear waves. For longitudinal ultrasonic waves, the measurement accuracy is determined primarily bythe maximum relative error of measurement of the sample thickness. Table 2 shows the results of optoacoustic measurements of the local Young's modulus and Poisson's ratioat the center (E_1, ν_1) andon the periphery (E_2, ν_2) of each CM sample. Figure 4 shows the results of calculations by formula (3) taking into account (4) and optoacoustic measurementsof the average values of Young's modulus *E* for different volume concentrations of SiC particles n_{SiC}. The dependence of the porosity P of the sample on the value of n_{SiC} is also shown in Fig. 4.Calculation results show that the addition of silicon carbide particles with a volume concentration of 14% should increase the Young's modulus by almost 6% compared to theinitial Young's modulus of the matrix, but this does not occur due to an increase in the porosity of the material.It can be seen that for a porosity of the sample $P \approx 5\%$, thedecrease in the Young's modulus relative to the calculated value is 8%.

Thus, to provide an increase in the Young's modulus of the CM based on AK12M2MgN silumin withincreasing volume concentration of the filler SiC during its mechanical mixing into the matrix melt, the porositythe final sample should not exceed 2%.

4. Conclusions

It was proposed and experimentally implemented an acoustic pulsed method using a laser source of ultrasoundfor quantitative evaluation of the combined effect of porosity and concentration of dispersed filler on the localYoung's modulus of isotropic metal matrix CMs. This method is suitable for nondestructive measurements ofYoung's modulus in samples 0.1–70.0 mm thick with a transverse size of the test area 1–2 mm. For the investigated CMs based on silumin with the addition of reinforcingmicroparticles of silicon carbide, SiC, we determined the maximum porosity of the prepared sample above whichthe effect of porosity on the reduction in Young's modulus is more significant than the effect of the filler on theincrease of this modulus.The developed laser ultrasonic method can be used for nondestructive testing of the local acoustic andmechanical properties of CM.

This method can also used in experimental studies of the effect of the composition, size and concentrationof reinforcing fillers on the elastic properties of composite materials.

5. Acknowledgement

The work was carried out with financial support from the Ministry of Education and Science of the Russian Federation *in the framework of* Increase Competitiveness Program of NUST «MISiS» (№ K1-2015-025)

References

1. Springer handbook of nanotechnology / Ed. B. Bhushan. Springer-Verlag, Berlin (2007), 1916 p.

2. A.D. Romig, jr, M.T. Dugger, PJ. McWhorter.ActaMater. 51, 5837 (2003).

3. Yu Golovin.Introduction to nanotech.Machinery, M. (2007).496 pp.

4. P.D. Tall, S. Ndiaye, A.C. Beye, Z. Zong, W.O. Soboyejo,H.-J. Lee, A.G. Ramirez, K. Rajan, Mater. Manuf. Processes22 (2007) 175-179.

5. E. Wibowo, C.Y. Kwok, J. Micromech. Microeng. 16 (2006)101-108.

6. A.K. Nanda Kumar, C.K. Sasidharan Nair, M.D. Kannan,S. Jayakumar, Mater. Chem. Phys. 97 (2006), 308-314.

7. M. Callisti, F.D. Tichelaar, B.G. Mellor, T. Polcar, Surf.Coat. Technol. 237 (2013) 261-268.

8. V. V. Polyakov and A. V. Golovin, "Elastic Characteristics of Porous Materials," Prikl. Mekh. Tekh. Fiz. **34** (5), 32-35 (1993) [J. Appl. Mech. Tech. Phys. **34** (5), 625-628 (1993)].

9. V. E. Gusev and A. A. Karabutov, *Laser Optoacoustics* (Nauka, Moscow, 1991) [in Russian]

10. S. Yu. Davydov, "Effect of Pressure on the Elastic Properties of Silicon Carbide," Fiz.Tverd. Tela46 (7),1169–1173 (2004).

Ultrasonic guided wave propagation and detection of high density core region in a honeycomb composite sandwich structure using embedded piezoelectric transducers

Shirsendu SIKDAR[1], Sauvik BANERJEE[2]

[1,2]Indian Institute of Technology Bombay, Powai, Mumbai-400046, India;
Phone: +91 22-25767343, Fax: +91 22-25767343;
Email: shirsendu@civil.iitb.ac.in[1], sauvik@civil.iitb.ac.in[2]

Abstract
An organized numerical and experimental study is carried out in order to understand ultrasonic guided wave (GW) propagation and interaction with a high density (HD) core region in a honeycomb composite sandwich structure (HCSS). Also the location of HD core region in a HCSS using embedded piezoelectric wafer transducer (PWTs) is investigated in this study. Due to complex structural characteristics, the study of guided wave (GW) propagation in HCSS with HD-core region inherently carries many challenges. Therefore, a three-dimensional (3D) numerical simulation of GW propagation in the HCSS with and without HD core region is carried out using embedded PWT network. It is observed that the presence of HD core significantly reduces the amplitude of the propagating GW modes. In order to verify the numerical results, experiments are conducted in the laboratory. A good agreement between the numerical and experimental results is observed, in all the cases studied. Finally, based on the change in amplitude of the received GW modes, the location of an unknown HD core region, within the PWT array is determined by applying a probability based signal difference coefficient (SDC) algorithm.

Keywords: Honeycomb composite sandwich structure (HCSS), guided wave (GW), high density core, piezoelectric wafer transducer (PWT)

1. Introduction

HCSS is a special kind of composite structure in which, thin fiber reinforced composite skins are bonded to the two faces of relatively thick and very lightweight aluminum honeycomb core using adhesive. This novel material is globally adopted in marine and aerospace industries as a specialized lightweight construction material [1]. The higher strength-to-weight ratio makes it suitable for construction of some of the major structural components such as flight wings, blades, fuselage, etc. and the high energy-absorption capability makes it attractive for impact mitigation and protection related applications. In some structural applications, the HCSS involves the use of various cores with different densities in the same sandwich element [2]. In order to introduce electronic and electrical devices, backing plates, fasteners, stiffeners for attachment or rigging purpose, etc., in these structures, it is suitable to use HD core sections. A part of the original core in sandwich structure is substituted by different core inserts like stiffeners, backing plates and many more [3,4]. Sandwich panels and beams with symmetric faces and cores of different stiffness were studied by Bozhevolnaya et al. [5]. In case of sandwich beam under 3-point bending, the closed form estimation of stress-strain field due to local effects was calculated. The faces close to the core junctions show significant rise in bending stress. Typical sandwich beams with glass fiber reinforced plastic face sheets and core junctions between polymer foams of different densities and rigid aluminum were tested under quasi-static and fatigue loading conditions by Johannes and Thomsen [6].

The Ultrasonic GW based techniques are proved to be an efficient and accurate procedure for nondestructive evaluation (NDE) and structural health monitoring (SHM) of composite structures [7, 8]. The general features of GW propagation that can be transmitted in isotropic and anisotropic media have been documented in many classical textbooks [9, 10]. Among the current methods, deployment of surface mounted broadband transducers or embedded PWTs [11, 12] are notable. Giurgiutiu et al. [13] reported review of the GW propagation technique for large area NDE of the laminated composite structures using surface bonded piezoelectric sensors. Maslov and Kundu [14] have shown that the tuning of GW modes in a received signal plays an important role for identification of hidden defects/discontinuities in laminated composites.

The dispersive properties of axially symmetric surface waves subjected to point loading on an isotropic elastic body with depth dependent elastic moduli and mass density was analyzed by Balogun and Achenbach [15] in the high frequency range [16]. The numerical and experimental studies of GW propagation in honeycomb sandwich structures (HSS) was conducted and reported by Hosseini and Gabbert [17]. Song and Huang [18] used a surface-bonded piezoelectric actuator/sensor system to investigate GW propagation mechanism in aluminum skin–hexagonal Nomex core sandwich structures, both numerically and experimentally. This study used a SDC to represent the differential features of debonding. Recently, Sikdar et al. [19] developed a baseline free damage index algorithm for detection of disbond in an HCSS using a PWT actuator/sensor network by considering time-frequency information of the received GW signals.

From the brief review of past literature, it appears that research dealing with GW propagation in HCSS is very few in number. Also no theoretical and/or numerical model is available to carry out elasto-dynamic analysis of a HCSS with joint core of varying densities. Therefore, the present research is motivated by the need to study the GW field produced by PWT sources on the surface of HCSS with a soft-dense-soft core for NDE/SHM applications. Towards this, the characteristics of the propagating GW are studied numerically, and the numerical results are successfully verified with the laboratory experiment. Also a SDC based algorithm is applied to identify the location and size of HD core region.

2. Experimental set-up and data acquisition

Experiments are carried out on a HCSS sample plate using embedded PWTs (20mm × 20mm × 0.4mm) and a NI set-up as shown in Fig. 1(a). The PWTs can serve as a transmitter as well as a sensor, and they are operated using the NI-instrumental set-up. The NI set-up is a PXI system that consists of an 8-channel oscilloscope (SCOPE), a multiplexer switch, an embedded arbitrary function/signal generator (FGEN) and a desktop monitor [19,20].

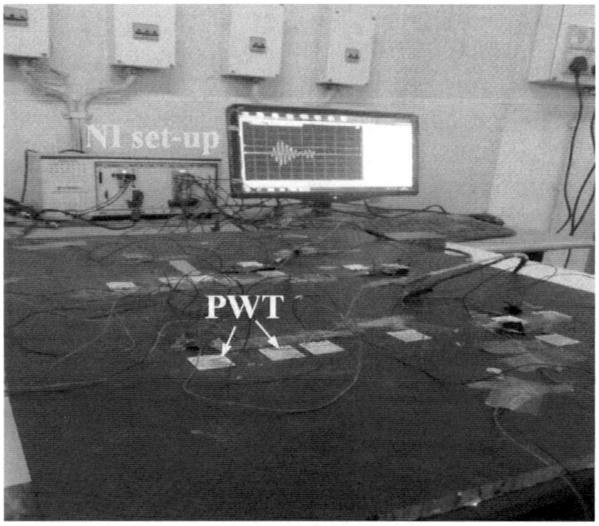

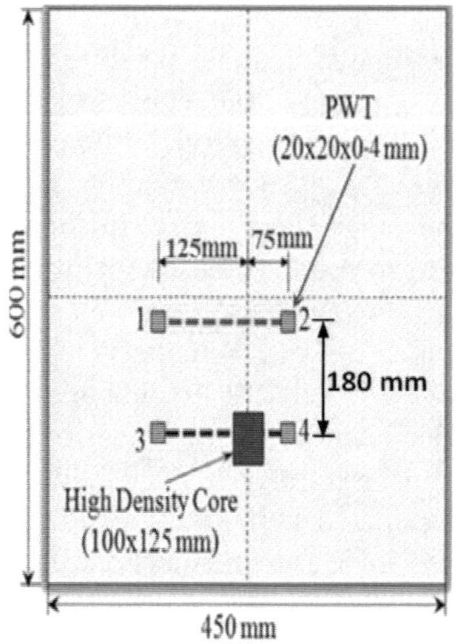

(a) (b)

Fig. 1 (a) NI experimental set-up and (b) Schematic representation of the HCSS with PWT locations.

A Hanning window modulated 5-cycle sine pulse is injected into the HCSS by the FGEN soft front-panel and a SCOPE soft front-panel is applied to collect the output GW signal. The PWTs are bonded on surface of the sample plate (600 × 450 × 13.5 mm) with a known HD core zone in order to generate GW. A schematic arrangement of PWTs against the HD core zone is shown in Fig. 1(b). In order to obtain the optimum frequency range of the PWTs, a frequency modulation is carried out by placing two PWTs (transmitter and sensor) at 200 mm distance on the surface of the sample plate, as shown in Fig. 2(a). The frequency vs. amplitude graph (also known as PWT calibration curve) is plotted in Fig. 2(b) for a range (0.05 to 0.230 MHz) of central frequencies, in order to identify the optimum excitation frequency. It is noticed that the PWTs show maximum response at 150 kHz frequency. Therefore, a 150 kHz 5-cycle sine pulse is selected for all the experiments and numerical simulations. The input signal and its frequency-spectrum are shown in Fig. 3.

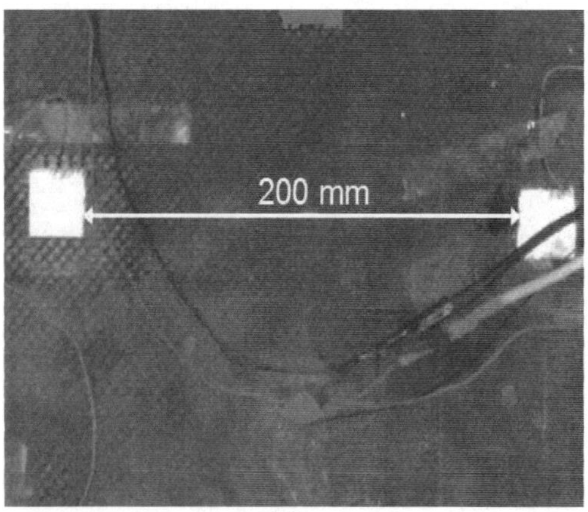

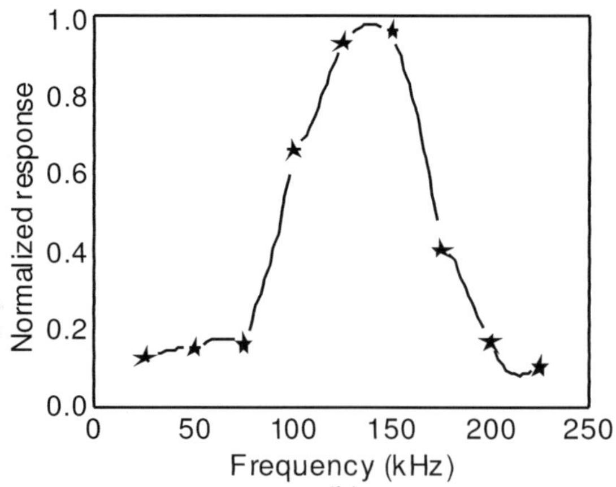

(a) (b)

Fig. 2 (a) Experimental set-up for frequency modulation and (b) corresponding calibration curve.

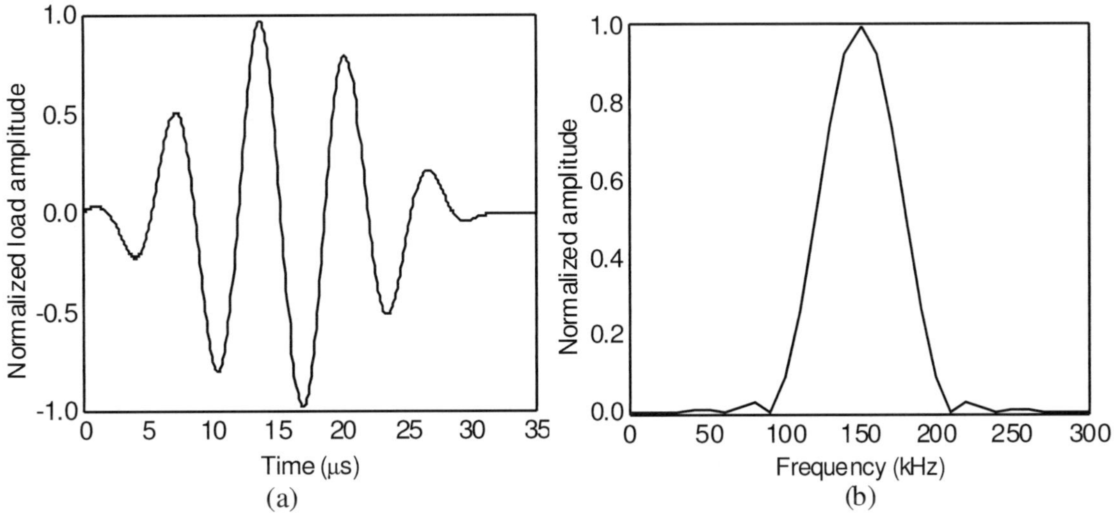

Fig. 3 Input signal of (a) 150 kHz five-cycle sine pulse in a Hanning window, (b) its frequency spectrum.

In the experiment, the FGEN of the NI set-up is applied to produce the 150 kHz input signal as shown in the Fig. 3(a). The output signal from PWT path: 1-2 can be assumed as baseline signal, as the transmitter-sensor path is sufficiently away (180 mm) from the HD-core region, as shown in the Fig. 1(b). Whereas, the output-signal from PWT path: 3-4 can be considered as affected signal due to HD core.

3. Finite Element Modeling

The finite element method is used as an alternative to numerically solve this class of problems. The finite element software ABAQUS/6.12 is used in this study to simulate the GW propagation in HCSS. In order to know the HD core effects on the propagating GW signal, a 3D numerical model (600 mm × 450 mm × 13.5 mm) of the experimental HCSS plate is made in ABAQUS, as shown in Fig. 5.

Table 1 Elastic properties of the HCSS

Material	E_1 (GPa)	E_2 (GPa)	E_3 (GPa)	G_{12} (GPa)	G_{23} (GPa)	G_{13} (GPa)	ν_{12}	ν_{13}	ν_{23}	ρ (kg/m^3)	t (mm)
CP-lamina	110.31	110.31	18.247	42.41	4.136	4.136	0.30	0.12	0.12	1.65	0.17
UD-lamina	60.212	60.212	10.252	18.20	3.611	3.611	0.20	0.03	0.03	1.42	0.08
Soft-core	0.0804	0.0804	1.6121	0.0321	0.0964	0.0964	0.25	0.025	0.025	0.032	12
HD-core	0.734	0.734	1.846	0.0763	0.4826	0.4826	0.34	0.042	0.042	0.128	12
Adhesive	0.0486	0.0486	0.0486	0.0174	0.0174	0.0174	0.40	0.40	0.40	1.25	0.01

The HCSS model consists of two 0.74 mm thin Graphite-epoxy fiber-reinforced composite skins, 0.02 mm thin adhesive layer (at the interface) and a 12 mm thick soft aluminium-core (Al5056) with a HD region (100 mm × 125 mm × 12 mm). The details of the elastic material properties of different layers in the HCSS are tabulated in Table 1. The PWTs (20 mm× 20 mm× 4 mm) are modeled on the surface of the HCSS, using the ABAQUS implicit code. The SP-5H PWT material properties are used [21] and the numerical simulations are carried out with the 150 kHz input signal. In ABAQUS, the C3D8R (8-noded linear brick element with reduced integration and hourglass control) are used for modeling the honeycomb core, adhesive, skin and electrode part, and the C3D8E elements (8-noded linear piezoelectric brick element) are used for piezoelectric part. In order to model the PWT excitations, the horizontal symmetric surface source excitation is applied on the top surface of the HCSS plate model.

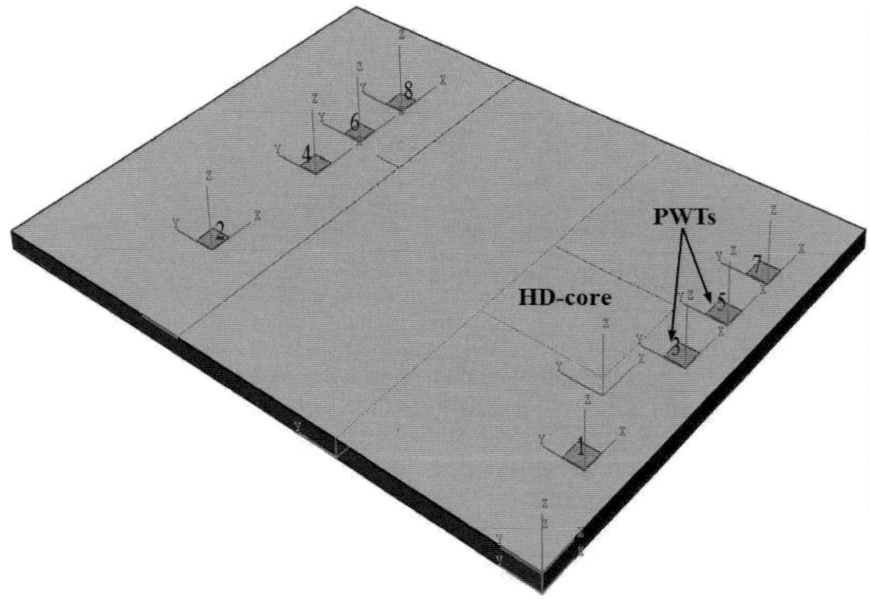

PWT coordinates:

PWT Number	X (mm)	Y (mm)
1	90	50
2	90	440
3	240	50
4	240	440
5	290	50
6	290	440
7	340	50
8	340	440

Fig. 4 Numerical model of HCSS with HD-core and PWT network in ABAQUS.

4. HD core detection algorithm

In order to locate the disbond region a disbond imaging algorithm is used, which works on the basis of an SDC. The Wavelet transform of the experimental sensor signals are used for the characterization of disbond in HCSS. The SDC based on the transformed GW signals in the time domain is used to capture the differential features of the disbond. To image the disbond region, the damage probability distribution is computed by using the extracted SDC as inputs, at each pixel. The quality of the final disbond image is improved by using the fused image at individual frequency. The damage localization probability, D_d, of any arbitrary position *(x, y)*, within the sensor network is expressed as (Zhao *et al* 2007)

$$D_d(x,y)=\sum_{i=1}^{N-1}\sum_{j=i+1}^{N}D_{ij}(x,y)=\sum_{i=1}^{N-1}\sum_{j=i+1}^{N}sdc_{ij}(x,y)\left[\frac{\beta-A_{ij}(x,y)}{\beta-1}\right] \tag{1}$$

where, $D_{ij}(x,y)$ represent the damage distribution probability, measured from actuator-sensor pair: *i-j* and $sdc_{ij}(x,y)$ is the signal different coefficient, which is the difference in amplitude area with disbond and without disbond for a particular GW mode. The SDC can be expressed as:

$$sdc_{ij} = \sqrt{\frac{\int_{t_1}^{t_2}(s^d - s^b)dt}{\int_{t_1}^{t_2}[s^d]^2 dt}} \quad (2)$$

where, s^d and s^b are the guided wave signal with disbond and without disbond, respectively, t_1 is the time arrival of signal for particular mode and $t_2 = (t_1 +$ bandwidth of signal), $\left[\frac{\beta - A_{ij}(x,y)}{\beta - 1}\right]$ is the spatial distribution function, which has contour in the shape of ellipse with a non-negative value, and

$$A_{ij}(x,y) = \begin{cases} P_{ij}(x,y), & P_{ij}(x,y) < \beta \\ \beta, & P_{ij}(x,y) \geq \beta \end{cases} \quad (3)$$

where,

$$P_{ij}(x,y) = [\sqrt{(x-x_i)^2(y-y_i)^2} + \sqrt{(x-x_j)^2(y-y_j)^2}\,]/p_{ij} \quad (4)$$

where, P_{ij} is the distance between actuator '*i*' and receiver '*j*', and the β is a small scaling parameter that reduces the size of the affected zone and it is independent of wave velocity. The value of β is determined empirically and in this study it is selected as 1.05 [22].

5. Results and discussions

In order to study the HD core effect on the propagating GW signal, numerical simulation is carried out for 3D HCSS models with a HD core region, as shown in Fig. 5. The plots in Fig. 5 indicate the presence of multiple GW modes. In order to clearly understand the multimodal behaviour, wavelet transform (WT) is performed with *AGU-Vallen Wavelet* [23] on the signals in Fig. 5(a), and are presented in Fig. 5(b).

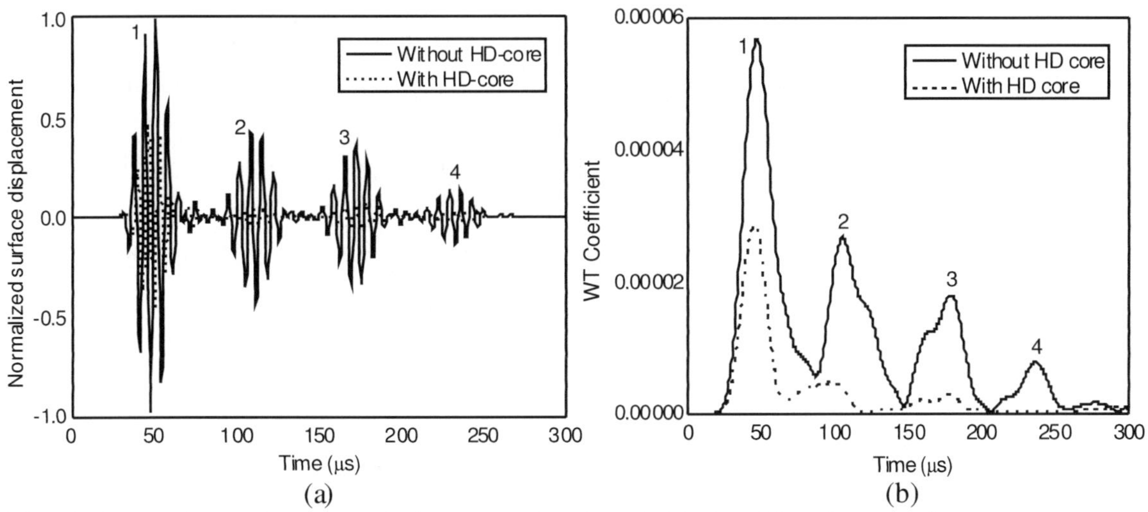

Fig. 5 Comparison of (a) numerical output signals and (b) their Wavelet transform.

Four individual wave mode are observed in both the cases and the modes are designated as 1, 2, 3 and 4. From the comparison in Fig. 5, it is observed that the presence of HD core significantly reduces the amplitude of the received signals.

In order to validate the numerical simulation results, the laboratory experiments are conducted on the HCSS sample plate for both the without HD core (actuator/sensor path: 1-2) and the with HD core (actuator/sensor path: 3-4) case (Ref. Fig. 1(b)). The output signals for both the cases are presented in Fig. 6, which confirms the presence of all four wave modes as observed in the numerical signals. Similar behaviour with simulation results is noticed in terms of considerable reduction in amplitude of the GW modes.

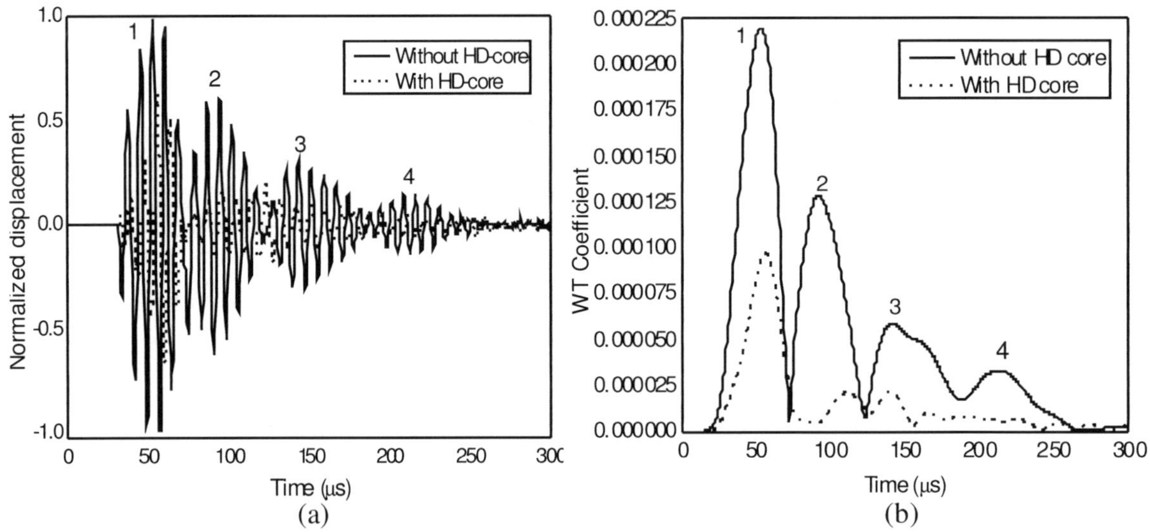

Fig. 6 Comparison of (a) experimental output signals and (b) their Wavelet transform.

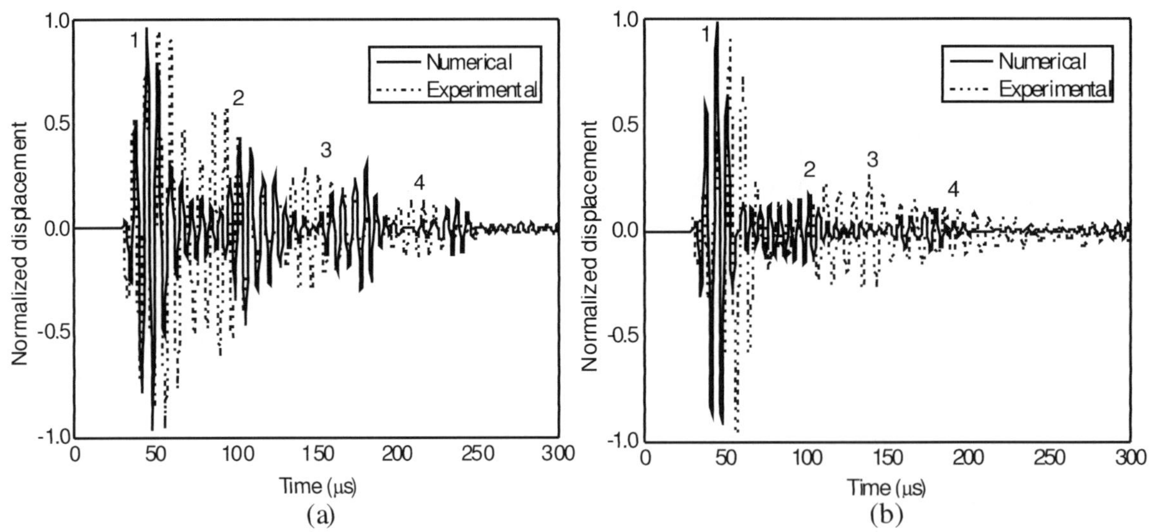

Fig. 7 Comparison of numerical and experimental response in the (a) 'without' and (b) 'with' HD-core case.

The comparison of numerical and experimental output signals of the 'without' and 'with' HD core cases, are shown in Fig. 7(a) and 7(b) respectively. A good agreement between the numerical and experimental results is found in both the cases. Therefore it can be stated that the numerical results are successfully validated by the laboratory experiments.

5.1 Detection of HD core in HCSS

In order to identify the exact location of the HD-core zone, the transformed received signals correspond to mode-1 along the path 1-2, 3-4, 5-6, 7-8, 1-8 and 2-7 are applied (Ref. Fig. 4) as input to the SDC based algorithm in *MATLAB*. The SDC map is shown in Fig. 8, which represents the maximum SDC intensity value (negative) close to the HD-core location in the HCSS.

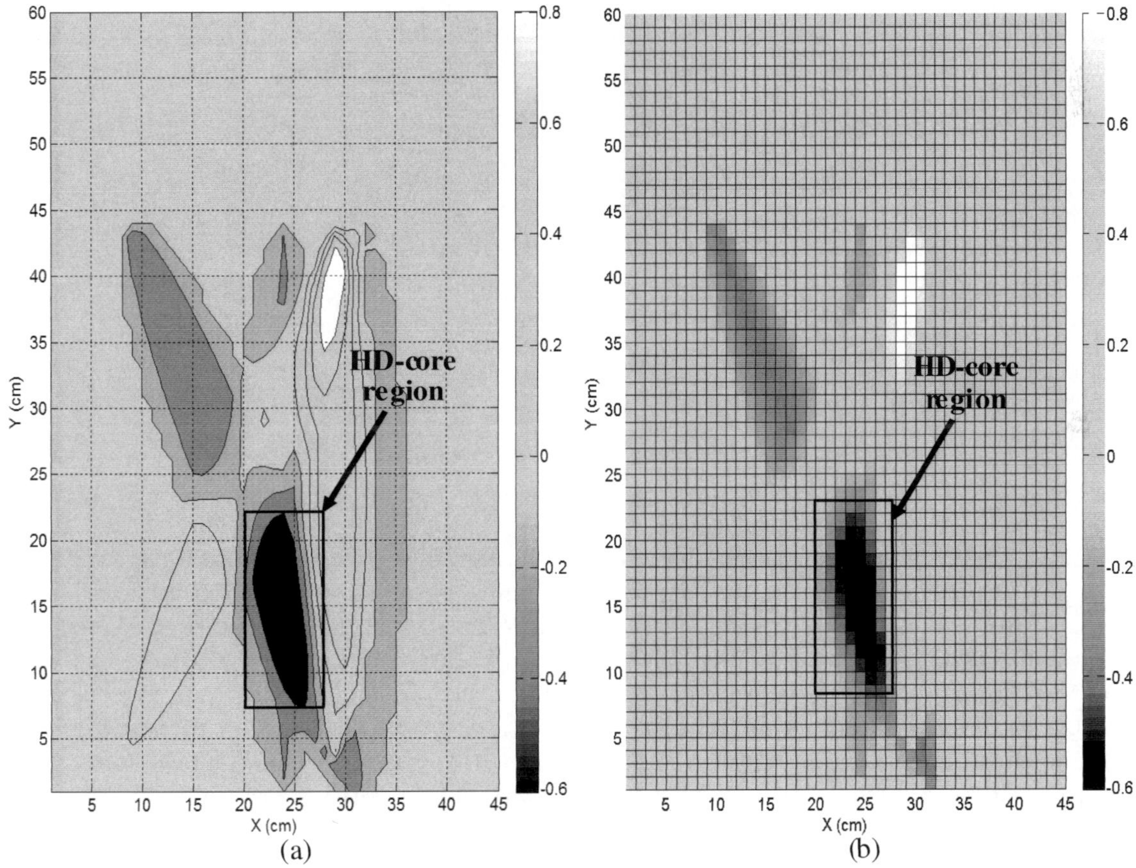

Fig. 8 SDC map in (a) contour and (b) grid pattern, showing the exact location of the HD core region.

In order to clearly understand the SDC behavior, a 3D representation of the SDC map is also obtained, as shown in the Fig. 9.

It is expected that the availability of baseline data will significantly improve the HD-core detection capability and possibly size it with some degree of confidence. However, it is found that the algorithm is capable to identify HD-core size and location using the change in modal amplitude of the received GW signal with a minimum number of transmitter/sensor paths available.

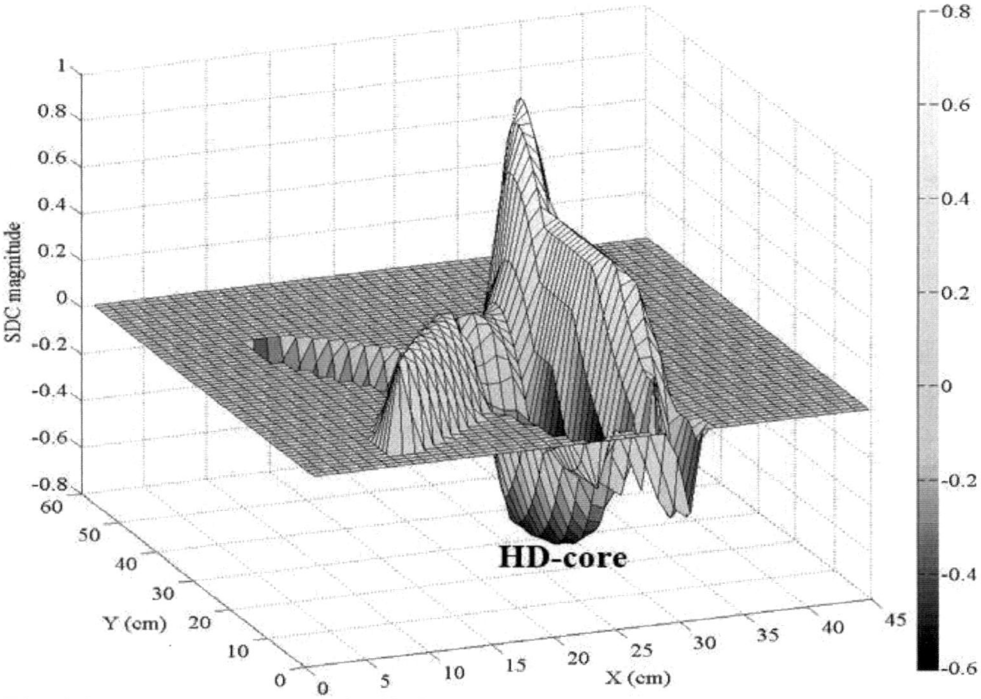

Fig. 9 3D representation of the SDC map.

6. Conclusions

In order to understand the mechanism of GW propagation in a HCSS in the presence of a HD core region, a combined numerical and experimental study is carried out. Significant variation in wave mechanism is noticed due to the increase in core density. A good agreement between the 2D numerical and experimental results is found in all the cases studied. Due to the presence of the HD core zone substantial reduction in amplitude of the GW modes in the received signal is observed. Therefore, it can be concluded that the increase in mass density of a particular region in HCSS plate leads to decrease in the amplitude of the propagating GW. The SDC based HD core identification technique shows its efficiency to accurately identify the HD core location and also the approximate size of the HD core zone.

Acknowledgement

The authors wish to thank the Indian Space Research Organization (ISRO) for supporting this work under grant 11ISROC001.

References

1. S T Peters, 'Handbook of composites', Chapman and Hall, Boca Raton, 1988.

2. D Gay, S V Hoa and S N Tsai, 'Composite materials: design and application', New York, CRC, 2003.
3. D Zenkert, 'The handbook of sandwich construction', London, EMAS Publishing, 1997.
4. O T Thomsen, 'Sandwich plates with 'through-the-thickness' and 'fully-potted' inserts: evaluation of differences in structural performance', Compos Struct, Vol 40, pp 159–74, 1998.
5. E Bozhevolnaya, O T Thomsen, A Kildegaard and V Skvortsov, 'Local effects across core junctions in sandwich panels', Compos B, Vol 34, pp 509–517, 2003.
6. M Johannes and O T Thomsen, 'Examination of the failure of sandwich beams with core junctions subjected to transverse shear loading', J Sandwich Struct Mater, Vol 12, No 2, pp 199–236, 2010.
7. A Raghavan and C E S Cesnik, 'Review of guided-wave structural health monitoring', Shock Vib Dig, Vol 39, No 2, pp 91–114, 2007.
8. Z Su, L Ye, and Y Lu, (2006) 'Guided lamb waves for identification of damage in composite structures: a review', J Sound Vib, Vol 295, pp 753–780, 2006.
9. J D Achenbach, 'Wave propagation in elastic solids', North-Holland, New York, 1978.
10. J L Rose, 'Ultrasonic waves in solid media', Cambridge University Press, Cambridge, 1999.
11. V Giurgiutiu, 'Tuned lamb wave excitation and detection with piezoelectric wafer active sensors for structural health monitoring', J Intell Mater Syst Struct, Vol 16, pp 291–305, 2005.
12. A K Mal, F Ricci, S Banerjee, and F Shih, 'A conceptual structural health monitoring system based on vibration and wave propagation', Struct Health Monit, Vol 4, pp 283–293, 2005.
13. V Giurgiutiu, A N Zagrai, and J J Bao, 'Piezoelectric wafer embedded active sensors for aging aircraft structural health monitoring', Structural Health Monitoring, Vol 1, pp 41–61, 2002.
14. K I Maslov, and T Kundu, 'Selection of lamb modes for detecting internal defects in laminated composites', Ultrasonics, Vol 35, pp 141–150, 1997.
15. O Balogun, and J D Achenbach, 'Surface waves on a half space with depth- dependent properties', J Acoust Soc Am, Vol 132, No 3, pp 1336–1345, 2012.
16. A L Shuvalov, 'The high-frequency dispersion coefficient for the Raleigh velocity in a vertically inhomogeneous anisotropic half-space', J Acoust Soc Am, Vol 13, pp 2484 – 2487, 2008.
17. S M H Hosseini and U Gabbert, 'Numerical simulation of the Lamb wave propagation in honeycombsandwich panels: A parametric study', Composite Structures, Vol 97, pp 189-201, 2013.
18. F Song and G L Huang (2012), 'Online guided wave-based debonding detection in honeycomb sandwich structures', *AIAA Journal* 50(2): 284-293.
19. S Sikdar , S Banerjee and S M Subhani (2013), 'Detection of disbond in a honeycombcomposite sandwich structure using ultrasonic guided waves and bonded PZT sensors', In: *Proceeding of 14th Asia Pacific Conference on Non-destructive Testing*, Mumbai, India, 18-22 November, 2013.
20. S Sikdar and S Banerjee (2014), 'Wave propagation in a honeycomb composite sandwich structure in the presence of high-density core using bonded PZT-sensors', In: *Vibration Engineering and Technology of Machinery, Mechanisms and Machine science* (ed JK Sinha), Manchester, UK, 9-11 November 2014, pp 619-628, Springer International Publishing Switzerland.
21. C R Bijudas, M Mitra and P M Mujumdar, 'Coupling effect of piezoelectric wafer transducers in distortions of primary Lamb wave modes', Smart Mater Struct, Vol 22, No 6, 065007, 2013.
22. X L Zhao, H D Gao, G F Zhang (2007), 'Active health monitoring of an aircraft wing with embedded piezoelectric sensor/actuator network:1, Defect detection, localization and growth monitoring', *Smart Materials and Structures* 16(4): 1208–1217.
23. M A Hamstad, A O'Gallaghar, and J Gary, 'A wavelet transformation applied to acoustic emission signals: part 1: source identification', J Acoustic Emission, Vol 20, pp 39–61, 2002.

A multichannel ultrasound pulser with arbitrary signal excitation unit for scanning acoustic microscopy

A. Juhrig[1], S. Kümmritz[1], M. Wolf[1], E. Kühnicke[1]
[1]Department of Electrical and Computer Engineering – Solid-State Electronics Laboratory, TU Dresden, Mommsenstraße 15, 01069 Dresden, Germany
Phone: +49 351 463 43781, E-Mail: Andre.Juhrig1@tu-dresden.de

Abstract
Ultrasound microscopy is used in non-destructive testing, especially for the inspection of very small test items like semiconductor chips and electrical circuits. For testing layered structures with curved and non-parallel interfaces by performing a single scan only, high frequency array technique is a smart solution.
This paper presents a functional model for signal excitation by a special circuit as well as preprocessing of the received signals. The unit emits rectangular pulse signals on four separate channels. Furthermore, to improve the noise reduction, four different receive circuits are tested consisting of amplifiers and different types of low pass filters processing the received echo signals.

Keywords: high frequency; high resolution; focusing and steering; arbitrary waveform; pulser for microscopy; annular array

Introduction

Scanning acoustic microscopes are used to examine hidden structures of opaque objects, especially to detect delaminations between two layers or inclusions inside these objects. In one hand, there are commercial microscopes utilising single element transducers with a centre frequency up to 240 MHz. On the other hand, there are modern ultrasound imaging techniques [1,2] utilising annular arrays for electronic focussing and steering of the ultrasound beam to gather additional information out of the reflected sound field by analysing the received signal of each element. Recent annular arrays, manufactured from composite materials, offer a centre frequency up to 20 MHz [3] and for PVDF up to 45 MHz [4]. Actually, there is no commercial multichannel electronic hardware available, making a further development of the control electronics indispensable.
A first approach is the pulser and receiver unit presented by Gust [5] that offers time delayed, fixed short time pulses to excite up to 16 elements of an annular array. Additionally, the received signals can be amplified. Another approach is an 8 channel arbitrary waveform generator for acoustic scanning microscopy developed by Assef [6]. However, this unit does not offer amplification for received signals. In order to combine the advantages of both solutions, a 4 channel functional model of a new pulser and receiver unit is constructed in a modular way to easily extend the system to a higher number of channels. Furthermore, it is important to implement an effective tool for noise reduction that acts as anti-aliasing filter to minimise superposing effects at the analogue to digital conversion.

Functional Model

The current functional model is intended to test an alternative way generating excitation signals for ultrasound pulses. Therefore, a special pulser circuit, controlled by high speed digital inputs, is used. The pulser circuit has 4 independent channels. The functional model works with upcoming ultrasound transducers that offer a centre frequency up to 20 MHz. Furthermore, the circuit should provide a good noise rejection and electromagnetic compatibility.

Ultrasound excitation

The existing solution by Gust [5] offers very short fixed time pulses to excite an ultrasound transducer. To perform the step targeting arbitrary signal this functional model offers an excitation signal with rectangular shape generated by H-bridges included in the pulser circuit. They are regulated by a 4-bit parallel interface for each of the four separate channels to control the time width as well as the voltage level of the rectangular pulse. With this two options it is possible to match the signal for excitation pulses to the centre frequency of the used transducer to achieve an optimal excitation by using lower voltages.

Preprocessing of received echo signals

Generally, received signals need a minimum amplification for a qualitatively good analogue to digital conversion, because their amplitude is too small compared to the effective size of a quantisation step of common used converters. The best results are achieved when the biggest width of the voltage reference range is filled by the signal.

Therefore, the amplification is realised in two steps. At first, the required minimum amplification of the received signal is realised by a fixed gain amplifier. In a second step a variable gain amplifier, consisting of an electrical circuit that includes all necessary functional components, is implemented to match the output signal to the range of the analogue to digital converter. All channels include a high pass filter to remove DC-components of the received signal. Additionally, two channels have a low-pass filter that acts as an anti-aliasing filter for the high-frequency noise suppression.

The small receive signal, consisting of the useful signal and the high-frequency noise, is amplified at first. Then, the amplified high-frequency noise, but as less as possible of the useful signal, is damped by the low-pass filter. Finally, the variable gain amplifier amplifies the useful signal with the remained but damped high-frequency noise, so that its negative influence on the conversion is reduced.

During development of the amplifiers and filters, the named technical requirements have to be concerned. Special care has to be taken at the design of the filters. A certain distance between cut-off frequency and centre frequency has to be implemented, acting as a kind of safety buffer, to avoid a loss of the useful signal. Figure 2 shows the frequency response characteristics of the three different amplifiers identified based on the simulation of the electronic circuits.

Channel 1 (blue curve) consists of a fixed gain and variable gain amplifier and a high-pass filter between. This channel shows the lowest damping of all presented. Channel 2 and 3 additionally contain a low-pass filter. In case of channel 2 it is designed as a first order filter and in channel 3 as a Butterworth [7] low-pass filter of third order. They show different frequency response characteristics regarding the range above 20 MHz and channel 3 shows the highest damping of high frequency noise of all. Furthermore, the frequency response characteristic demonstrates that the security-buffer gets smaller with the rising order of the implemented filter.

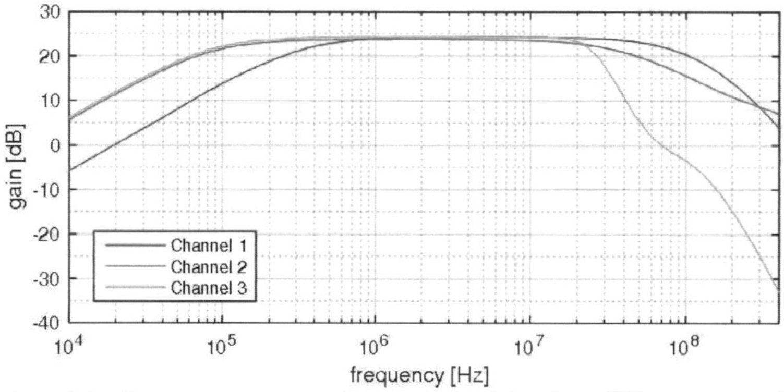

Figure 1: Simulation of the frequency response characteristic of the three different types of amplifier and filter.

The last of the offered four channels has an identical structural and functional construction like channel 1. It just differs by the usage of an other variable gain amplifier, so that the results of channel 1 are the same as for channel 4.
Finally it can be assumed, that channel 3 will offer the best performance regarding the recommendation of noise rejection, because of its highest damping ratio at high frequency.

Characterisation of the functional model

Testing method

The most critical part of the functional model is the transmission behaviour of the receiver. To examine the transfer behaviour the receiver is tested with a function generator instead of using ultrasound echos.
To analyse the transmit behaviour of each channel, the signals at the output of the filter (before the variable gain amplifier) are evaluated.

Test results

By the usage of a digital oscilloscope the transmitted sinusoidal signals are recorded and their frequency spectrum is deduced. Figure 2 compares the circuit simulation with the extracted frequency spectrum of the measurement for each of the three different types of filter and amplifier combinations.
Each of the three types shows a significant peak at the frequency of 20 MHz in the simulation as well as in the measurement. This represents the useful signal that is applied by the signal generator. In the following range between 30 MHz and 200 MHz they differ to each other.
Channel 1 and channel 2 are showing a similar frequency spectrum. A quantitative comparison shows that the low-pass filter, implemented in channel 2, is damping the amplitude of higher frequencies with respect to channel 1.

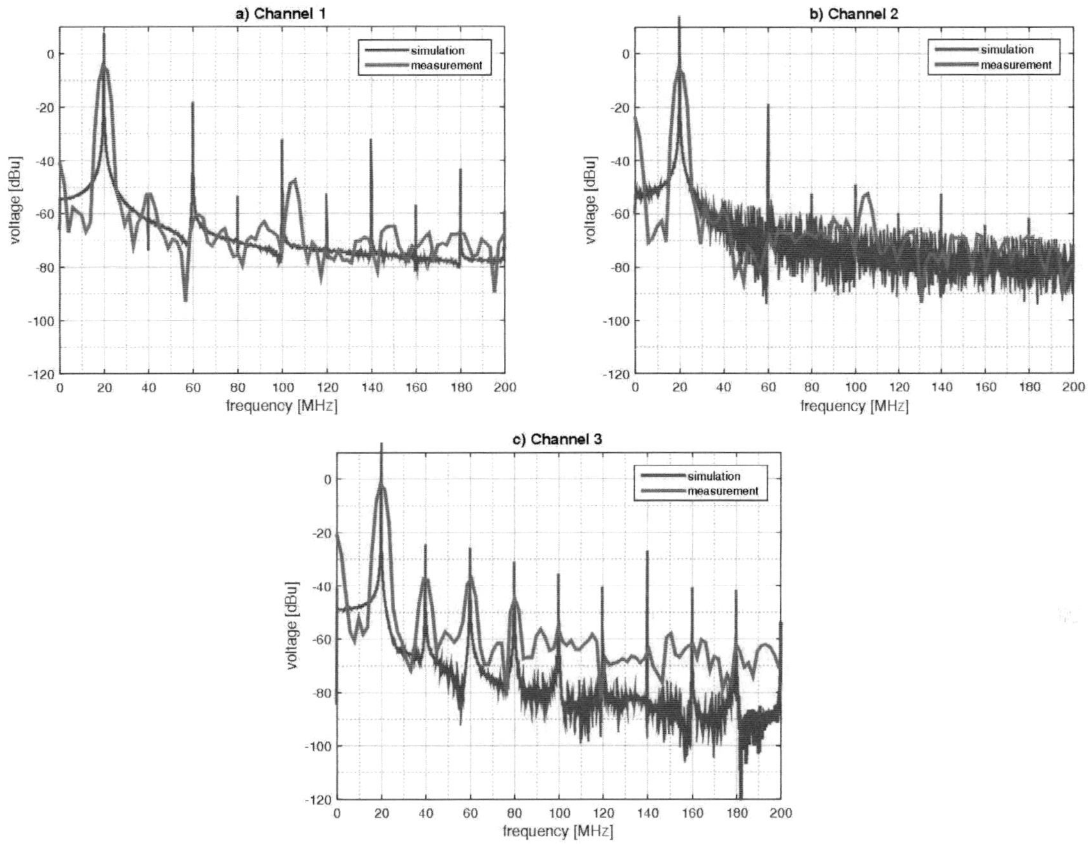

Figure 2: Comparison of the result of the circuit simulation and the extracted frequency spectrum of the measured signal for each of the three different types of filter and amplifier combinations. A sinusodial function of 250 mV and 20 MHz is applied as test signal.

In comparison to the other two channels, the simulated frequency spectrum of channel 3 shows more peaks with higher amplitudes. Hereby the measurement confirms the simulation. These peaks arise when the circuit is oscillating. This is caused by violating the stability criterion of active components. In this case the measurement of echo signals received by the transducer becomes impossible. To solve this problem a complete redesign of the implemented active low-pass filter is necessary. A possible solution is for example the replacement of the active filter by a passive one.

Finally, the measured frequency spectra of each channel are compared to each other to find an applicable solution for the further development (figure 3).

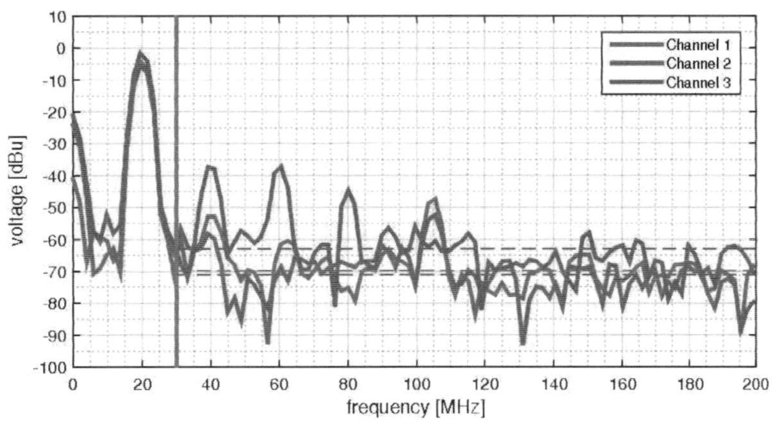

Figure 3: Comparison of all measured frequency spectra.

The comparison shows that channel 3 presents the worst behaviour of the three different types in the functional model. With a mean damping of -62,9 dBu in the range from 30 MHz to 200 MHz channel 3 has the highest noise level of all channels. The two remaining channels have a similar characteristic in their high frequency behaviour. To find a difference between them, the mean values of their high frequency damping are quantified in the range specified above. Hereby, channel 2 has the highest damping with -71,1 dBu with respect to channel 1 with -69,8 dBu. Therefore, the design of channel 2 is the preferred design for signal preprocessing in upcoming functional models and prototypes.

Conclusion

The presented functional model contains a pulser circuit that offers 4 independent channels to emit and receive ultrasound signals. On each of the four channels there is a different receiver consisting of a fixed gain and variable gain amplifier as well as different filters between.
The characterisation of the receiver shows that a first order low-pass filter, acting as an anti-aliasing filter has improved the damping of high frequency noise compared to the channel without an applied low-pass filter. Contrary to the assumption, the third order filter has shown the worst result.
The functional model is a modular setup and can be simply extended to more channels. For further development of the ultrasound pulser and receiver, the implementation of a first order low-pass filter is advantageous with respect to the signal to noise ratio. For higher operating frequencies of transducers, a redesign of the filters is necessary.

References

[1] M. Wolf, E. Kühnicke, "Non-invasive and Locally Resolved Measurement of Sound Velocity by Ultrasound", Sensors & Transducers, Vol. 184, Issue 1, pp. 53-59, January 2015

[2] S. Kümmritz, M. Wolf, E. Kühnicke, "Simultaneous Determination of Thickness and Sound Velocity in Layered Structures", ECNDT 2014, October 6-10, 2014, Prague

[3] J. Zhang, W. Ren, X. Jing, P. Shi, X. Wu, "Deep reactive ion etching of PZT ceramics and PMN-PT single crystals for high frequency ultrasound transducers", Ceramics International, Vol. 41, pp. 656-661, 2015

[4] J.A. Brown, C.M.E. Démoré, G.R. Lockwood, "Design and Fabrication of Annular Arrays for High-Frequency Ultrasound", IEEE Trans. Ultrason. Ferroelectr. Freq. Control, 51:8, 2004, 1010-1017

[5] N. Gust, F. Schellhorn, "High Frequency 16 channel ultrasonic microscope for annular arrays", IEEE Ultrasonics Symposium (IUS) 2012, 2012

[6] A.A. Assef, J.M. Maia, F.K. Schneider, V. LSN. Button, E.T. Costa, "A reconfigurable arbitrary waveform generator using PWM modulation for ultrasound research", BiomiMedical Eng. OnLine, vol. 12, no. 1, pp. 1-13, 2013

[7] R. Kories, H. Schmidt-Walter, "Taschenbuch der Elektrotechnik", Verlag Harri Deutsch, Vol. 7, 2006, pp. 421-429

Complementary near field technique for assessment of fiber reinforced composites

Adriana SAVIN[1*], Rozina STEIGMANN[1,2], Nicoleta IFTIMIE[1], Maria PAPEZOVA[3], Dagmar FAKTOROVA[3]
[1]Nondestructive Testing Department, National Institute of R&D for Technical Physics, Iasi, Romania
Phone +40232430680; fax: +40232231132 email asavin@phys-iasi.ro
[2]Faculty of Physics, University Al.I.Cuza, 11 Carol I Blvd, Iasi, 700506, Romania
[3] Faculty of Electrical Engineering, University of Žilina, Univerzitná 1, 010 26 Žilina, Slovak Republic

Abstract
The paper present the possibility to use a sensor with metamaterial lens for the nondestructive evaluation of metallic strip gratings and carbon fiber reinforced plastics. The sensor has enhanced spatial resolution due to the apparition of evanescent waves in the space between strips and respectively between carbon fibers at the excitation with a transversal magnetic along z axis polarized electromagnetic field. The evanescent waves can be manipulated by a lens made from two conical Swiss rolls that act as a field concentrator. The detection is made with spatial resolution better than λ/2000.

Keywords: metamaterials lens, metallic strip gratings, fiber reinforced plastic composites, evanescent waves

1. Introduction

The electromagnetic (EM) nondestructive evaluation (eNDE) of materials consists in the application of an EM field with frequencies ranging from tens of Hz to tens of GHz, to the examined object and evaluating the interaction between the field and the eventual material discontinuities and can be schematic represented as in Figure 1.

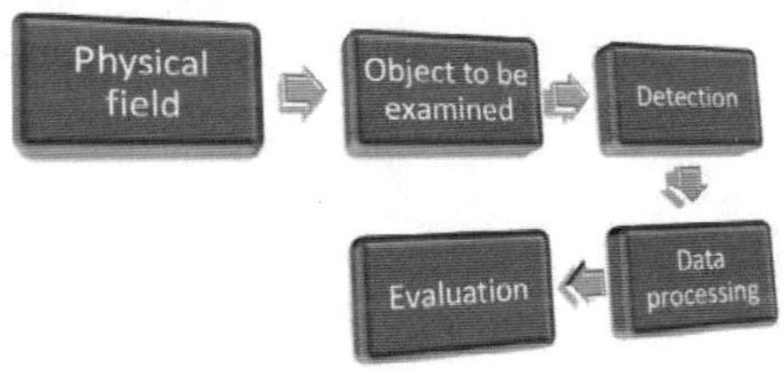

Figure 1. Principles of nondestructive evaluation operation

If the object to be examined is electrical conductivity, under the action of the incident EM field, eddy currents will be induced in the material, according Faraday's Law [1].
If the incident EM field has low-frequency, this is usually created by coils circulated with alternating current [2], [3] or current impulse [4] and the coils of different shapes assure the detection of the secondary EM field. It is necessary to use the smallest possible value of lift-off (distance between the EM transducer and the controlled piece) for obtain better signal/noise ratio [5]. This leads to the necessity of working in the near field due to the fact that the generated and scattered EM waves are evanescent waves (waves that are rapidly attenuated with distance [6]) and are difficult to be focalized using classical materials for sensors.
The interaction of the EM field with periodical metallic structures constitutes an interesting domain, both from theoretical point of view as well as experimentally [7],[8]. Metallic strip gratings (MSG) can act as filters and polarizers, representing also the basic elements in printed circuits boards, serving both to supply as well as to the transmission of the signals.
MSG presents a series of special properties when they interact with a transversal electric/magnetic along z axis (TE_z/TM_z) polarized EM field. Extending their study to the composite materials, carbon fiber reinforced plastics (CFRP) with uniaxial symmetry [9], as well in the case of reinforcement with carbon fibers woven [3], [10], the possibility to manipulate the evanescent waves that appears in the space between fibers indicate the

necessity to improve the spatial resolution of the sensors, in order to emphasize microscopic flaws.
The composite materials are involved in modern technologies due to the development of industries, from an engine valve to an aircraft wing. The advantages of fiber reinforced polymer composite (FRPC) compared with metallic materials are well known. FRPC are classified, in general, as advanced and present continuous fiber reinforcement (for example, carbon, glass or aramide) of high modulus or high strength embedded in a thermoset or thermoplastic polymeric matrix. Despite the fact that thermoplastic composites are more expensive, they have been preferred, due the advantages of thermoplastic matrix as recyclability, aesthetically finishing, high impact resistance, chemical resistant, hard crystalline or rubbery surface option, eco-friendly manufacturing, making the entire process costless. Woven carbon fibers are most suitable for applications requiring a high strength-to-weight ratio. Polyphenilene sulfide (PPS) has excellent properties [11], framing into the advantages described above. Woven carbon fibers/PPS laminates are characterized by reduced damages but are susceptible to impacts with low energies, leading to delamination's, desbonding of carbon fibers and/or matrix cracking. The aerospace industry has the requirement for the highest quality control and product release specifications [12]. The raw material standard, prepreg materials require mechanical property testing (i.e. interlaminar strength and tensile strength) from specimens. The presence of different types of defects such voids, inclusions, desbonding, improper cure and delamination are almost common during the manufacture and use of composite materials.
This paper proposes a possibility to enhance the spatial resolution of eNDE methods applied to MSG and to composite materials reinforced with carbon woven fibers, using a sensor with metamaterials (MM) lenses [13] with conical Swiss rolls, allowing the manipulation of evanescent waves created in slits and respectively in the dielectric insulating the carbon fibers and reaching a spatial resolution for visualization of carbon fibers layout and eventually flaws as delamination created by impact.

2. Metamaterial sensor for eNDE and theory

The MM, EM structures with distinguished properties have started to be studied especially in the last few years. EM MM belongs to the class of artificially engineered materials, can provide an engineered response to EM radiation that is not available from the class of naturally occurring materials. These are often defined as the structures of metallic and/or dielectric elements, periodically arranged in three or two dimensions [14]. The size of the structure is typically smaller than the free space wavelength of incoming EM waves. Nowadays, multitudes of MM structural elements type are known, conferring special EM properties. Depending on the frequency of the incident EM field, the type and geometrical shape MM may have a high relative magnetic permeability either positive or negative. Also, the MM lenses allow the evanescent waves amplification [15]. These properties strongly depend on the geometry of MM rather than their composition [16] and experimentally demonstrated [17], MMs have attracted intensive research interest from engineers and physicist in recent years because of their wide application in perfect lens [18], slow light [19], data storage [20], etc. All researchers reported until now, capitalize, in different manners the possibility to realize perfect lens.
In order to find the effective permittivity and permeability of a slab of MMs from the S parameters data, it must be characterized as an effective homogeneous slab.
The connection between ε and μ for a medium as well as the wave propagation through, can categorize the medium into the following classes [21]

a) Double positive medium (DSP) when $\varepsilon > 0$ and $\mu > 0$; Only propagating waves;

b) Single negative medium-electric negative (ENG), when $\varepsilon < 0$ and $\mu > 0$; Only evanescent waves;

c) Double negative medium (DNG) when $\varepsilon < 0$ and $\mu < 0$; Propagating waves and evanescent waves;

d) Single negative medium-magnetic negative (MNG), when $\varepsilon > 0$ and $\mu < 0$; Only propagating waves.

When the effective electrical permittivity ε_{eff}, and the effective magnetic permeability μ_{eff}, of a MM slab are simultaneously -1, the refractive index of the slab is $n=-1$ [22]. Therefore, the surface impedance of such MM is $Z=1$, such that there is no mismatch and consequently no reflection at the interface slab – air [17]. This MM slab forms a perfect lens [15] and is not only focusing the EM field, but also focuses the evanescent waves [15].

Due to experimental difficulties in obtaining a perfect lens, the manipulation of the evanescent modes can be made with an EM sensors with MM lenses that have, at the operation frequency, either ε_{eff}=-1and electric evanescent modes can be manipulated, either μ_{eff}=-1, and the lens can focus magnetic evanescent modes [14]. Moreover, working at frequency that assures $\mu_{eff} = -1$ for the same lens, the magnetic evanescent modes can be focalized [6],[13], [23].

According to the above classes, as shown in [9], the electric evanescent modes can be manipulated with a sensor made from a special MM, named Conical Swiss Roll (CSR) functioning in a frequency range that assure that μ_{eff} is maximum [13], that did not accomplish the conditions imposed for a perfect lens but shall lead to the substantially enhancement of spatial resolution. The spatial resolution of the system (=the distance between two distinctively visible points) was verified according to [13] and the analysis of data obtained shows that the realization of MM lenses in the RF range is possible using the CSR, whose distortions are minimal and whose calculation is made based on Fourier optic principles [24].

EM sensors with MM lenses have been made using two CSRs (Figure 2), the operation frequencies depending both by the constitutive parameters of MM as well as by the polarization of the incident EM field (TE_z or TM_z). CSR consists in a number of spiral-wound layers of an insulated conductor on a conical mandrel [13].

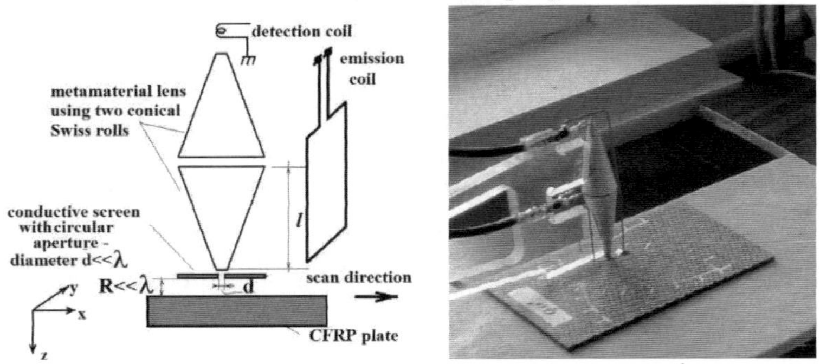

a b

Figure 2. Sensor with MM lens; a) schematic representation, b) photo

The EM sensor used is absolute send-receive type has the emission part made from one turn rectangular coil with 35x70mm dimension, using Cu wire with 1 mm diameter. When an EM TM_z polarized waves acts, at normal incidence, the magnetic field being parallel with the y axis such that $H_x=H_z=0$ and $H_y \neq 0$, in very near field, between carbon fibers or MSG's strips can appear evanescent waves. As shown in [21], the electrical evanescent modes can be manipulated with the lens realized with CSR, functioning in the range of frequencies so that

μ_{eff} is maximum. The reception coil of the sensor has one turn with 1 mm average diameter made of Cu wire with 0.1 mm diameter, placed in the focal image point, converting the localized energy into electromotive force (e.m.f.). An optimized work frequency of 476MHz assures a magnetic effective permeability of 22, verifying that CSR acts as an alternative magnetic flux concentrator.
The calculus of the MM lenses based on CSR was presented in [10], the field in the focal plane of the lens is given by [24]

$$H(x, y, z_0 + 2l) = \frac{H_0}{\pi^2} e^{j\left(\frac{k_b+k_a}{2}x + \frac{k_b+k_a}{2}y\right)} \frac{\sin\left(\frac{k_b - k_a}{2}x\right)}{x} \frac{\sin\left(\frac{k_b - k_a}{2}y\right)}{y} \quad (1)$$

where $z_0 = R$ (Figure 2a) with $R << \lambda$, H_0 is amplitude of incident magnetic field.
The spatial resolution of the sensor is enhanced by a conductive screen having a circular aperture with d = 100 μm diameter, made from perfect electric conductor (PEC), serving for the diffraction of the evanescent waves that can occur on slits. This assure paraxial incident beam. The diameter of focal spot provided by MM lens is given by [24]

$$D = \frac{4\pi}{k_b - k_a} \quad (2)$$

and is equal with the diameter of the small basis of the conical Swiss roll, *i.e.*, 3.2 mm. The MM lens with CSR will be displaced along the x-axis (Figure 2a). From this reason $k_a = 0$ and inserting this value in Equation (4), k_b is obtained and the field in focal plane is calculated. The detection principle is similar with the one of near-field EM scanning microscopy (NFESM). NFSEM imaging is a sampling technique, *i.e.*, the sample (in our case MSG or carbon fibers is raster scanned point by point with the sensor over the sample surface and energy image pixel a corresponding EM signature is recorded.
Selecting a region with dimensions (x_c+x_d) x (x_c+x_d), (where x_c and x_d are coordinates for conducting/dielectric material), using the Fourier optics methods [24],[25], an object *O(x,y)* that can represent the eigenmodes in function of the polarization of incident EM field, has, while passing through the circular aperture and the lens, an image *I(x',y')* given by [21].
From experimental point of view, in RF domain, the most convenient way is to measure S parameters for the MM that fill a waveguide or in free space, using an emission and a reception antennas [13]. The relation between S parameters (S_{11} and S_{21}) and applying the effective method [26] using a 4395A Network/Spectrum/Impedance Analyzer Agilent (Agilent Technologies, Santa Clara, CA, USA) coupled with an Agilent 87511A S Parameters Test kit and effective refractive index n is given by [27],[28].
The focal distance of the lens using a MM is [10] $f \cong l$ where l is the height of a CSR. Assuming that the passive MM slab has an effective refractive index n and Z impedance, according to [26] that effective permittivity ε_{eff} and permeability μ_{eff} are directly $\mu_{eff} = nZ$ and $\varepsilon_{eff} = n/Z$.

2. Studied samples and experimental set-up

The functioning of MM sensor has been verified using two types of materials metallic strip gratings (MSG) and fiber reinforced polymer composite (FRPC).

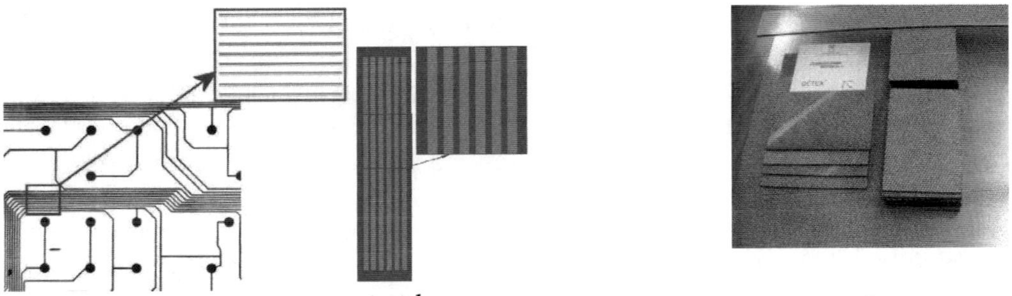

a b c

Figure 3. Studied samples: a) and b) MSG; c) FRPC

3.1. Metallic strip gratings (MSG) - we have into study two type of MSG:

- *flexible printed circuit* with transparent polyimide support, 80 μm thickness, with silver conductive strip of 10 μm thickness, the width of strips being x_c=0.6 mm and the width of slits being x_d=0.4 mm, and (Figure 3a)
- *silver strips realized* with polyimide support 65 μm thickness made from silver strip having 14 μm thickness, the width of strips being x_c=1.2 mm and width of slits being x_d=0.8 mm; are analyzed (Figure 3b).

For the MSG having the width of strips being x_c=0.6 mm, the study is focused on the appearance of abnormal and/or evanescent modes for the cases where various dielectric fluids fill the gaps between the strips, and also improvement the EM images so that we will obtains a better spatial resolution in order exceeded the limit imposed by diffraction.

For the strips having x_c=1.2 mm, we have taken into account the need for not presented interruptions as well as non-alignments. Once a metal coating has been applied to a surface, it is critical to determine the adhesion properties of the deposit. Silver strips were realized by successive deposition of silver paste, using an adequate stencil by screen printed method. The adhesion of silver on polyimide has been done with a thin film of resin. The conductive silver past made from microparticules with concentration >80%, density 10.49 g/cm^3, resistivity $1 \div 3 * 10^{-5}$ Ω cm was used (SPI supplies web site). Silver paste exhibit, in addition to a reasonably fast drying rate even at room temperature, good adhesion to most substrates and high conductivity. Around of 500 MHz, the permittivity of silver [29] is $\varepsilon_m = -48.8 + j \cdot 3.16$.

The use of polyimide implies a mechanically flexible substrate which can be easy manipulated into unique three-dimensional designs. So it provides an ideal surface for the selective attachment of various important bioactive species onto the device accomplished.

3.2. Plates from FRPC composite materials - the study involved quasi-isotropic FRPC plates made by Tencate [11] (Almelo, The Netherlands), having 150 × 100 × 4.2 mm^3, containing 12 layers of five harness satin carbon fibers woven with balanced woven fabric [30] (Figure 3c). The matrix is made of polyphenylene sulphide (PPS), a thermoplastic polymer consisting of aromatic rings linked with sulphide moieties, resistant to chemical and thermal attack, and the amount of gas released due to matrix ignition is substantially low. The carbon fibers are T300JB type and their volume is 0.5 ± 0.03 and the density is 1460 kg/m^3.

The composites carbon fiber woven – PPS have increased strength to impact. The FRPC has the transverse electric conductivity between 10 S/m and 100 S/m and longitudinal conductivity ranging between 5 × 10^3 S/m and 5 × 10^4 S/m and is paramagnetic allowing eNDE. The sample were impacted with energy of 8J using FRACTOVIS PLUS 9350-CEAST-Instron USA with a hemispherical bumper head with 20 mm diameter and 2.045 kg weight, in order to induce delamination's. Under impact, the FRPC suffer delamination that is usually accompanied by a dent, deviation and/or breaking of the carbon fibers. In all cases appears a reduction in the spacing between fibers in the thickness direction and this causes an increase in fiber contact leading to decrease of electrical resistance in the thickness direction and modifies the electrical conductivity local both in the plane of the fibers and perpendicularly on fibers [31]. The energy absorbed by the composite serves at the plastic deformation of the composite in the contact zone, being dissipated through internal friction between the matrix's molecules, carbon fibers, matrix-carbon fibers as well as at the creation of delamination's. Typical records of force *vs.* time during impact can give information about the FRPC status (delaminated or not).

The generation and detection of evanescent waves from slits/fibers has been made using a sensor with MM lenses [32] and the equipment shown in Figure 4. The distance between screen aperture and the surface to be examined has been maintained at 20±1 μm. The EM sensor with MM lenses, presented above is connected to a Network/Spectrum/Impedance Analyzer type 4395A Agilent USA. During the measurements, the transducer was fixed and the samples is mounted on a XY displacement system, type Newmark USA that assures the displacement in plan with ±10 μm precision and rotation with $\pm 2''$. That assures the scanning of sample with established steps in both directions.

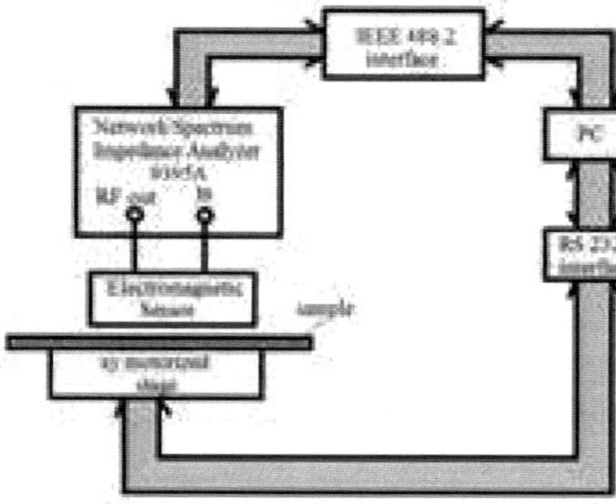

Figure 4. Experimental set-up

The measurement system is commanded by PC through RS232 interface for displacement system and IEEE 488.2 for analyzer. The acquisition and storage of data are made by software developed in Matlab 2012b. The e.m.f. induced in the reception coil of the measurement system is the average of 10 measurements in the same point in order to reduce the effects of the white noise, the bandwidth of the analyzer being set-up to 10 Hz, also to diminish the noise level.

The measurements of S parameters were carried out with Agilent S Parameters Test kit 87511A coupled to Agilent 41395 Analyzer.

3. Results and discussions

The EM sensor used in NDE must accomplish two roles:

- shall induce eddy current into the conductive material to be examined
- shall emphasize their flow modifications due to material degradation.

To emphasize the induced eddy current and effect of material degradation over their propagation, sensors sensitive to variation of magnetic field can be used, in our case sensor with MM lenses. The theoretical and experimental study of eigenmodes [13] that appear in this type of MSG, open new domains of applications in eNDE of stratified structures.

The method and the developed sensor can serve for eNDE of conductive strips, in order to detect eventually interruptions as well as non-alignments of silver strips or short-circuits between the traces. In the case of MSG with silver strips and sub-wavelength features, excited with TM_z polarized incident EM field, it has been shown that only a single evanescent mode appears in slits; this mode disappears when water is inserted in slits (ε_{water}=81) [32]. This mode could be detected and visualized using sensor described above. For MSG with features compatible with λ of incident EM field, TM_z polarized; the structure presents known selective properties of transmission and reflection [33]. When an EM TM_z polarized waves acts, at normal incidence, for MSG with $x_0 << \lambda_0$, according to [34] the reflection coefficient is practically 1.

The apparition of evanescent and abnormal modes in the space between MSG's strips is experimentally confirmed at TM_z polarized EM wave with λ=0.6m.

The images present an increasing of the amplitude of the signal induced in the reception coil of the evanescent and abnormal modes created in slits at the scanning of 1×1 mm^2 region with 10 μm step on both directions. When the slits are filled with air, only evanescent mode will be generated, the amplitude of the signal induced in the reception coil has the shape presented in Figure 5a. In central zone of the slits the amplitude is maximum followed by an accentuated decreasing towards the flanks of metallic strips. The existence of a single evanescent mode, theoretical foreseen [3] is experimentally confirmed by the existence of a local maximum in the middle zone of the slits, with maximum amplitude on the middle of the slits, followed by an accentuated decreasing, symmetrically on the flanks of the strip. The two secondary maxima with smaller amplitude are at ±42μm from vertical wall of the strip.

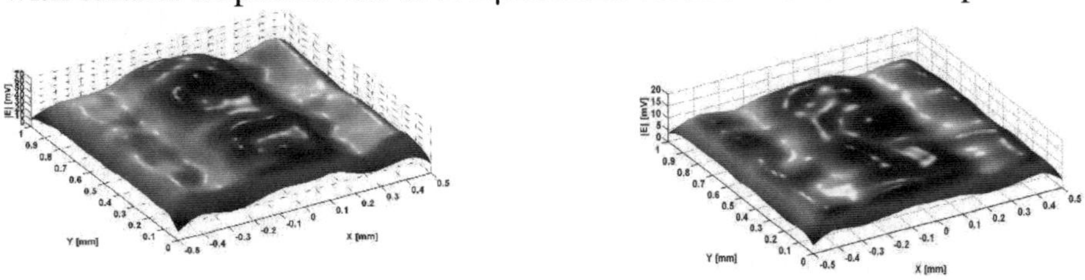

a b

Figure 5. Image of evanescent and abnormal modes generated in slits for TMz polarized excitation at frequency of 476 MHz: (a) slits filled with air; (b) slits filled with isopropyl alcohol.

In slits are generated abnormal modes, in the case of excitation with TM_z polarized wave, for large values of the liquid dielectric constants larger than 10, when ε_d = 17.9, (isopropyl alcohol (Figure 5 b). Because the real component of the propagation constant β_v for isopropyl alcohol is smaller (β_v=4.2 + i·104.321) than the imaginary component, abnormal modes will be generated in slits, the EM image of these modes shows a similar behavior like in the case of air in the slits. The amplitude of the signal is smaller and has a central maximum more flat (Figure 5b). This opens a large perpective for the use of MSG in sub-subwavelength regime as a sensors (including biosensors based on the evanescent modes generated in slits and extremely low frequency plasmons).

In order to obtain silver strip with different widths and thickness in range of micrometers, different masks/stencil with different microstrip widths were used. Once a metal coating has been applied to a surface, it is critical to determine the adhesion properties of the deposit. The presence of the low frequency, of evanescent waves in slits [3], having maxima of the field in the interior of the trace, opens the perspective of using the layered mesostructures in sub-wavelength regime as sensors and biosensors.

The case of a MSG made from silver strip having 14 μm thickness, the width of strips being x_c=1.2 mm and width of slits being x_d=0.8 mm is analyzed, considering that the wavelength of incident field is λ=0.6 m (corresponding to 500 MHz fequency).

A region of 16 × 16 mm^2 from MSG has been scanned with 0.1 mm steps in both directions. The scanning along x direction was made so that shall correspond for few periods of gratings, $x_0=x_c+x_d$. The working frequency was 476 MHz.

Figure 6 shows the dependency of e.m.f. amplitude induced in the reception coil of sensor at the scanning of the MSG taken into study, the image showing that type of sensor allows the correct emphasizing of 14 μm thickness conductive strips and eventual interruptions.

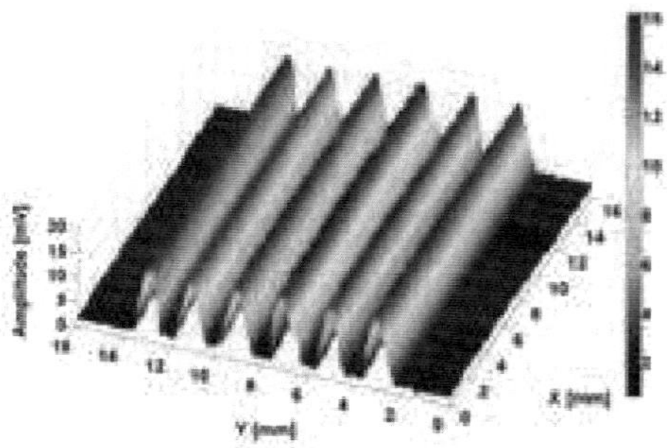

Figure 6 Amplitude of e.m.f. induced in the reception coil at the scanning of silver strip grating

The results are in good concordance with theoretical estimations, that confirms good adhesion silver paste on polyimide, also good alignments of strips. It is known that biosensing characteristic is strongly dependent on the deposition condition that affects the physical properties of the thin film. Using the procedure and the sensor described earlier, for eNDE of MSG, an interruption of a strip stop the propagation of evanescent waves in the nearest slit; so that the amplitude of e.m.f induced in the reception coil practically decreases to zero when the aperture's sensor is in the respective region of slits. The roughness feature is emphasized by the propagation of surface polaritons [35]. It is known that bio sensing characteristic is strongly dependent on the deposition condition that affects the physical properties of the thin film.

For the second type of material taken into study (FRPC), the EM behavior of composite was simulated by FDTD software, the samples being CAD designed following textiles features, and compared with eNDE tests. One cell of carbon fiber woven has been designed in TexGen -Textile Geometric modeler software (Figure 7a) and exported as CAD format in order to be used in FDTD software - XFDTD Remcom USA [36]. The 5x5 tows were represented in different colors in order to easily follow their intersection as woven. Cell dimension in FDTD simulation is 0.04 mm, the grid size being 146 x 146 x 45 cells. The perfectly matched layer boundary condition was applied at the grid boundaries.

The result of the simulation presented in Figure 7b is a snapshot from a field sequence, showing the H_y field progress at a particular slice of the geometry. In this case, the roll of conductive strips is taken by carbon fibers, which act as MSG [13] and the apparition of evanescent waves can be emphasized.

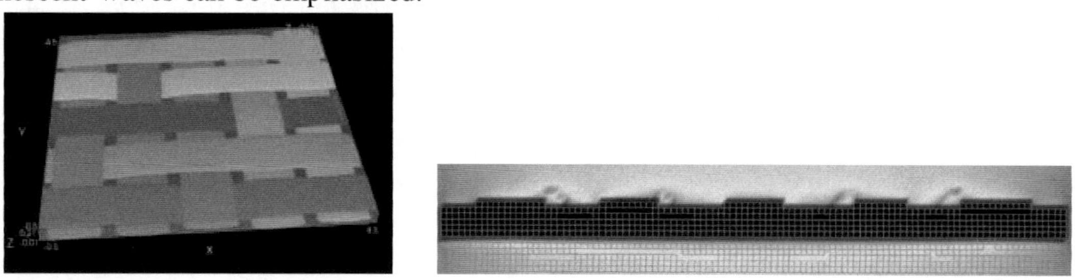

a b

Figure 7. One cell carbon woven simulation: a) TexGen – different colors show the tows intersections in the 5HS woven; b) XFDTD – H_y propagation in a plane orthogonal at composite

The detection of eventual delamination or the characterization of carbon fibers woven structure using eNDE method and the sensors with MM lenses is an emerging ND technique which combines the advantages of conventional eddy current testing and evanescent waves detection giving a higher resolution than classical eddy current. Figure 8 presents a scan with 1 mm step in both directions of a 60 x 60 mm^2 region from the FRPC sample impacted with 8J .

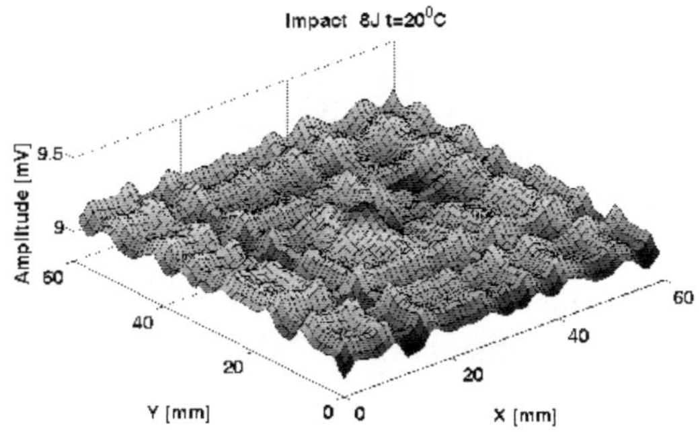

Figure 8. The answer of the sensor with MM lenses for 8 J impact

It can be shown that the MM lenses allow the enhanced of spatial resolution, the layout of the woven being emphasized. The proposed method can be thus extended not only for the evaluation of MSG but also for the eNDE of FRPC in order to evaluate delaminations as well as the woven layout.

4. Conclusions

Monitoring the complex structures as FRPC and MSG has become a requirement, the trend consisting in using of special sensors with high performances. The special properties of EM MM recommend their use in the eNDE methods.
The structures conductive grating allow, by extension, the estimation of results from eNDE of few real situation as evaluation of metallic strips in printed circuit, metallic strip grating in sub-wavelength regime as a biosensor and FRPC.
The use of evanescent waves and sensors with MM lenses allows the manipulation of evanescent waves can reach a spatial resolution, by approximately of $\lambda/2000$.

Author Contributions

All authors have equal contributions to this paper.

Conflicts of Interest

The authors declare no conflict of interest.

Acknowledgements

This paper is partially supported by a grant of the Romanian Ministry of Education, Research and Innovation CNCS – UEFISCDI, project number PN-II-ID-PCE-2012-4-0437.

References

1. V Bladel , 'Electromagnetic fields', 2nd Ed., IEEE Press, 2007.
2. T P Theodoulidis, E Kriezis, 'Impedance evaluation of rectangular coils for eddy current testing of planar media', NDT&E International, Vol 35, No 5, pp 407-414, 2002.
3. R Grimberg, A Savin, R Steigmann, 'Electromagnetic imaging using evanescent waves', NDT&E International, 46, pp 70-76, 2012.
4. L Wang, S Xie, Z Chen, Y Li, X Wang, T Takagi, 'Reconstruction of stress corrosion cracks using signals of pulsed eddy current testing', Vol 28, No 2, pp 145-154, 2013.
5. B Stephen Wong, Non-Destructive testing-Theory, 'Practice and Industrial applications', Lap-Lambert Publishing GmbH KG, 2014.

6. A Gribic, G V Eleftheriades, ‘Growing evanescent waves in negative refractive index transmission line media’, Applied Physics Letters, Vol 82, No 12, pp 1815-1817, 2003.
7. R Petit, L C Botten, ‘Electromagnetic Theory of Gratings’, Springer-Verlag, 1980.
8. C A Balanis, ‘Antenna Theory: Analysis and Design’, 3rd Edition, ed.J.Wiley, NY, 2005.
9. R Grimberg, ‘Electromagnetic metamaterials’, Mater. Sci. Eng. B, 178, pp 1285–12, 2013.
10. R Grimberg, A Savin, R Steigmann, B Serghiac, A Bruma, ‘Electromagnetic non-destructive evaluation using metamaterials’, INSIGHT, Vol 53. No 3, pp 132-137, 2011.
11. www.tencate.com
12. Standard AS 9100/2009, ‘Quality sistems –aerospace-model for quality assurance in design , development, production, installation and servicing’
13. A Savin, R Steigmann, A Bruma, R Sturm, ‘An electromagnetic sensors with a metamaterial lens for nondestructive evaluation of composite materials’, Sensors, 15, pp 15903-15920, 2015.
14. J Pendry, A J Holden, D J Robbins, W J Stewart, ‘Magnetism from conductors and enhanced non-linear phenomena’, IEEE Trans.Microw Theory Tech., 47, pp 47-58, 1999.
15. J B Pendry, ‘Negative Refraction makes a perfect lens’, Phys.Rev.Lett., Vol 85, No 18, pp. 966-3969, 2000.
16. W Cai, U K Chettiar, A V Kildishev, V M Shalaev ‘Optical cloaking with metamaterials’ Nature Photonics, Vol 1, No 4, pp 224-227, 2007.
17. D R Smith, J B Pendry, M C K Wiltshire, ‘Metamaterials and negative refractive index’ Science, 305, pp 788–792, 2004.
18. R A Shelby, D R Smith, S Schultz, ‘Experimental Verification of a Negative Index of Refraction’ Science, 6, 292, pp 77-79, 2001.
19. Q Bai, C Liu, J Chen, C Cheng, M Kang, ‘Tunable slow light in semiconductor metamaterial in a broad terahertz regime’, J. Appl. Phys., 107, 093104:1–093104:8, 2010.
20. M Wuttig, N Yamada, ‘Phase-change materials for rewriteable data storage’, Nat. Mater., 6, pp 824–832, 2007.
21. N Engheta, R W Ziolkowski, ‘Electromagnetic Metamaterials: Physics and Engineering Explorations’, Wiley. NY, 2006.
22. V G Veselago, ‘The electrodynamics of substances with simultaneously negative values of ϵ and μ’, Physics-Uspekhi, Vol 10, No 4, pp 509-514, 1968.
23. R Grimberg, G Y Tian, ‘High Frequency Electromagnetic Non-destructive Evaluation for High Spatial Resolution using Metamaterial’ Proc. R. Soc. A, 468, pp 3080–3099, 2012.
24. M Born, E Wolf, ‘Principle of Optics, 5th ed. Pergamon Press, Oxford, UK, 1975.
25. J W Goodman, ‘Introduction to Fourier Optics’, 3rd ed.; Roberts and Company, Englewood, CO, USA, 2005.
26. J A Kong, ‘Electromagnetic Wave Theory’, EMW, Cambridge, MA, 2000.
27. R A Shelby, D R Smith, S C Nemat-Nasser, S Schultz, ‘Microwave transmission through a two-dimensional, isotropic, left-handed metamaterial’, Appl. Phys. Lett. 78, pp 489–491, 2001.
28. X Chen, T M Grzegorczyk, B I Wu, J Jr Pacheco, J A Kong, ‘Robust method to retrieve the constitutive effective parameters of metamaterials’ Phys. Rev. E, 70, 016608, 2004.
29. E D Palik, ‘Handbook of Optical Constants of Solids’; Academic Press: London, UK, 1985.
30. R Akkerman, ‘Laminate mechanics for balanced woven fabrics’, Compos. Part B Eng. 37, pp 108–116, 2006.
31. H Menana, M Féliachi, ‘Electromagnetic characterization of the CFRPs anisotropic conductivity: Modeling and measurements’ Eur. Phys. J. Appl. Phys., 53, 2011.

32. A Savin, R Steigmann, A Bruma, 'Metallic Strip Gratings in the Sub-subwavelength Regime', Sensors, Vol 14, No 7, pp 11786-11804, 2014.
33. C Balanis, 'Advanced Engineering Electromagnetics', Wiley Eds, NY, 1989.
34. A E Peterson, S L Ray, R Mittro, 'Computational Methods for Electromagnetics', IEEE Press, NY, 1989.
35. A Kolomenski, A., Kolomenskii, J Noel, S Peng, H Schuessler, 'Propagation length of surface plasmons in a metal film with roughness'. Applied Optics, 48(30), pp 5683-5691, 2009.
36. K S Kunz, R J Luebbers, 'The Finite Difference Time Domain Method for Electromagnetics', Boca Raton, FL: CRC Press, 1993.

OLYMPUS

Your Vision, Our Future

SCREW AND CRITICAL PART INSPECTION BY EDDY CURRENT ARRAY.

PATRICK CABANIS

Senior Product & Application Specialist , NDT Level 3.

OLYMPUS INDUSTRIAL SYSTEMS EUROPA

patrick.cabanis@olympus-europa.com

tel : 33 1 45 60 68 40

mobile :+ 33 6 77 88 46 00

R/D TECH PANAMETRICS-NDT™ NDT engineering corporation NORTEC SONIC

Presentation plan

- What is ECA
- Critical parts inspections
- SCREW applications

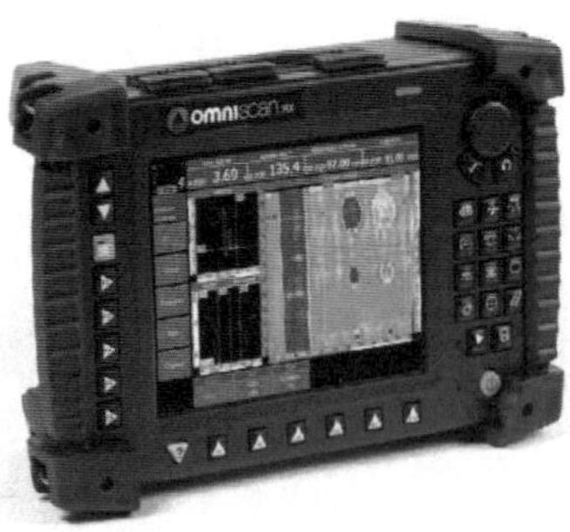

Omniscan ECA

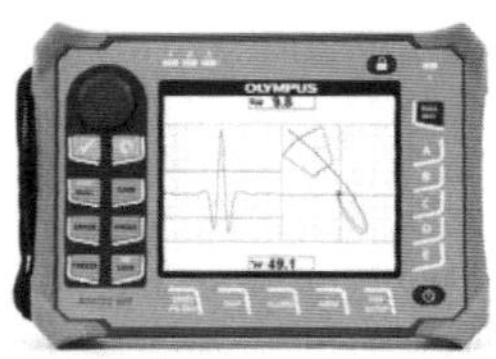

Nortec N600

2

History

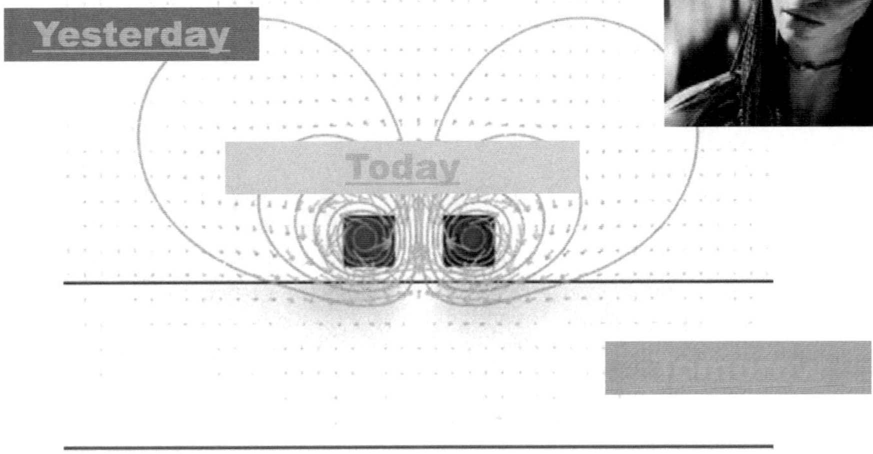

Historique théorie

$$e = |\Delta\Phi| / |\Delta t| \quad \text{or} \quad \Phi = \mathrm{Li} \quad \text{donc} \quad |\Delta\Phi| = L\,|\Delta i|$$

$$J_x = J_0 e^{-\left(x\sqrt{\pi\mu\sigma f}\right)} \sin\left(2\pi f t - x\sqrt{\pi\mu\sigma f}\right)$$

$$\theta = 57\frac{x}{\delta} \qquad J = 100\cdot\exp\left(-x/\delta\right) \qquad \delta = 50\sqrt{\frac{\rho}{f\mu_r}}$$

$$U^2 = R^2 I^2 + L^2\omega^2 I^2 = (R^2 + L^2\omega^2) I^2$$

D'où $\quad U = \mathrm{I}\sqrt{\mathrm{R}^2 + \mathrm{L}^2\omega^2}$

Donc $\quad Z = U / I = \sqrt{\mathrm{R}^2 + \mathrm{L}^2\omega^2}$

Eddy Current Array – Basic Concept

- In ECA, several eddy current coils are placed in the same probe assembly to:
  - Allows larger coverage in a single probe pass while maintaining a high resolution.
  - Reduces the need for complex robotics to move the probe; a simple manual scan is often enough.
  - Improves flaw detection and sizing with C-scan imaging.
  - Inspect complex shape with a probe made with to the profile of the parts

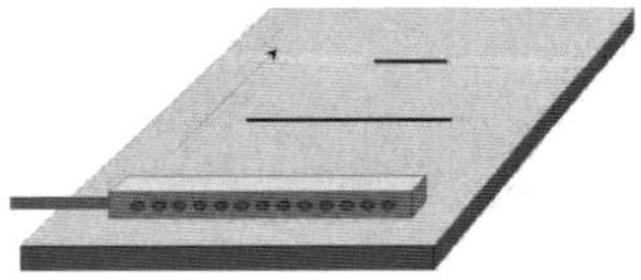

What is Eddy Current Array ?

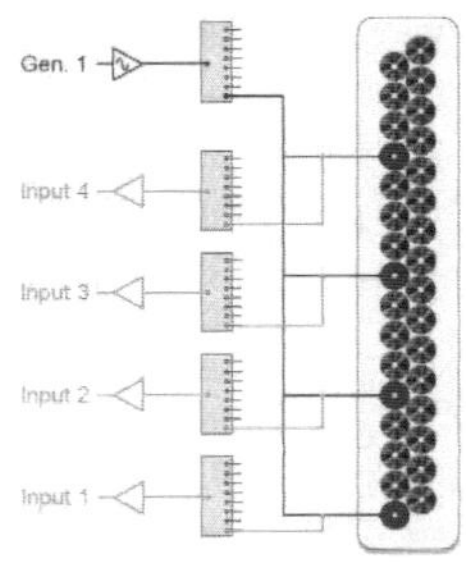

- Eddy current array technology is the ability to ***electronically drive several eddy current sensors*** .
- Data acquisition is performed by ***multiplexing the eddy current sensors*** .
- Most eddy current probes and techniques for flaw detection can be used.

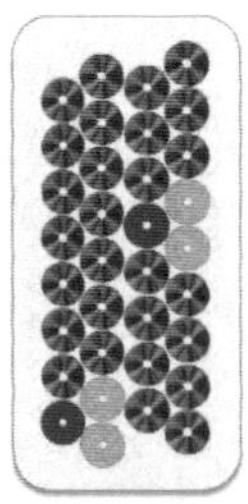

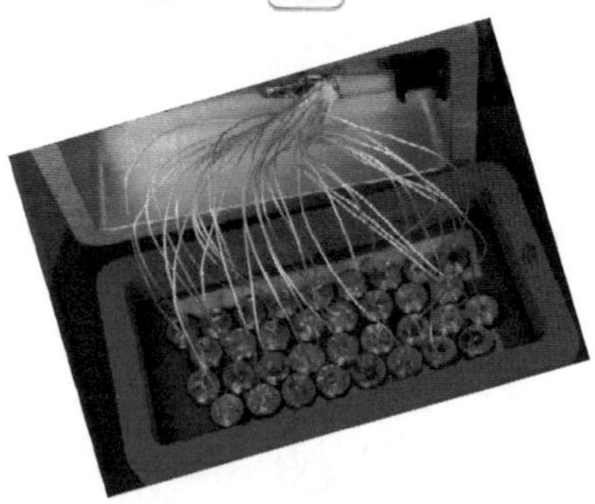

OLYMPUS Innovation in NDT™

Equipment

1CH	BNC connector used for one single-coil eddy current probe using conventional eddy current technique.
4CH/MUX	19-pin connector used for eddy current probes that have up to four single coils.
EC ARRAY	R/D Tech® connector used for eddy current array inspection technique.

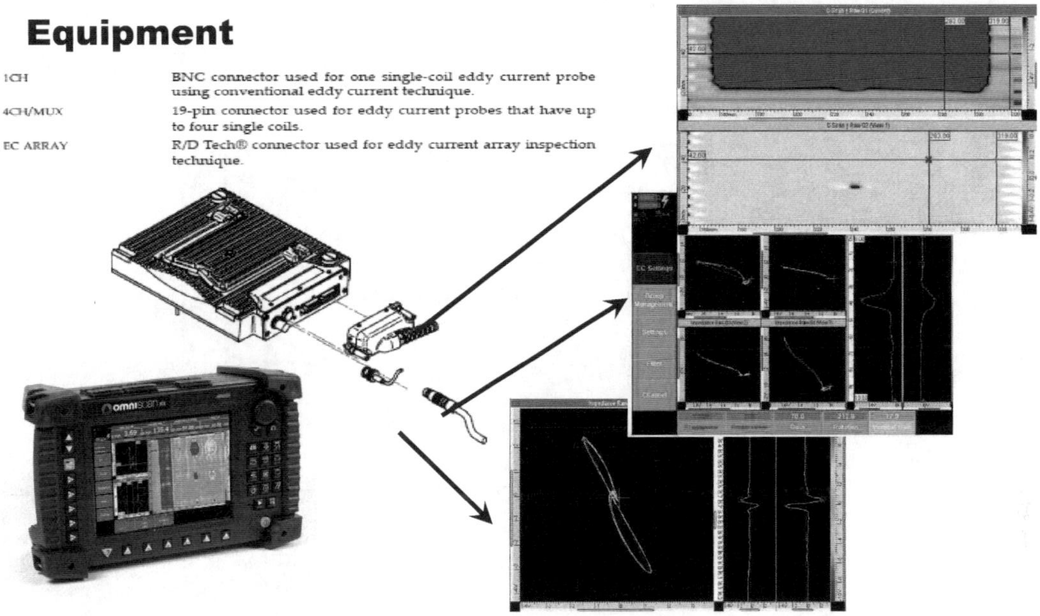

7

OLYMPUS Innovation in NDT™

ECA – Signal Representation

- The C-scan can only show one component of the signal at a time (X or Y component).
- In the example below, the C-scan shows the vertical component of the signal (Y axis) and this is why the lift-off, which has been rotated horizontally, is almost not visible.

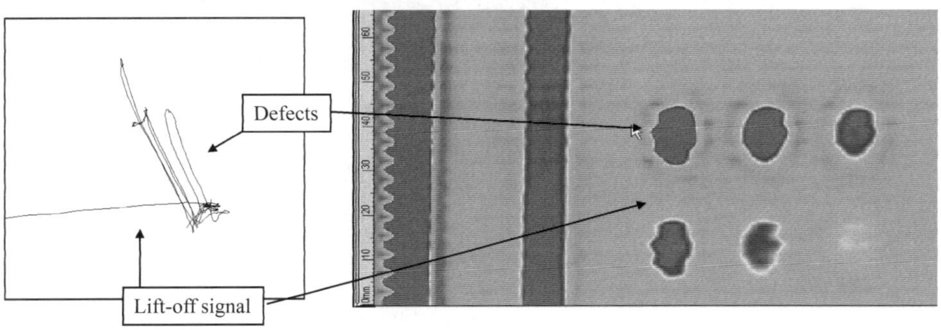

This document and the information contained are OLYMPUS France OISE property and shall not be copied or disclosed to any third party without OLYMPUS France OISE prior written authorization

8

ECA Prepare the Future

OLYMPUS Innovation in NDT™

ECA industrial application

- **Dovetail**
- Aircraft Corrosion
- **Gaz Turbine**
- Tubes inspection
- **Pipe**
- Rails
- Train axis
- **Doubler edge**
- FSW (weld)
- Blade
- **SCREW**

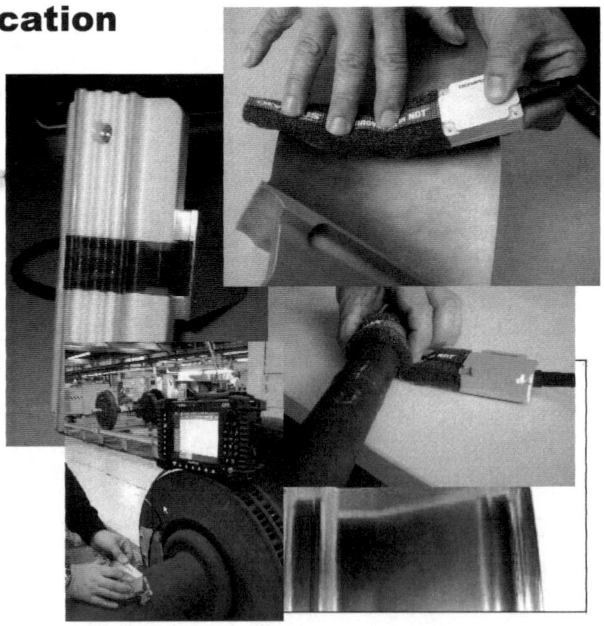

This document and the information contained are OLYMPUS France OISE property and shall not be copied or disclosed to any third party without OLYMPUS France OISE prior written authorization

OLYMPUS Innovation in NDT™

Engine disk dovetail slot Inspection

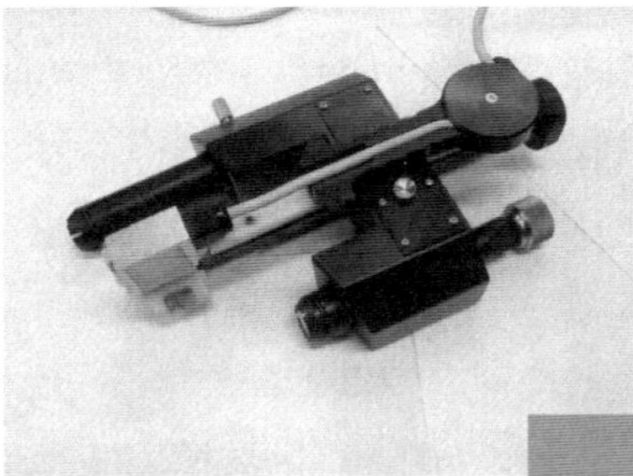

movie

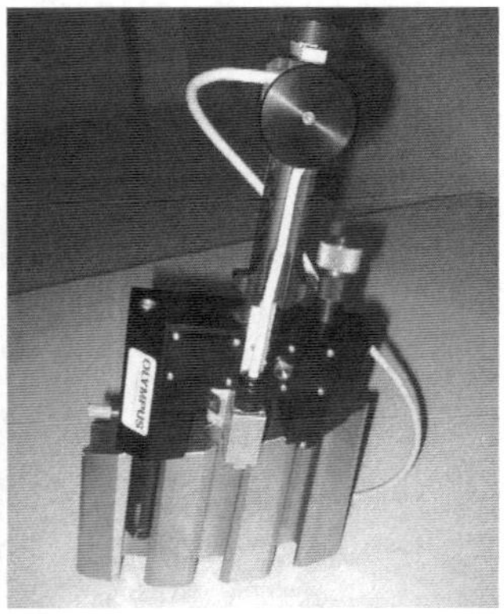

- **ECA probe**
- **32 elements (2 x 16)**

OLYMPUS Innovation in NDT™

ECA HF Inspection method

Inspection in the maintenance shop

This document and the information contained are OLYMPUS France OISE property and shall not be copied or disclosed to any third party without OLYMPUS France OISE prior written authorization

OLYMPUS Innovation in NDT™

ECA - Flexible Probe

Flex probe FBB-051-500-032
Frequency 50KHz to 4MHz

This document and the information contained are OLYMPUS France OISE property and shall not be copied or disclosed to any third party without OLYMPUS France OISE prior written authorization

13

OLYMPUS Innovation in NDT™

ECA - Flexible Probe

- **Two mode**

Absolut
Need a encoder

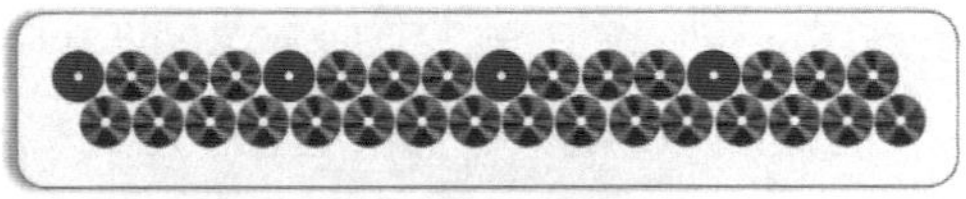

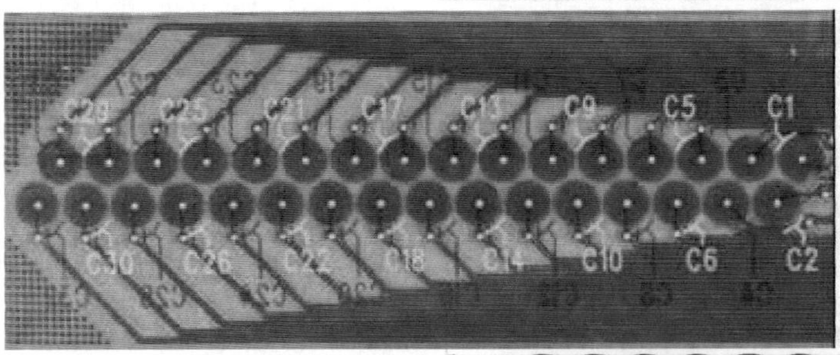

Reflection

This document and the information contained are OLYMPUS France OISE property and shall not be copied or disclosed to any third party without OLYMPUS France OISE prior written authorization

14

OLYMPUS Innovation in NDT™

Gas turbin ECA Flexible probe and holder

- Special probe holder
- Flex probe 32 elements

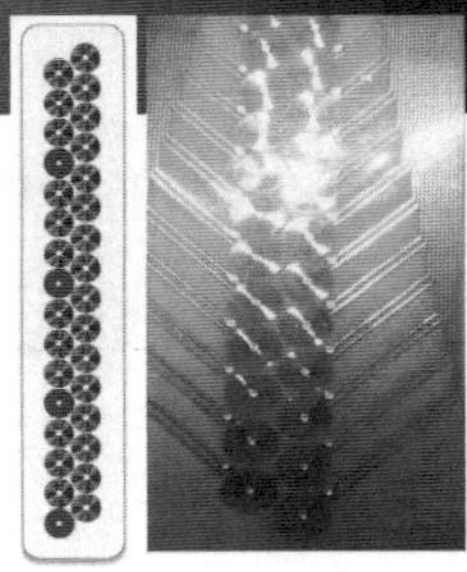

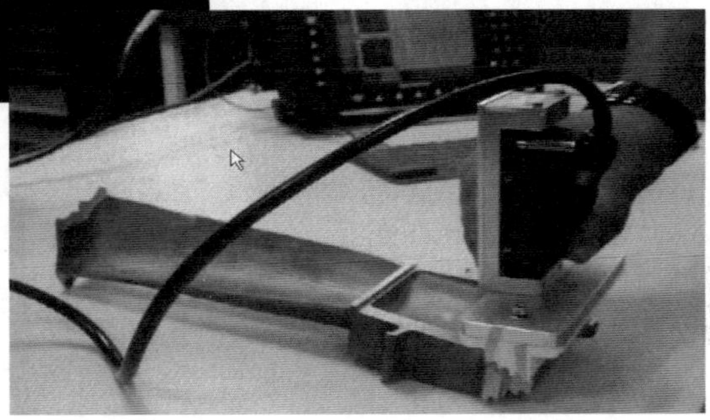

15

OLYMPUS Innovation in NDT™

Gaz turbin

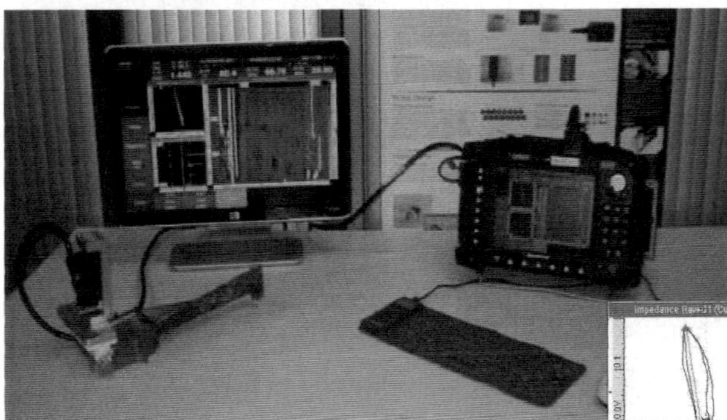

•One pass inspection
•Notch detection from 0.06 to 1 in to the edge
•Reliable inspection (100% full coverage)
•Easy interpretation on Cscan
•No need for encoder

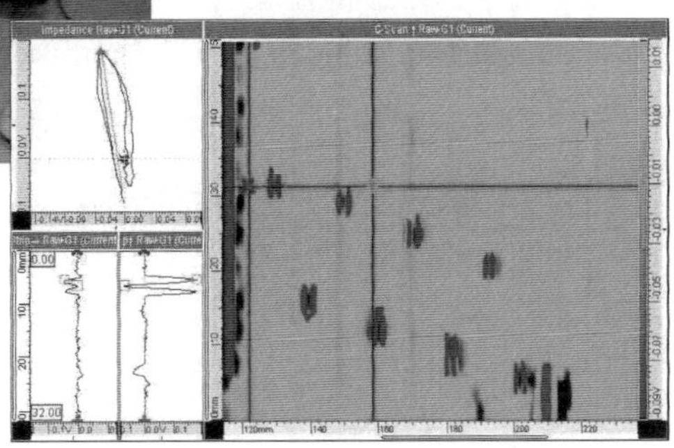

16

TUBE INSPECTION

- Reference standard
- Banc

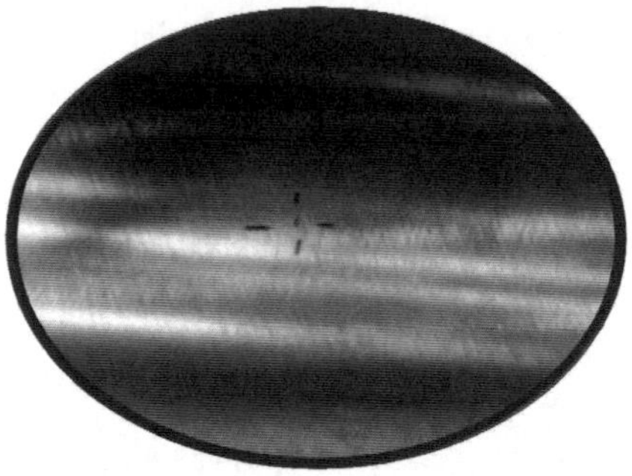

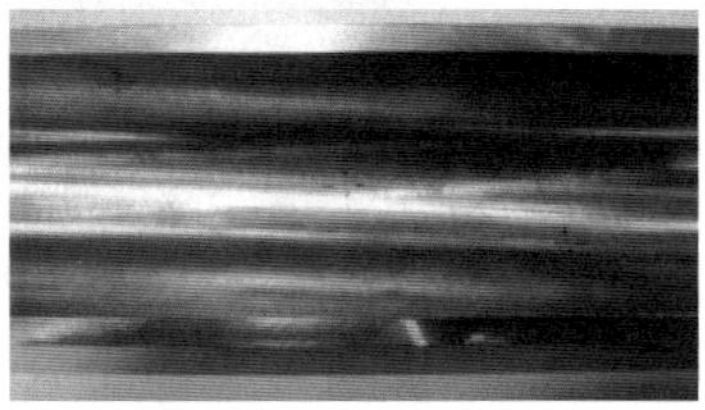

TUBE INSPECTION

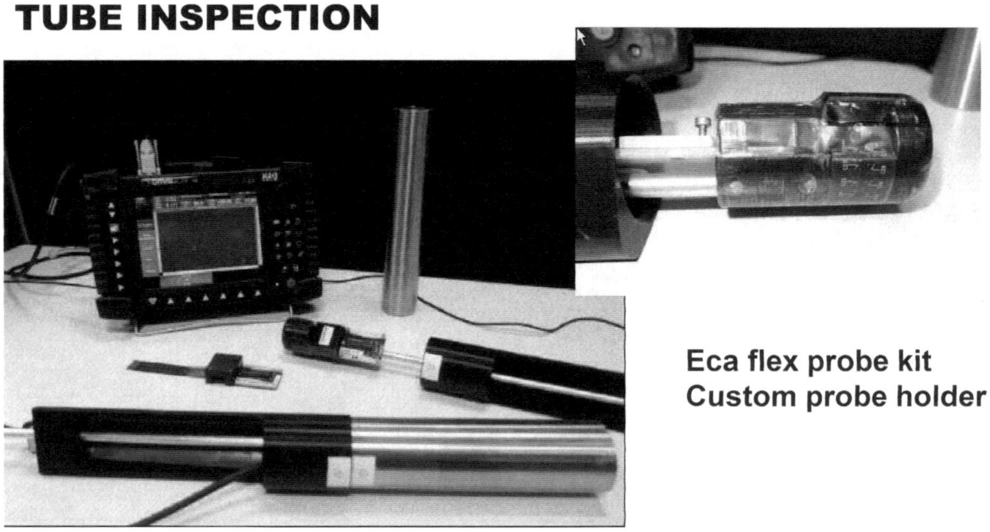

Eca flex probe kit
Custom probe holder

Method of ECA High-frequency inspection of the lower skin for scribe mark

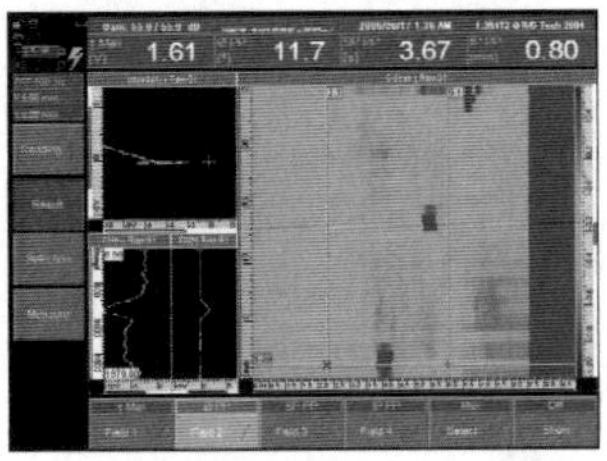

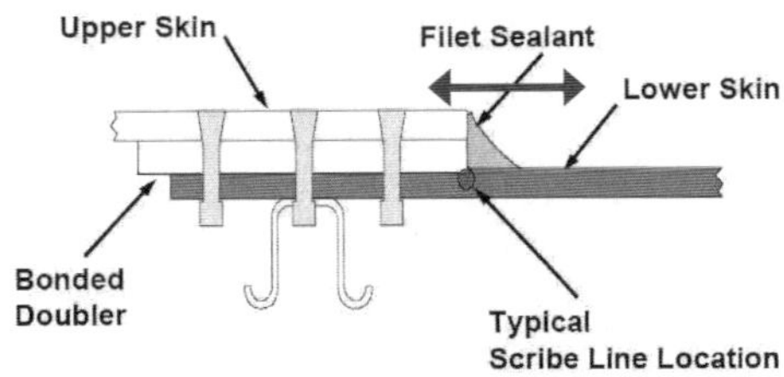

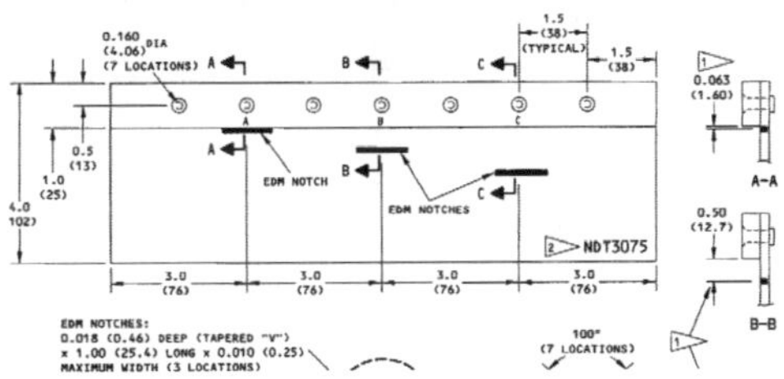

0.06 in to 1 in to the edge

Method of ECA high-frequency inspection of the lower skin for scribe mark

- **One pass inspection**
- **Notch detection from 0.06 to 1 in to the edge**
- **Reliable inspection (100% full coverage)**
- **Easy interpretation on Cscan**
- **No need for encoder**
- **General procedure**

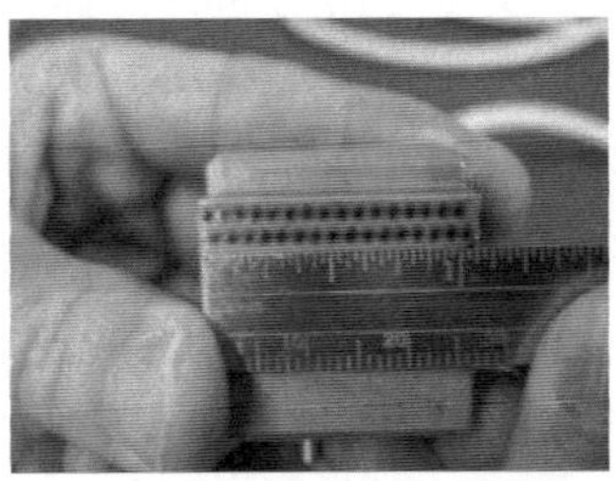

film

BOEING 757 LAP SPLICE C-SCAN INSPECTION

- inspection procedure to Part 6, 53-30-06.
- Has at least a 32 channel probe head that can scan an area greater than 0.87 inch (22 mm) but less than 1.5 inches (37 mm) in width.
- Probe SBBR-026-300-032
- Operates at a frequency range of 200 kHz to 400 kHz.
- Has a linear position encoder.
- Has a C-Scan display mode

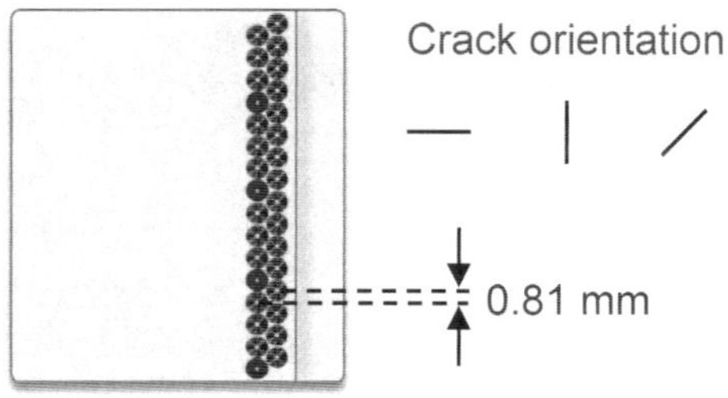

BOEING 757 LAP SPLICE C-SCAN INSPECTION
Calibration Sample

NDT 588
5/32 and 6/32 rivet hole

Typical cross-section

EDM notch:
length .1 in from rivet shank
Thickness: through 1st skin

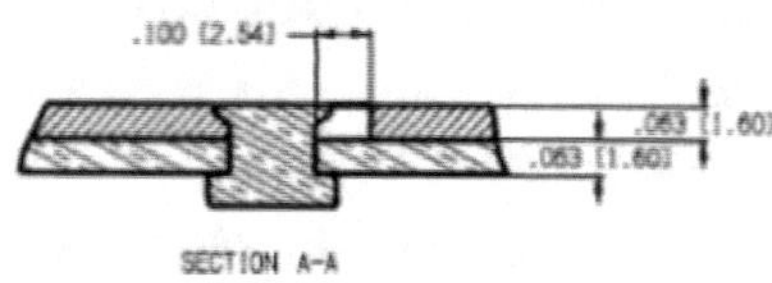

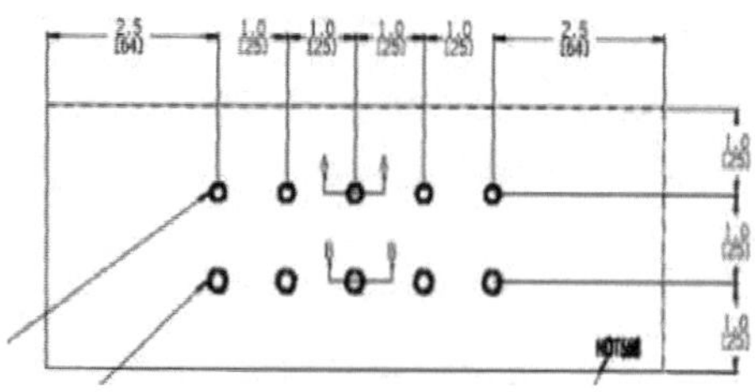

OLYMPUS Innovation in NDT

Results

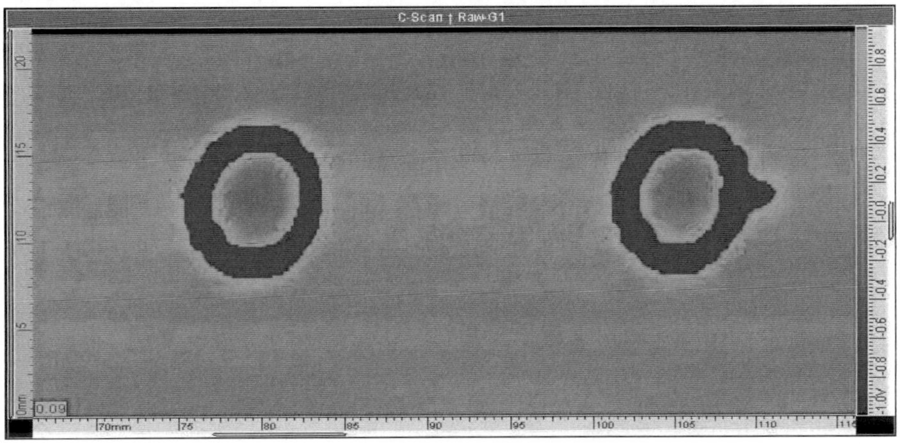

Good rivet hole Rivet hole with notch

OLYMPUS Innovation in NDT

ECA – Pipe inspection

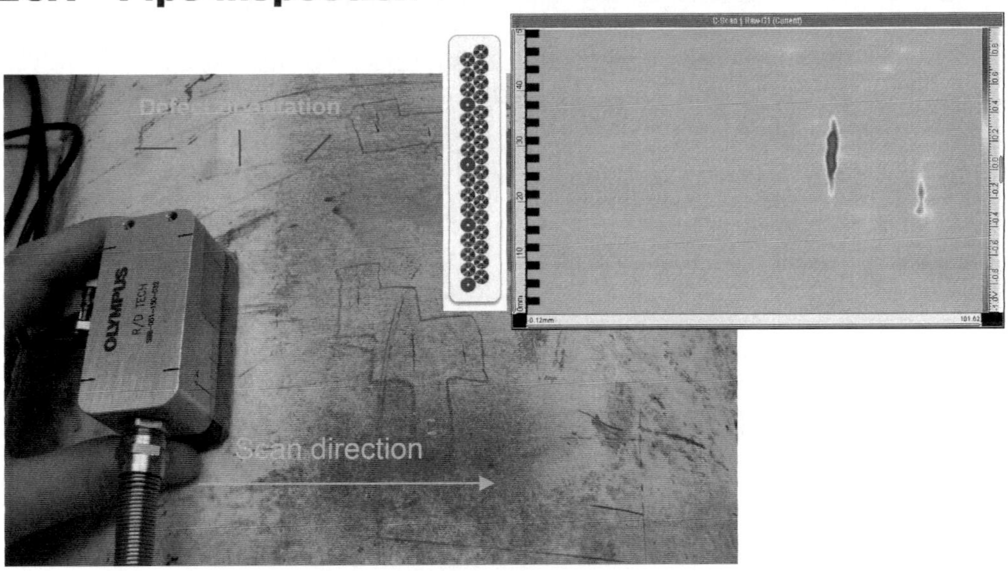

Low-frequency EC array application:
Boeing 737 Skin crack detection at doubler edge

- Probe SAB-067-005-032 referenced in the NTM737 NDT 53-30-25 part 6, dec 2004.
- Description of the application:
  - Shear and compression loading cause cracking at the doubler edge.
  - Cracks need to be detected at their initial stage during the "C" Check,
  - If the crack reach the surface, the aircraft must be removed from service

Boeing 737 Skin crack detection at doubler edge

- The inspection is done from the outside and crack as small as 0.240" long by 0.010" deep located at the edge of the doubler need to be detected.
- Benefits:
  - Simple manual inspection
  - C-Scan allows easy location of the doubler edge for fast and simple detection of the initiating cracks
  - Better reliability
  - Better reproducibility
  - Time saving:
    - » Normal time: 200 hours
    - » With ECA: 48 hour

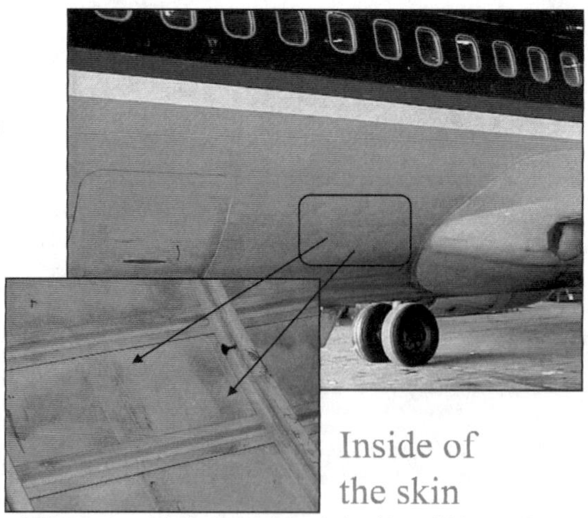

Inside of the skin

Boeing 737 Skin crack detection at doubler edge

- The user can see very well the doubler edge represented by the light to dark green color transition.

- Fastener will show up in light green.

- Defect above the rejection level are in red like shown in this picture

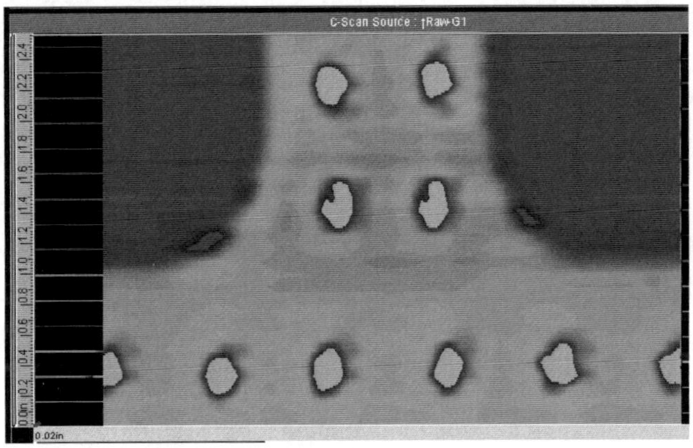

Weld Inspection

OLYMPUS Innovation in NDT™

Notches identification on the C-Scan view

length: 6mm, depth: 1,2mm, width: 0.2mm

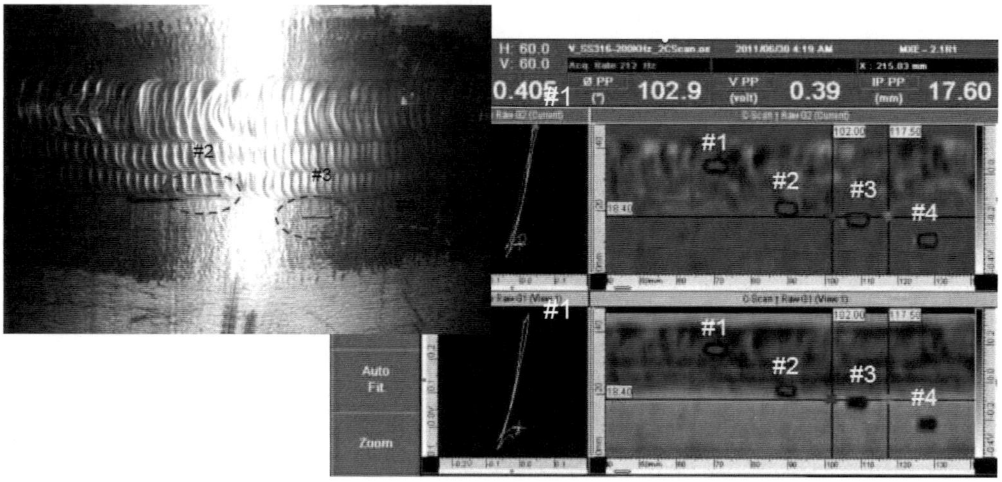

This document and the information contained are OLYMPUS France OISE property and shall not be copied or disclosed to any third party without OLYMPUS France OISE prior written authorization

29

OLYMPUS Innovation in NDT™

SCREW INSPECTION

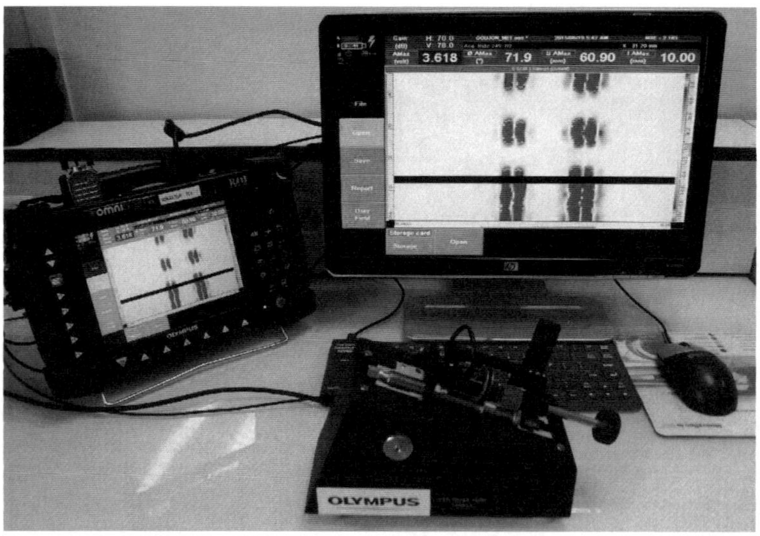

This document and the information contained are OLYMPUS France OISE property and shall not be copied or disclosed to any third party without OLYMPUS France OISE prior written authorization

30

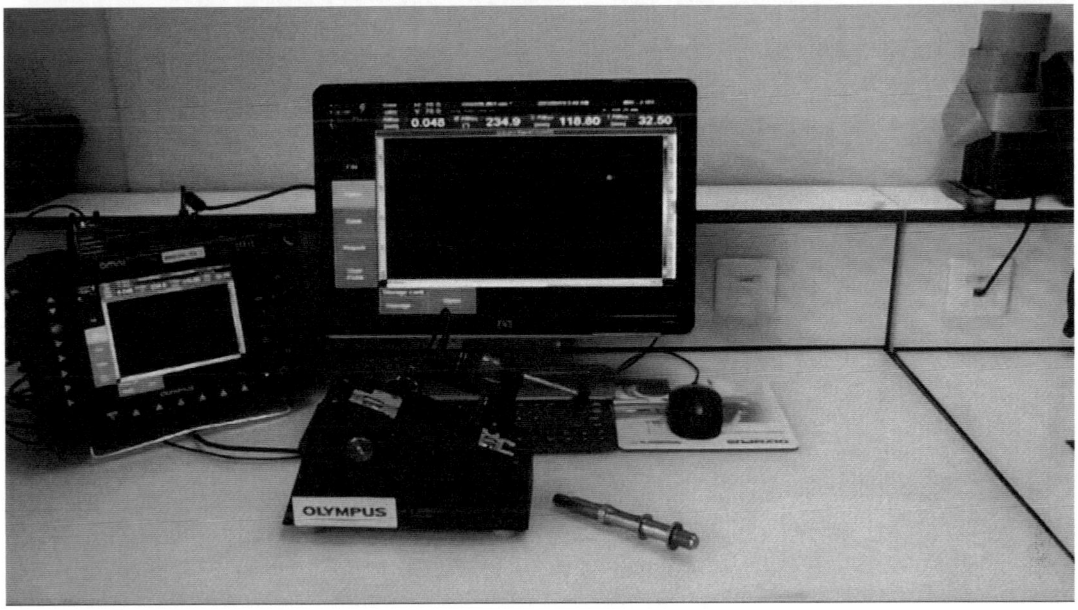

31

SCREW INSPECTION

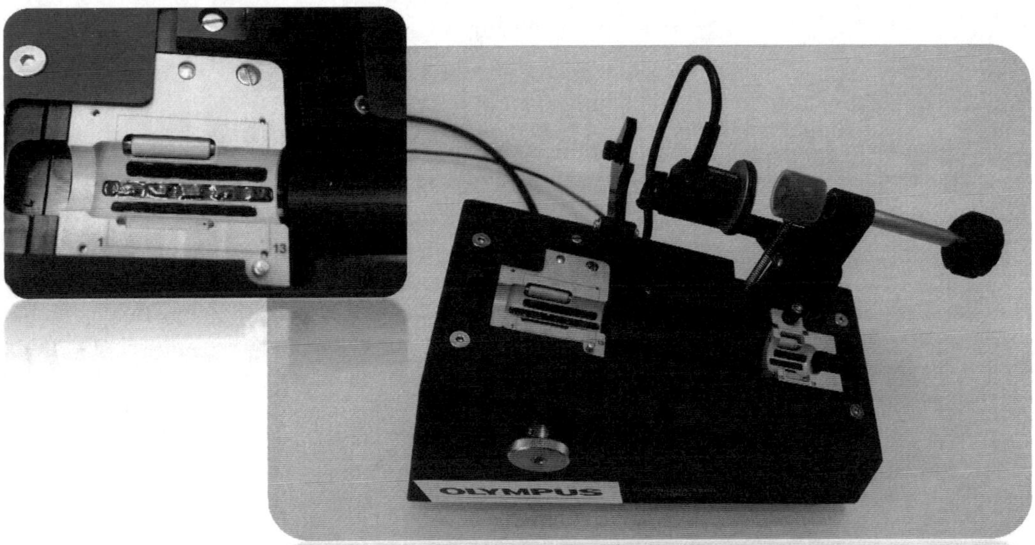

32

Conclusions about ECA

- Deployed in the field for many applications
- Referenced in procedures for aircraft and engine maintenance
- Fast (large area)
- Reliable (Full coverage and Cscan imaging)
- Many ECA standard probes
- Omniscan ECA module compatible with Bondmaster C-Scan.
- More to come…

OLYMPUS

Your Vision, Our Future

BEST REGARDS

RD TECH PANAMETRICS-NDT™ NDT engineering corporation NORTEC SONIC

Material Characterisation of Layered Structures by the Simultaneous Determination of Sound Velocity and Thickness

Sebastian KÜMMRITZ[1], Mario WOLF[1], Elfgard KÜHNICKE[1]
[1]Department of Electrical and Computer Engineering – Solid-State Electronics Laboratory,
TU Dresden, Mommsenstraße 15, 01069 Dresden, Germany
Phone: +49 351 463 32167, E-Mail: sebastian.kuemmritz@tu-dresden.de

Abstract
This paper presents a method for the characterisation of two-layered structures in one measuring process with ultrasound. This is, in particular, necessary if the dimensions of the specimens are not known, and the layers are not accessible. The main approach is the simultaneous determination of thickness and sound velocity for both layers. This is achieved by evaluating the amplitude beside the time of flight (TOF). The amplitude of a reflected sound wave depends on the position of the interface in the sound field and is maximal, if it is located in the focus. By the use of an annular array, the focal position is varied and the amplitude as a function of focusing is evaluated. The determination of density and shear wave velocity is performed, with known layer thicknesses and pressure wave velocities.

Keywords: Focusing, annular array, material characterisation

1 Introduction

Ultrasound is widely used for material characterization, for thickness measurements or for the localization of defects in nondestructive testing. In all cases either the specimen dimensions (thickness *d*) or the sound velocity *c* has to be known. All these methods can not (or only under strong limitations) be performed, if neither sound velocity nor thickness are known.

Gust [1] determined sound velocity and thickness of one layer simultaneously with a scanning acoustic microscope via moving a probe vertically to the surface in water. The echo of each interface is highest if the focus is located at the interface. So the displacement between the positions, where the echoes of the first and the second interface become strongest, and the TOF between the interfaces are used to determine sound velocity and thickness simultaneously. Moving the focus can also be realized by an annular array at a fixed position via driving the elements with time lags. In this case, the amplitude of the focused signal is analysed as a function of focusing [2].

In this contribution steel, aluminium and brass plates after a 25 mm water delay line are inspected. The first layer is the water layer, the second layer the metal plate. Since the presented method of simultaneous determination of sound velocity and distance acts very well for the first layer, and the sound velocity of water does not change in all performed measurements, it is assumed as known and in this work only the second layer is considered. First, the sound velocities of longitudinal waves and thicknesses of all plates are determined. If these parameters are known, the determination of shear wave velocity is possible. Additionally, the densities were determined with a multiple reflection approach.

2 Methods

2.1 Simultaneous determination of sound velocity and thickness

2.1.1 Configuration

In this work, an annular array is used to vary the focal position without moving the probe. The main advantage of annular arrays in contrast to linear or matrix arrays is, that only a few elements are needed for good focusing in a wide range. This reduces the requirements on control- and evaluation electronic. Furthermore, with annular arrays a circular focus in the plane parallel to the transducer surface can be realised, whereas the focus of a linear array is highly expanded in one direction.

The presented results are achieved with a self-developed pulser system [3]. It allows do drive up to 16 channels simultaneous or time shifted (time lag increment: 0.25 ns) with pulses (duration: 4 or 6 ns, amplitude: 10-350 V). For digitizing a PCIe digitizer of with 8 chanels, 14 bit resolution and a sampling rate of 125 MS/s was used.

Since the sound velocities of the investigated specimens are not known, focusing has to be realized after the measurement (synthetic focusing). Therefore, each and any element is excited with pulses and each and every element receives the echo signals. The used array has a centre frequency of about 6 MHz and 6 ring elements (fig1, Table 1), so that 36 different sending-receiving combinations can be recorded.

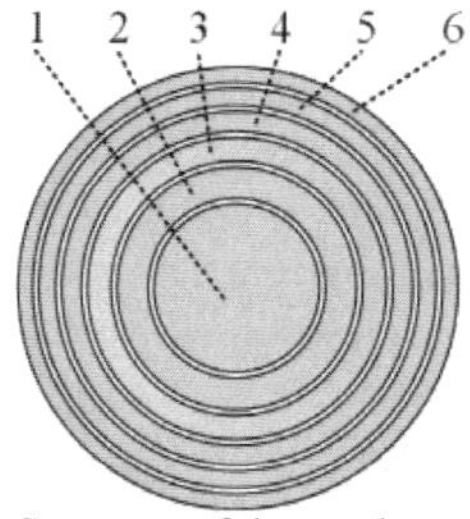

Figure 1. Structure of the used annular array

Table 1. Correct inner- (r_i) and outer radii (r_a) of the ring elements

Element	1	2	3	4	5	6
r_i [mm]	0,6	2,4	3,39	4,16	4,83	5,43
r_a [mm]	2,3	3,29	4,06	4,73	5,33	5,87

2.1.2 Calculation of time delays

The delay times Δt_n (n = 1, 2, ..., 6) are required, for focusing in an arbitrary depth along the acoustic axis. For that, the sound travel paths from each sending element to the focal point and back to each receiving element are calculated. The origin of each element *n* is

$$y_\mathrm{n} = \frac{1}{A}\int_{r_\mathrm{i,n}}^{r_\mathrm{a,n}} r(2\pi r)dr = \frac{2\pi}{3A}\left(r_\mathrm{a,n}^3 - r_\mathrm{i,n}^3\right) \text{ ....(1)}$$

with the active sensor area A and the inner- $\boldsymbol{r}_\mathrm{i,n}$ and outer $\boldsymbol{r}_\mathrm{a,n}$ radii of transducer element n.

The sound propagation time from an element to a focal point in the second layer under consideration of diffraction (Figure 2) can be calculated with:

$$t_n = \frac{\sqrt{(d_1{}^2 + y_p{}^2)}}{c_1} + \frac{\sqrt{(y_n - y_p)^2 + (z_\mathrm{fok} - d_1)^2}}{c_2} \text{ ....(2)}$$

Here, d_1 denotes the thickness of the first layer, c_1 and c_2 the sound velocities of the first and the second layer, z_fok the position of focal point and y_p the point there the sound travel path transmits the interface between the two media. This point can be calculated with FERMAT's principle:

$$\frac{dt_n}{dy_p} = 0 \text{ ....(3)}$$

The required delay time Δt_{n} from a specific element n to the focal point results from

$$\Delta t_{\mathrm{n}} = t_n - t_1.$$

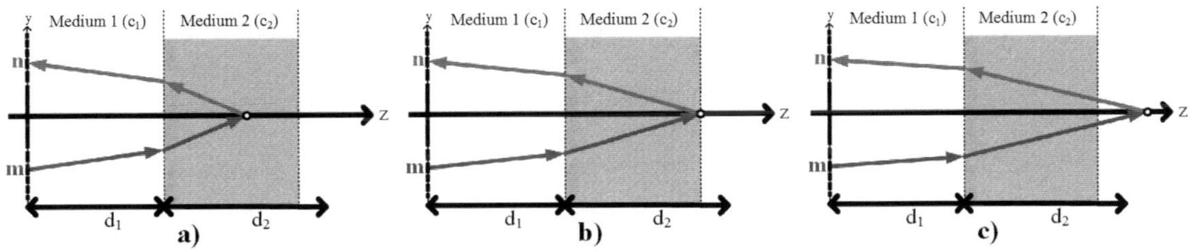

Figure 2: Sound propagation paths for focusing in front of (a), on (b) and behind the backwall of the second layer (c)

2.1.3 Determination of Layer thickness and sound velocity

In [1] the echo amplitude from the interface was evaluated in dependence of the transducer position. Equivalent to this, the amplitude in dependence of the used delay time regime has to be evaluated. A delay time regime means the set of delay times

$$\Delta T(z_{\mathrm{fok}}, c) = \begin{pmatrix} t_1 + t_1 & \cdots & t_N + t_1 \\ \vdots & \ddots & \vdots \\ t_1 + t_N & \cdots & t_N + t_N \end{pmatrix} - 2t_1 \quad (4)$$

for focusing in a specific depth z_{fok} with an assumed sound velocity c (control sound velocity). Every element of this matrix is composed by the sound propagation time from an emitter to the focal point and the sound propagation time from the focal point back to a receiver.

The presented method uses an iterative algorithm, which varies sound velocity (control sound velocity) and focus depth for focusing. Figure 3 demonstrates the first step of the algorithm for the example: steel plate (c = 5920 m/s, d = 14 mm) behind a water layer (c = 1488 m/s, d = 25 mm). In the shown case, the chosen control sound velocity is 3000 m/s. In Figure 3a) the focused signals for focusing in three different depths behind the interface between the two media are shown. Figure 3b) shows the amplitude in dependence of the adjusted focal position, the so called focus curve. The focus curve is a measure for the quality of the adjusted focusing. In the example the maximum can be seen at $z_{\mathrm{fok}} = 30$ mm. This means, if the specimen has a sound velocity of 3000 m/s, it would have a thickness of about 30 mm.

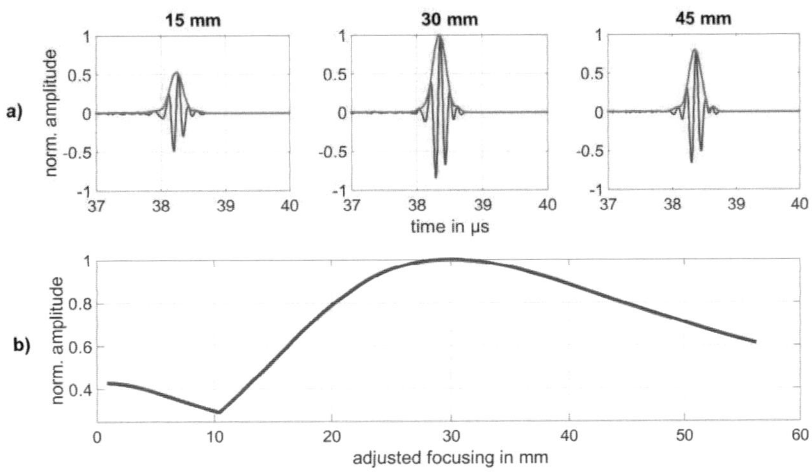

Figure 3: Principle of the focus curves - a) signals for focusing in different depths with c = 3000 m/s; b) signal amplitude in dependence of the adjusted focusing

In the second step the control sound velocities are varied and the focus curves are determined (Figure 4a). Every curve gives a distance which is related to the chosen sound velocity. In Figure 4b the reflector distance is shown as a function of the control sound velocity (red crosses). These points are interpolated with a polynomial (blue dotted curve). The TOF provides additional information (black curve), so that sound velocity and thickness can be determined from the intersection point of both curves.

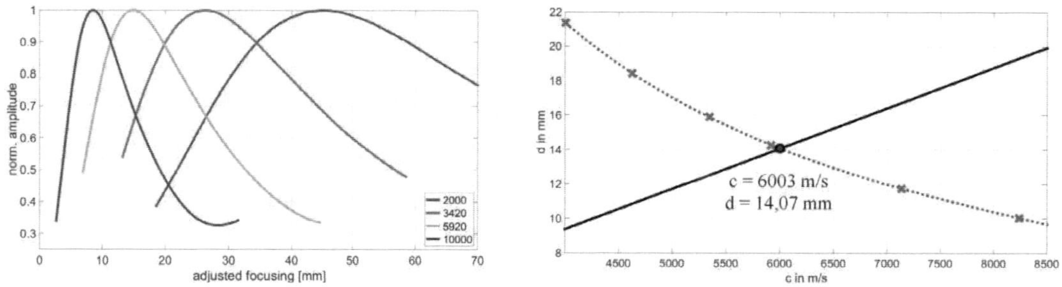

a) b)

Figure 4. a) Focus curves for different sound velocities; b) determination of sound velocity and distance - red crosses: maxima of focus curves for different control sound velocities; blue dotted line: fit; black curve: TOF; red point: intersection point

2.2 Density

The density was determined with a multiple reflection method [5]. The sound pressure p of the echoes signals can be expressed with

$$p_{\mathrm{SE}} = p_{\mathrm{inc}} R_{12} \quad (5)$$

for the surface echo (SE),

$$p_{\mathrm{BWE1}} = p_{\mathrm{inc}} T_{12} R_{23} T_{21} e^{-2d\alpha} \quad (6)$$

for the first backwall echo (BEW1) and

$$p_{\mathrm{BW2}} = p_{\mathrm{inc}} T_{12} R_{23}{}^{2} R_{21} T_{21} e^{-2d\alpha} \quad (7)$$

for the second backwall echo (BWE2), with the transmission

$$T_{12} = \frac{2Z_2}{Z_1+Z_2} \quad (8)$$

and reflection coefficient

$$R_{12} = \frac{Z_2-Z_1}{Z_1+Z_2} \quad (9)$$

with the acoustic impedance Z of the respective layer, and the sound pressure p_{inc} of the incident wave. The coefficient denotes the number of layer. Equation 5 to 9 can be formed to

$$\frac{p_{\mathrm{BW2}}/p_{\mathrm{BW1}}}{p_{\mathrm{BW1}}/p_{\mathrm{S}}} = -\frac{(Z_2-Z_1)^2}{4Z_2Z_1} \quad (10)$$

so that the acoustic impedance of the second layer can be determined. The density of the second layer can be calculated with the formula of the characteristic acoustic impedance:

$$Z = \rho c \quad (11)$$

This method acts on the assumption of plane waves.

2.3 Shear Wave Velocity

To determine the shear wave velocity, a generation of shear waves is required. This can be realized by mode conversion, which appears by inclined sound incident. Therefore, the ring elements of the annular array are used. Figure 5a) shows a principle drawing of different sound propagation paths and Figure 5b) the corresponding signals. Since the surface reflection is much higher, only the first and the second backwall reflections are shown for a better illustration. The shear wave velocity can be calculated, if the following parameters are known: thicknesses and longitudinal wave velocities of both layers, TOF of the surface echo, TOF of the shear wave echo, TOF of the first backwall and of TOF to the second backwall echo.

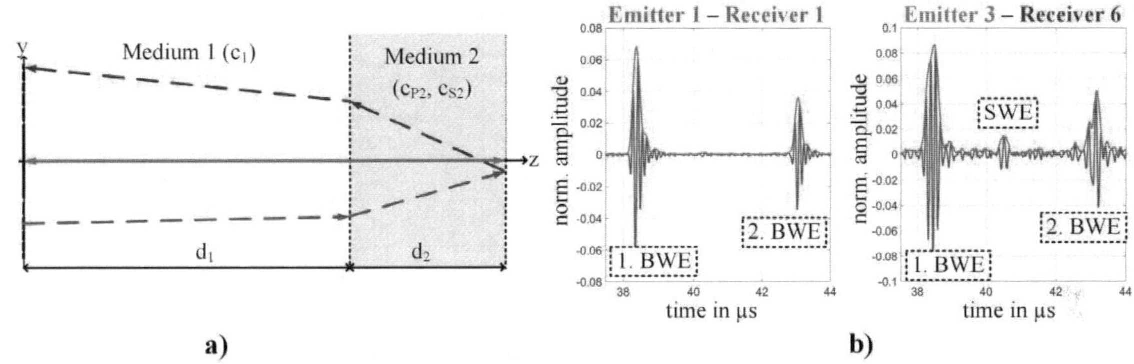

Figure 5. Exciting of shear wave by the use of ring elements. a) Sound travel paths for sending and receiving with the central element (green) and sending with element 3 (blue) and receiving with element 6 (red); b) corresponding signals

3 Results

Plates of steel, aluminium and brass behind a 25 mm water layer were investigated. For all plates a c-scan with 61x61 measurement points was performed. The algorithm for the simultaneous determination of sound velocity and thickness was applied for every measurement point. Figure 6 shows exemplarily the determined sound velocity for each point of the 6 mm steel plate. From all determined sound velocities within the plate, a mean value and a standard deviation was calculated.

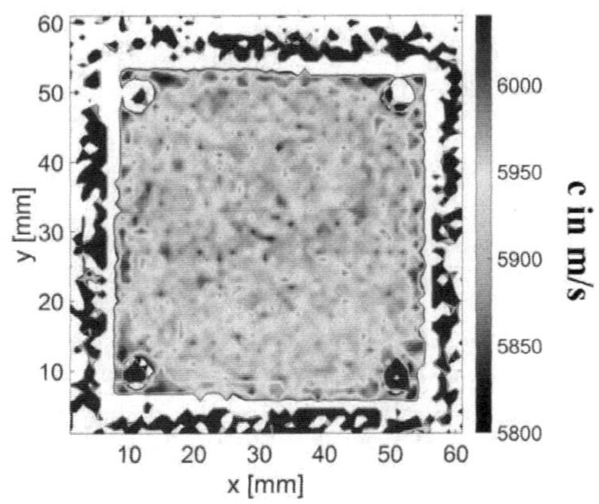

Figure 6. Sound velocity map of 6 mm steel

The results for all plates are summarised in Table 2 - Table 4. The second row shows the thickness d_{con} of the plates, determined with a calliper gauge (accuracy ± 50 µm). In the third and fourth row, the simultaneous determined thickness d_{sim} and sound velocity of the longitudinal wave c_{sim} are shown. Row five shows the shear wave velocity c_{S} and row six the measured density ρ. In line two of each table, densities, longitudinal and shear wave velocities given in literature [5], [6] are shown.

Table 2. Results steel

Plate No.	d_{con} [mm]	d_{sim} [mm]	c_{sim} [m/s]	c_s [m/s]	ρ [g/cm^3]
Literature	-	-	5790[5]-5920[6]	3100[5]-3230[6]	7.8[5]-7.85[6]
1	3.98	3.95 ± 0.03	5955 ± 39	-	7.0 ± 0.2
2	5.88	5.77 ± 0.03	5913 ± 34	3209 ± 77	7.3 ± 0.2
3	7.89	7.84 ± 0.04	5943 ± 33	3291 ± 38	7.5 ± 0.3
4	9.82	9.97 ± 0.05	6004 ± 31	3360 ± 40	7.6 ± 0.3

Table 3. Results aluminium

Plate No.	d_{con} [mm]	d_{sim} [mm]	c_{sim} [m/s]	c_s [m/s]	ρ [g/cm^3]
Literature	-	-	6380[5]-6420[6]	3040[5]-3100[6]	2.7[5,6]
1	3.80	3.77 ± 0.03	6488 ± 29	-	1.9 ± 0.0
2	6.10	5.84 ± 0.04	6251 ± 34	3095 ± 33	2.1 ± 0.1
3	7.89	7.57 ± 0.04	6253 ± 33	3074 ± 45	2.1 ± 0.1
4	9.59	9.32 ± 0.06	6281 ± 37	3112 ± 31	2.2 ± 0.1

Table 4. Results brass

Plate No.	d_{con} [mm]	d_{sim} [mm]	c_{sim} [m/s]	c_s [m/s]	ρ [g/cm^3]
Literature	-	-	4430[6]-4700[5]	2100[5]-2140[6]	8.4[6]-8.6[5]
1	4.12	4.24 ± 0.14	4593 ± 142	-	7.6 ± 0.7
2	5.98	6.14 ± 0.19	4603 ± 131	-	8.2 ± 1.0
3	7.81	8.12 ± 0.28	4605 ± 147	-	9.9 ± 2.0
4	9.84	10.59 ± 0.79	4753 ± 107	-	-

4 Discussion

All determined sound velocities deviate from the values from the literature at most of about 3 percent. Due to the simultaneous determination of sound velocity and thickness, the same deviation occurs for the thicknesses. All determined densities are partially much smaller than the values from the literature. This can be explained, that for the discussed multiple reflection method plane waves are assumed. For better results a more longer water delay line is necessary.

The shear wave velocity for the smallest plates was not able to determine, since no shear wave echo could be identified. Here, the shear wave echo may be superposed from different signal components, which occurs due to different sound propagation paths caused by receiving with rings [7]. Also, no shear wave echo could be found for all brass specimens. This can be explained with the high attenuation of brass. The attenuation is also an explanation for the high standard deviation of the simultaneous determined values of thickness and sound velocity of brass.

5 Summary

This contribution presents a method for the simultaneous determination of sound velocities and thicknesses of two-layered structures. It acts with an annular array, no additional transducers or reflectors at known positions are necessary. The method is fast and suitable for monitoring of production processes, since no mechanical displacement of the probe is required. If the longitudinal wave velocities and thicknesses of the layers are determined with the method, for material characterization, shear wave velocity and density can be estimated with conventional methods.

The simultaneous determination of thickness and sound velocity is demonstrated for the second layer of a fluid-solid system, the implementation for the first layer is analog. It is also applicable to solid-solid systems, however, since additional problems like bad coupling, multiple reflections and mode conversions occurs, in this paper water was used for the first layer. The benefit of this approach is the determination of material parameters and thickness of both layers, which means that an inspection of hidden objects is possible.

Deviations between the determined values and values from literature occurs due to the geometric approach, no wave properties are considered. For the calculation of diffraction at the interface between the first and the second layer, SNELLs law is used. This is only approximately valid and, therefore, a model based on sound field simulations for the diffraction correction is developed currently.

References

[1] N. Gust, 'Improvement of signal analysis for the ultrasonic microscopy', PhD Thesis, TU Dresden, 2011

[2] S. Kümmritz, M. Wolf, E. Kühnicke, 'Simultaneous determination of thickness and sound velocities of layered structueres', Technisches Messen, Vol 82, No 3, pp. 127-137, 1991

[3] N. Gust, F. Schellhorn, 'High frequency 16 channel ultrasonic microscope arrays', IEEE IUS proceedings, 2012

[4] J. C. Adamowski, F. Buiochi, C. Simon, E. C. N. Silvay, R. Siegelmann, 'Ultrasonic measurements of density of liquids', J. Acoust. Soc. Amer., Vol 97,pp354-361, January 1995

[5] Onda Corp., 'Acoustic Properties of Solids', http://www.ondacorp.com/tecref_acoustictable.shtml, October 2015

[6] J. L. Rose, 'Ultrasonic Waves in Solid Media', Cambridge University Press, 2004, ISBN 978-0-521-54889-2

[7] M. Wolf, S. Kümmritz, E. Kühnicke, 'Advanced signaly analysis for the examination of multilayered structures using annular arrays', 11th ECNDT, proceedings, October, 2014

Przedsiębiorstwo Badawczo-Produkcyjne
OPTEL Sp. z o.o.
ul. Jerzego Kowalskiego 5
PL-52-428 Wrocław
tel.: +48 (071) 329 68 53
fax.: +48 (071) 329 68 52
e-mail: optel@optel.pl
http://www.optel.pl

OEM opBox ver 2.3

OEM USB Ultrasonic device with integrated 2-channel pulser&receiver

Overview

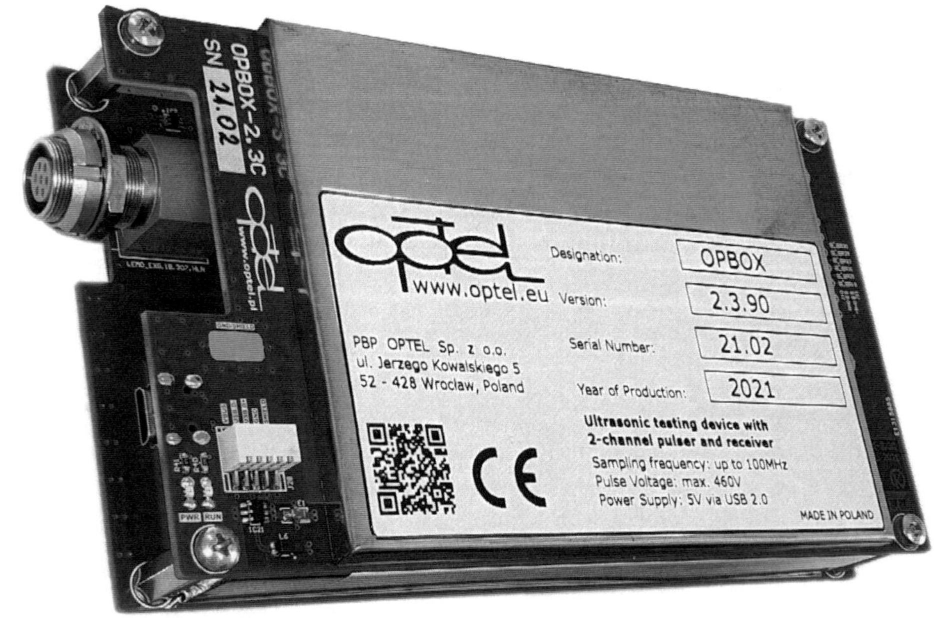

Valid for hardware version: 2.3.90

Document version: 1v3

Last changes: 2021-07-06

SpecificationElectrical specification

Parameter of analog receiver:	
Adjustable amplifier:	-31dB to 65dB (step 0.5dB)
Switchable pre-amplifier:	+24dB
Switchable attenuator:	-20dB
Input voltage:	± 275mV, ± 2.0V with attenuator -20dB.
Bandwidth (T/R switch):	0.8 MHz - 25 MHz (-3dB)
Switchable hardware filters (-3dB):	0.5 - 6MHz, 0.5 - 10MHz, 0.5 - 15MHz, 0.5 - 25MHz, 1 – 6MHz, 1 - 10MHz, 1 - 15MHz, 1 - 25MHz, 2 - 6MHz, 2 – 10MHz, 2 – 15MHz, 2 - 25MHz, 4 - 6MHz, 4 - 10MHz, 4 - 15MHz, 4 – 25MHz.

Pulsers specification (2-channels):	
Type:	Step-up, short-circuits pulser; 1. charging-up within PulseTime phase 2. short-circuit discharge
Voltage adjustment:	0V - 460V (positive pulse);
Loading time:	Adjustable from 0 do 6.3μs step 0.1μs;
Short circuit time:	<= 20 ns;
Bandwidth:	Up to 50MHz.

A/D converter:	
Resolution:	10 bit (8-bits are stored);
Maximum input voltage:	±0.5V;
Sampling frequency - switchable (MHz):	100; 50; 33.3; 25; 20; 16.7; 14.3; 12.5;
	11.1; 10; 9.1; 8.3; 7.7; 7.14 and 6.67;
Data buffer:	1 - 262090 (256k) samples;
Measuring frame delay:	0 - 65535 sample periods.

Hardware data processing:	
Data representation:	RF, Absolute;
Measurements in defined frames:	3 peak detectors;
	3 level comparators – modes of transition detection: Level, Rising, Falling, Transition.

DAC (TGC) with arbitrary function generator:	
Sampling frequency	100MHz;
Resolution	8 bit;

Trigger:	
Software	Software command;
Internal	Programmable timer;
*External (not available on standard assembly)	2 incremental encoder modules; 2 TTL inputs.

Ultrasonic channels:	
PE1 (sending&receiving)	Lemo EXG.1B.307 pin 6
PE2 (sending&receiving)	Lemo EXG.1B.307 pin 1

LEMO connector:	
Type:	EXG.1B.307.HLN
Pinout:	front view: real view: 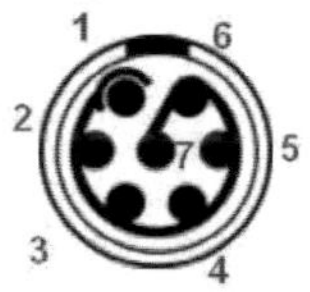 

Pin	Description	Type	Description
1	PE2_signal	HV/RF	PE2: HV pulse output, RF signal input
2	PE2_return	GND	
3	LED1_pos	OC with Rs	Driver: USB 5V, max 100mA Ser. res. installed (R55): 220ohm
4	LED2_pos	OC with Rs	Driver: USB 5V, max 100mA Ser. res. installed (R55): 220ohm
5	PE1_return	GND	
6	PE1_signal	HV/RF	PE1: HV pulse output, RF signal input
7	LEDs_return	GND	

USB interface:	
Connector type:	USB micro-B socket **or** Molex PicoSpox 5-pin socket ("wired-USB") **Note!** Only one USB connection can be used! Connecting USB micro-B and wired USB can lead to communication errors and USB interface failure or damage!
Standard:	USB-2.0 High Speed (480Mbps)
Vendor ID:	0x0547
Product ID:	0x1003
Hardware version:	2.3 (0x0302)
Power supply:	Bus powered <500mA
USB device class:	Vendor
Available endpoints:	
Endpoint 0 (control register)	Direction: IN/OUT, Type: Control, Packets: max. 64Bytes;
Endpoint 2 (writing TGC curves)	Direction: OUT, Type: Bulk, Packets: max. 512Byte;
Endpoint 6 (reading measurement data)	Direction: IN, Type: Bulk, Packets: max. 512Bytes.

J2B "wired-USB" connector:			
Type:	Molex Pico-SPOX, 5-pin, 1.5mm pitch, single row, low profile, friction lock		
Components:	PCB socket: part no: 874380543 Cable plug: part no: 874390500 Crimp terminals: part no: 874210000		
Pin	Description	Type	Description
1	VBUS	power	USB 5V power supply
2	USB_D-	signal	USB D- data line
3	USB_D+	signal	USB D+ data line
4	GND	ground	Return for data and power lines
5	SHIELD	shield	USB cable shield (shorted to GND)

***CONTROL connector (not available on standard assembly, only as a special request):**			
Type:	DB15 female		
DB15 pin	Description	Standard	Description
1	GPO 0	TTL 5V	Digital input GPO0 or synchronization output SYNC_OUT
2	GPO 2	TTL 5V	Digital output GPO2
3	GND		Ground
4	GPI 1	TTL/LVTTL	Digital input GPI1 or trigger input EXT_Y or CHB input for encoder module ENC1
5	GPI 3	TTL/LVTTL	Digital input GPI3 or CHA input for encoder module ENC2
6	GPO 5	TTL 5V	Digital output GPO 5
7	Reserved		Reserved
8	GPI 5	TTL/LVTTL	Digital input GPI5 or IDX input for encoder module ENC2
9	GPO 1	TTL 5V	Digital output GPO1
10	GPO 3	TTL 5V	Digital output GPO3
11	GPI 0	TTL/LVTTL	Digital input GPI0 or trigger input EXT_X or CHA input for encoder module ENC1
12	GPI 2	TTL/LVTTL	Digital input GPI2 or IDX input for encoder module ENC1
13	GPO 4	TTL 5V	Digital output GPO4
14	GND		Ground
15	GPI 5	TTL/LVTTL	Digital inputGPI4 or CHB input for encoder module ENC2

Mechanical specification

Mechanical parameters	
Overall dimensions:	(LxWxH)
boards stacked with 15mm spacers:	135.0 x 76.5 x 29.0mm (with shields)
boards placed flat with 5mm space:	135.0 x 158.0 x 19.0mm (with shields)
Digital board dimensions:	
board LxW	135.0 x 76.5mm
board thickness	1.6mm
top shield height	5.0mm
bottom B2B connector height	4.3mm (open), 7.7mm (mated)
USB connector height	2.9mm
bottom SMD components height	<2.0mm
Analog-board dimensions:	
board LxW	135.0 x 76.5mm
board thickness	1.6mm
top shield height	14.0mm (pulser shield can)
bottom SMD components height	<3.0mm
Lemo connector offset: (beyond PCB outline)	about 7.5mm
USB connector offset: (beyond PCB outline)	about 0.9mm
B2B Flat cable offset: (beyond PCB outline)	about 3.0mm
Mounting holes:	4x holes for M3 spacers/screws, centered on square 125x70mm
Weight (both boards with shields):	215g

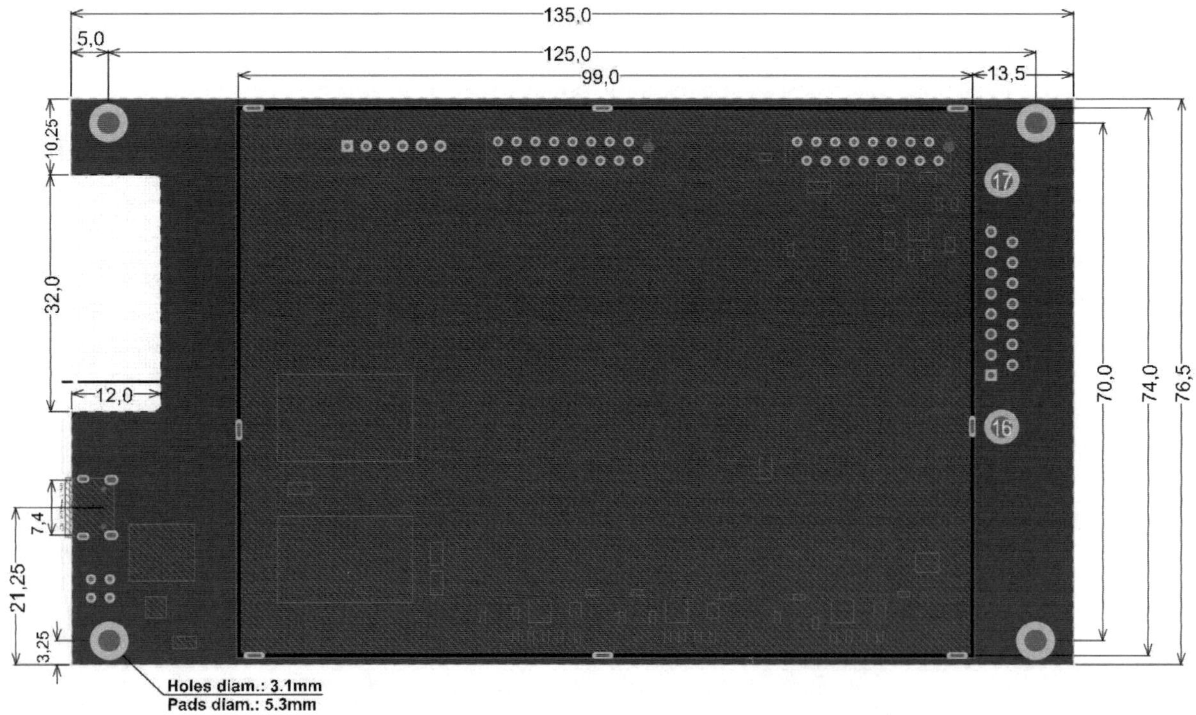

OPBOX-2.3 Digital Board Dimensions
(rev.1, date: 2021-05-18)

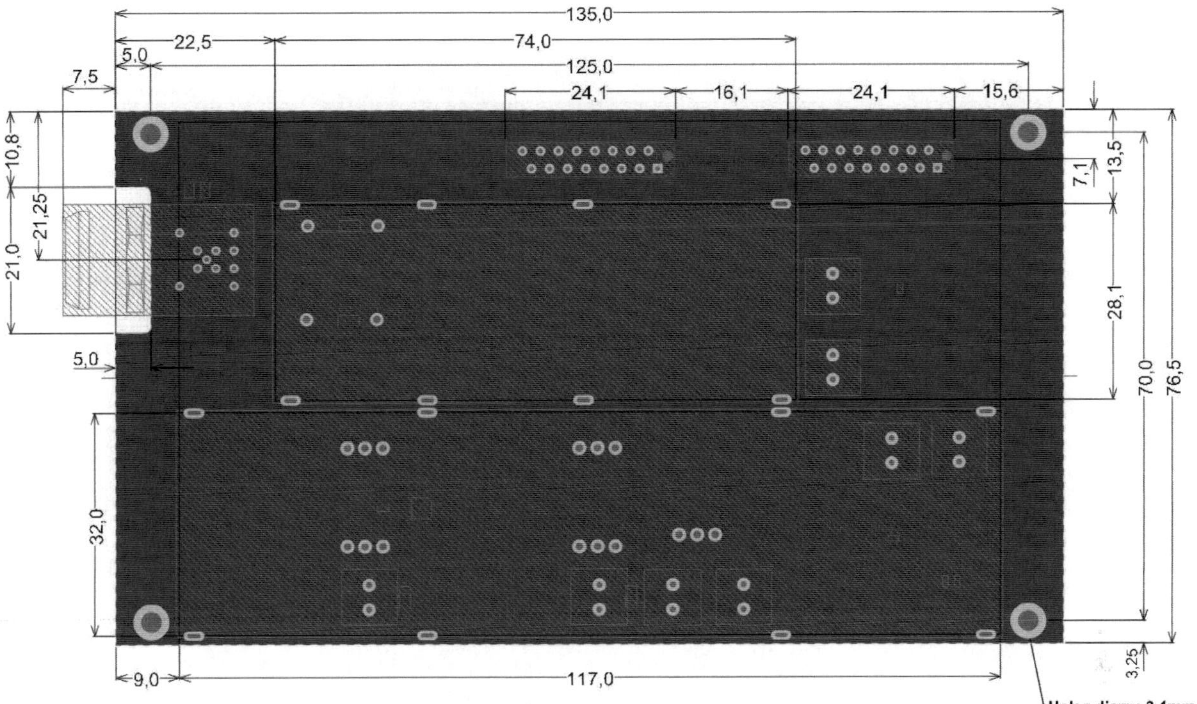

OPBOX-2.3 Analog Board Dimensions
(rev.1, date: 2021-05-18)

Internal LEDs locations on Digital board:

PWR LED (red):
blinking - powered down or power failure (INITIAL/SLEEP mode);
solid on - powered up, ready for measurements (READY mode).

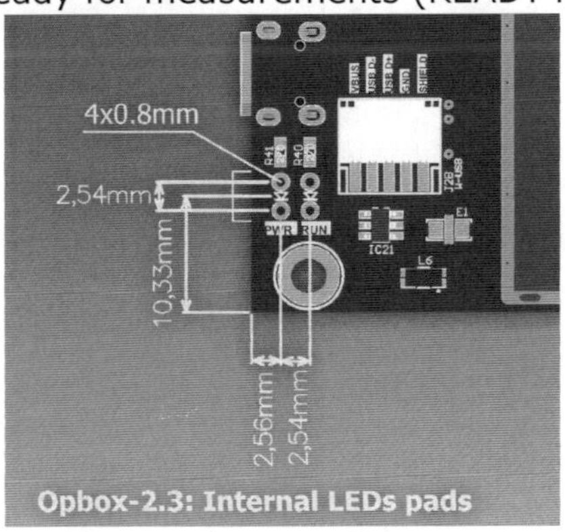

Opbox-2.3: Internal LEDs pads

RUN LED (yellow):
ON when acquisition is in progress and/or there is a data available in the FIFO;
OFF otherwise.

AUTHOR INDEX

Aliabadi, M. 145
Angelo, J. 76
Banerjee, S. 170
Bokůvka, O. 1
Brockmann, R. 61
Bukacek, V. 18
Busch, M. 11
Caliez, M. 126
Carbol, L. 46
Chaari, M. 111
Cherepetskaya, E. 165
Chlada, M. 32
Dos Santos, S. 126
Dráb, M. 71
Dvořáková, Z. 126
Faktorová, D. 1, 185
Faupel, B. 11
Grum, J. 53
Hajek, K. 118
Han, K. 38
Hanak, M. 71
Hančič, A. 53
Iftimie, N. 185
Jana, M. 18
Juhrig, A. 180
Jung, I. 38
Kek, T. 53
Kharitonov, I. 106
Khodaei, Z. Sharif 145
Kim, D. 38
Kim, S. 28
Kim, Y. 28, 38
Kober, J. 66
Komenda, L. 18
Kusic, D. 53
Kusák, I. 46
Kühnicke, E. 180, 213
Kümmritz, S. 180, 213
Lee, D. 28
Lints, M. 126
Liou, R. 155
Martinek, J. 46
Mazal, P. 18
Michaels, J. 155
Noh, Y. 28
Nový, F. 1
Oh, J. 38
Patrick, C. 196
Petrosyants, K. 106
Poirot, N. 126
Pápežová, M. 1, 185
Převorovský, Z. 32, 66, 84, 126, 136
Salmanpour, M. 145
Salupere, A. 126
Savin, A. 1, 185
Sikdar, S. 170
Son, M. 28
Steigmann, R. 1, 185
Tokar, D. 136
Triantafillou, T. 112
Vlašic, F. 18
Vojkůvková, P. 46
Weyrich, D. 11
Witoś, M. 91
Wolf, M. 180, 213